Understanding Land Warfare

Understanding Land Warfare provides a thorough grounding in the vocabulary, concepts, issues and debates associated with modern land warfare.

The book is a thematic, debate-driven analysis of what makes land warfare unique, how it interacts with the other environments, the key concepts that shape how it is executed, the trade-offs associated with its prosecution, and the controversies that continue to surround its focus and development.

Understanding Land Warfare contains several key themes:

- the difficulty of conducting land warfare
- the interplay between change and continuity
- the growing importance of co-operation
- the variety of ways in which land warfare is fought
- the competing theoretical debates.

This book will be essential reading for military personnel studying on cadet, intermediate and staff courses. In addition, it will also be of use to undergraduate and postgraduate students of military history, war studies and strategic studies.

Christopher Tuck is a Lecturer with the Department of Defence Studies, King's College London, based at the Defence Academy of the United Kingdom. Prior to this, he was a lecturer at the Royal Military Academy, Sandhurst.

Understanding Land Warfare

Christopher Tuck

LONDON AND NEW YORK

First published 2014
by Routledge
2 Park Square, Milton Park, Abingdon, Oxon OX14 4RN

and by Routledge
711 Third Avenue, New York, NY 1001

Routledge is an imprint of the Taylor & Francis Group, an informa business

British Library Cataloguing in Publication Data
A catalogue record for this book is available from the British Library

Library of Congress Cataloging in Publication Data
Tuck, Christopher.
Understanding land warfare / Christopher Tuck.
pages cm
Summary: "This textbook provides a thorough grounding in the vocabulary, concepts, issues and debates associated with modern land warfare"– Provided by publisher.
Includes bibliographical references and index.
1. Military art and science. 2. Military art and science–Case studies. I. Title.
U102.T94 2014
355.4–dc23
2013046492

ISBN: 978-0-415-50753-0 (hbk)
ISBN: 978-0-415-50754-7 (pbk)
ISBN: 978-1-315-88225-3 (ebk)

Typeset in Times New Roman
by Taylor & Francis Books

Printed and bound by CPI Group (UK) Ltd, Croydon, CR0 4YY

Contents

List of illustrations

Figures

Tables

Boxes

Acknowledgements

This book has its origins in a multi-authored work, *Understanding Modern Warfare*, to which I contributed a chapter on land warfare. I had struggled at the time to find an accessible introduction to land warfare that I could use for teaching purposes and so I wrote one myself. At the time, I felt that a single chapter (rather long though it was) hardly did the topic justice and so this is the result: an entire volume dedicated to the subject. Inevitably, however, having believed originally that a single chapter was insufficient to say what needed to be said, I have found that having a whole book has hardly remedied this problem: but then again, I would hardly be the first academic to say that.

Sincere thanks go to Dr Deborah Sanders, who not only provided valuable feedback on what I had written but, as my partner, has also provided the love and support to complete what has seemed often to be a Sisyphean task. I owe also a great debt of thanks to Dr Ian Speller, who has provided a great deal of support in the form of invaluable advice and comment delivered, as always, with brutal disregard for my feelings. Thanks are due to my colleagues at the Joint Services Command and Staff College (JSCSC), who, after making the fatal mistake of asking me what I was working on, were polite enough to feign interest.

Thanks also go to the editorial team at Routledge, especially Andrew Humphrys, for continuing to prod me to complete this work which was, unsurprisingly for a book written by an academic, delivered late. Thanks also go to the Graphics Department of the JSCSC, who produced some marvellous diagrams from some decidedly dodgy originals that I provided.

It should be noted that the analysis, opinions and views expressed or implied in this book are those of the author and do not necessarily represent the views of the Defence Academy of the United Kingdom, the UK Ministry of Defence, or any other government agency.

Dr Christopher Tuck

Glossary

4GW: *Fourth Generation Warfare.*

AirLand Battle: The first US operational level doctrine, published as *Field Manual* (FM) 100-105 in 1982.

Asymmetric Warfare: Warfare between dissimilar adversaries. A contested concept, asymmetric warfare is often associated with conflict against such unconventional adversaries as terrorists and insurgents.

Attrition: The reduction of the effectiveness of a military formation caused by the loss of equipment and/or personnel.

Battalion: A military unit usually consisting of 500–1,200 troops divided into four to six companies. Two or more battalions may be grouped into a *brigade*, or one or more may constitute a *regiment*.

Battlegroup: A flexible combined arms force usually created around the nucleus of a *battalion/regiment*-sized infantry or armoured unit and augmented with other arms.

Battlespace dominance: A US concept that encompasses the ability to dominate the three-dimensional battlespace by establishing zones of superiority around deployed forces.

BCT: Brigade Combat Team.

***Blitzkrieg*:** Second World War German *Operational level* mobile warfare.

Brigade: A military unit consisting of two to five *battalions* or *regiments*. Two or more brigades may be grouped into a *division*.

C2: An acronym to represent the military functions of *command* and *control*. The process by which a commander exercises *command* and *control* of their forces in order to direct, organise and co-ordinate their activities to maximum effect.

C3I: An acronym to represent the following military functions: Command, Control, Communications and Intelligence.

C4ISR: An acronym to represent the following military functions: Command, Control, Communications, Computers, Intelligence, Surveillance and Reconnaissance.

C4ISTAR: An acronym to represent the following military functions: Command, Control, Communications, Computers, Intelligence, Surveillance, Target Acquisition and Reconnaissance.

Campaigns: A connected series of *tactical* engagements within a *theatre* designed to achieve a *strategic* objective.

Centre of gravity: Characteristics, capabilities or locations that are key to the ability of an organisation to function effectively.

CIA: Central Intelligence Agency.

Close Air Support: Air attacks launched against enemy targets in close proximity to, and with the aim of helping the operations of, friendly forces.

COIN: Counterinsurgency operations. Military, paramilitary, political, economic, psychological and civic actions taken by a government to defeat insurgency.

Combined arms: The integration of different arms into a single system to achieve a complementary effect. *Battlegroups*, for example, are designed to exploit the benefits of combined arms.

Combined operations: Traditionally, a term used to describe what would now be called *joint operations*. In modern nomenclature, the term is used to describe multi-national operations.

Command: The authority to *control* the activity and organisation of armed forces.

Company: A military unit consisting of 100–200 troops, divided into two to five *platoons*.

Conflict prevention: Diplomatic and other actions taken in advance of a predictable crisis to prevent or limit violence, deter parties, and reach an agreement short of conflict.

Control: The process by which a commander directs, organises and co-ordinates the activities of the forces under their *command*.

Corps: A military unit consisting of two or more *divisions*. Derived from the French term *corps d'armée* ('body of the army'), traditionally a corps was a *combined arms* force capable of fighting independently for a limited period of time.

Culminating point: The point at which the power of an attack ceases to overmatch that of the defence.

DDR: Disarmament, demobilisation and reintegration activities conducted with regard to local armed groups.

Deep operations: Large-scale, simultaneous attacks against enemy reserve and rear areas as well as the front line. Deep operations can be performed by remote firepower (air, artillery and missile attack) and by the rapid exploitation of breakthroughs by mobile forces.

Division: A large, usually *combined arms*, military unit consisting of two or more *brigades*. Two or more divisions may be grouped together into a *corps*.

Doctrine: Guidance, mandatory or discretionary, on what is believed officially to be contemporary best military practice.

Double envelopment: Simultaneous movement around the flanks of the enemy designed to achieve, or to threaten to achieve, the encirclement of the enemy.

Echelon: As a formation, describes the use of multiple reserve formations to create successive waves of attack. Generically, the term 'second echelon' is sometimes used to describe all of the forces that follow on from the initial attacking forces.

EBAO: Effects-based approach to operations. A 'way of thinking' about the relationship between military activity and political effect, and which prioritises the need for multi-agency approaches to operations. A less rigid form of *EBO*.

EBO: Effects-based operations. Rather than prioritising the destruction of enemy forces through the application of military power, EBO is an approach that focuses first on establishing the broad end states that need to be delivered (the 'effect'), and then choosing the most appropriate range of instruments to achieve it (of which military power may be only one).

Envelopment: Movement around or over the enemy.

Expeditionary operations: Operations where military force is projected and sustained at some distance from the home bases, usually across the seas.

FCS: Future Combat System.

Fire team: The building block of modern tactics, the fire team is, in general, a four-man grouping of soldiers. Two or more will comprise a *squad* or section.

FM: Field Manual. Standard designation for US army military doctrine publications, e.g. FM 3-24, *Counterinsurgency.*

Fourth Generation Warfare: A perspective that sees future warfare as an evolved form of networked insurgency.

Friction: The accumulation of those factors in warfare such as chance, complexity, human emotion, geography, weather and so forth that make war in practice more difficult than war in theory.

Globally Integrated Operations: US defence concept that aims to develop 'a globally postured Joint Force to quickly combine capabilities with itself and mission partners across domains, echelons, geographic boundaries, and organizational affiliations'.

Grand strategy: The application of national resources to achieve national/alliance policy objectives.

Hybrid Warfare: Warfare characterised by a fusion of regular and irregular techniques by the same organisation.

IDF: The Israeli Defence Forces, Israel's armed forces.

Insurgency: An organised movement aimed at the overthrow of a constituted government through the use of subversion and armed conflict. Like terrorism, however, insurgency is a contested concept. Its essence is the combination of irregular warfare techniques and political ideology.

Integration: The development of closer, complementary interaction between elements within a military system. Generically, integration is required for effective performance in all aspects of warfare, whether *command* and *control*, *logistics*, *combined* *arms* or *joint operations*. The term has acquired specific importance in relation to contemporary debates regarding the creation of a *system of systems*.

Interdiction: Actions to divert, disrupt or destroy the enemy before they reach the area of battle.

ISR: Intelligence, Surveillance and Reconnaissance.

JDAM: Joint Direct Attack Munitions. Ordinary bombs augmented with additional technology to turn them into precision-guided munitions (*PGM*).

Joint operations: Operations involving the integration of land, air and/or naval forces.

JSTARS: Joint Surveillance Target Attack Radar System.

Land power: The ability to exert influence on or from the land.

Linear tactics: A tactical system based upon the use of troops deployed in line, designed to maximise firepower and minimise problems of *command* and *control*. Characteristic of warfare in the nineteenth and early twentieth centuries.

Logistics: The art of moving armed forces and keeping them supplied.

Manoeuvre: Movement in relation to the enemy to occupy advantageous positions.

Manoeuvre Warfare: The application of the *manoeuvrist approach* to fighting war.

Manoeuvrist approach: A philosophy of war based upon the principle of defeating the enemy by attacking intangibles such as cohesion and will to fight rather than focusing on destroying the enemy's materiel. Despite its label, manoeuvre is not necessarily a prerequisite for a manoeuvrist approach to war.

Mechanised forces: Mobile forces where the transport is usually composed of armoured, armed and often tracked vehicles. The transport is designed to contribute to combat operations by providing protection and fire support.

Military strategic level: The application of military resources to help grand strategic objectives.

Mission command: A command philosophy based on the principles of decentralisation of responsibility and the use of initiative. Classically, subordinates are informed of what they must achieve (the mission), but are allowed to use their own judgement on how best this should be achieved given local conditions.

Motorised forces: Mobile forces where the transport is usually composed of wheeled and unarmoured vehicles. The transport is not usually designed to take part in combat.

MTR: Military-Technical Revolution.

Multi-Dimensional Peacekeeping: Sometimes known as 'New Peacekeeping' or 'Second Generation Peacekeeping'. Complex operations with wider remits, multiple participants and complex intra-state contexts.

NATO: North Atlantic Treaty Organization.

NCO: Non-Commissioned Officer, such as a Corporal, Sergeant or Warrant Officer. A military officer without a commission (i.e. they are still enlisted soldiers). NCOs provide the crucial interface between commissioned officers and the men and women under their command.

NCW: Network-Centric Warfare. A concept in which intense networking creates a *system of systems* which will, in turn, deliver information superiority over an adversary.

NEC: Network-Enabled Capability. A less ambitious, British version of *NCW*.

'New Wars': A perspective on contemporary and future war that argues that a paradigm shift has occurred in warfare reflected in such characteristics as new participants, new objectives, an increased blurring between war and criminality, and the obsolescence of much of traditional military thinking.

NGO: Non-governmental organisation.

OMG: Operational Manoeuvre Group. Manoeuvrable Soviet *combined arms brigades* created to exploit and raid into the depth of *NATO* forces.

OODA loop: The Observe-Orient-Decide-Act (OODA) process believed by Col. John Boyd to underpin effective decision making.

OOTW: Operations Other Than War.

Operational level: The level at which campaigns and major operations are planned, sequenced and directed.

Outflank: Movement around the enemy flanks. Although often used synonymously with the term '*envelopment*', envelopment often implies a much deeper outflanking movement designed to pin and encircle the enemy.

Paradigm: A worldview, conceptual model, or framework of thinking.

Peacekeeping: A term sometimes used generically to mean *Peace operations*, but more properly an operation undertaken with the consent of all major parties to a dispute, and one designed to monitor and facilitate implementation of an agreement to support diplomatic efforts to reach a long-term political settlement.

Peace building: A long-term process designed to create a self-sustaining peace, to address the root causes of conflict, and to prevent a slide back into war.

Peace enforcement: *Peace operations* that are coercive in nature and rely on the threat of the use of force.

Peace making: A diplomatic process aimed at establishing a ceasefire or an otherwise peaceful settlement of a conflict that is already ongoing.

Peace operations: Crisis response and limited contingency operations to contain conflicts, redress the peace and shape the environment to support reconciliation and re-building, and to facilitate the transformation to legitimate governance. Sometimes also termed peace support operations.

PGM: Precision-guided munitions such as laser-guided bombs and cruise missiles.

PLA: The People's Liberation Army, the Chinese armed forces.

Platoon: Comprising two or more *squads*, the platoon is a sub-division of a *company*.

PMC: Private military contractor.

Power projection: The ability to project military power overseas.

Regiment: A tactical or administrative grouping of one or more *battalions.*

RMA: Revolution in Military Affairs. *An* RMA is a paradigm shift in war: 'an observable breaking point between two recognisably different types of warfare.' *The* RMA refers to a concept that emerged in the 1990s that future war would be revolutionised by combinations of *PGM*, sensors and networking.

Scouting: Activity designed to locate the enemy.

Screening: Activity designed to protect your forces from enemy observation and enemy action.

Shock: A psychological state marked by fear and disorientation. Shock is a desirable state to induce in an enemy because it may result in slowed, uncoordinated and irrelevant activity. Translated through to military organisations, 'systemic shock' undermines the capacity of a military force to operate effectively and may result in heavy defeat without the infliction of large amounts of physical destruction.

Squad: A small infantry unit comprising two or more *fire teams.* In some armies the squad is known as a section. Two or more squads will be combined to form *platoons.*

Squadron: A small formation of mobile ground forces such as cavalry or tanks. This term is also used by navies and air forces.

SSR: Security sector reform. Activities designed to build capacity in the target nation's military, police and judicial structures.

Stability operations: From a US perspective, tactical activities designed to maintain or re-establish a safe or secure environment, provide essential government services, emergency infrastructure reconstruction and humanitarian relief. However, other states have alternative definitions.

Strategic level: The level of war at which a nation determines national or multinational (alliance or coalition) strategic security objectives and guidance, and then develops and uses national resources to achieve those objectives.

Strategy: The use of military power for political purpose. Strategy comprises the 'ways' that link 'means' to 'ends'.

Stryker: A US wheeled armoured fighting vehicle introduced in 2002.

Suppression: The neutralisation, rather than destruction, of forces through the psychological effect of firepower. Typically, 'suppressive fire' forces an enemy to remain in cover, pins them, and prevents them from firing themselves by rendering them unable (through *shock* or disorientation) or unwilling (through fear) to expose themselves.

Synergy: Interaction between two or more elements that produces combined effects that are greater than the sum of the parts.

System: A set of interrelated parts that together constitute a whole. Military organisations are made up of a multitude of systems, such as administrative systems, logistic systems and training systems.

'System of systems': A concept in which very high degrees of *integration* between different military systems, facilitated by improvements in data transfer and *C4ISR*, produce something more resembling a single *system*. In theory, this should dramatically improve the co-operation, co-ordination, speed and flexibility of a military organisation.

Tactical level: The level of war at which battles and engagements are planned and fought.

Tactics: The conduct of battles and engagements.

Tempo: The rhythm or rate of activity on operations, relative to the enemy.

Terrorism: A basic definition might be 'the of use of violence, or the threat of violence, to create fear'. Terrorism is a strategy in which violence is important for the psychological effect that it produces. However, terrorism is a contested concept, not least because it has associated moral, political and pejorative connotations, and a multitude of different definitions exist.

Theatre: Large geographically defined areas of land, water and air that bound the conduct of military operations.

TNC: Trans-national corporation.

TRADOC: Training and Doctrine Command. Established by the United States in 1973, focusing on knitting together the often disparate army training and doctrine programmes, TRADOC's purpose essentially was to change how the army trained and fought.

Transformation: A US vision of defence reform embodying a radical vision of the demands of future warfare, including the need for smaller, modular, rapidly deployable and highly networked forces.

UAV: Unmanned Aerial Vehicle, also known as a drone.

UNAMID: African Union-United Nations Mission in Darfur.

UNFICYP: United Nations Peacekeeping Force in Cyprus.

UNOSOM: United Nations Operation in Somalia.

UNPROFOR: United Nations Protection Force, deployed in the Former Yugoslavia.

UNTAC: United Nations Transitional Authority in Cambodia.

Vertical envelopment: Moving over the enemy in order to threaten their flanks or rear. This may be executed through air-mobile (helicopter-borne) or airborne (parachute or glider) forces.

War: Organised violence conducted by political units against each other.

Warfare: The military conduct of war; it is concerned centrally with fighting.

Introduction

Understanding land warfare

The purpose of this book is to provide an accessible introduction to the theory and practice of modern land warfare. *Understanding Land Warfare* provides the reader with a thorough grounding in the vocabulary, concepts, issues and debates associated with modern land warfare. This book is not a history of warfare: it is instead a thematic analysis of what makes land warfare unique; how it interacts with the other environments; the key concepts that shape how it is executed; the trade-offs associated with its prosecution; and the controversies that continue to surround its focus and development. *Understanding Land Warfare* is designed to provide a foundation of understanding of warfare on land that will better enable the reader to explore the issues further.

Particularly with regard to theories and concepts, an examination of land warfare requires some engagement with military doctrine, especially (but not exclusively) the doctrine of Western land forces. It is important that non-military students understand how modern militaries articulate the problems of land warfare and their solutions, and this book aims to make the vocabulary used by modern militaries digestible, particularly in terms of historical precedents. This will enable the non-specialist to decipher the often opaque vocabulary of modern warfare. This book intends to make the complex theory and obscure language of contemporary military doctrine accessible to the layman, but another objective is to test the assumptions and question the conclusions contained within such doctrine. For military students, often inculcated into 'presentist' notions that all of today's challenges and solutions are unique, it is important to understand that modern concepts are often a recapitulation of long-standing historical themes; that these themes are subject to debate; and, indeed, that they are sometimes spurious. A great deal of military doctrine displays a tendency towards certainty, presenting as objective fact that which is open to debate. Concepts developed by the US military will, inevitably, represent a significant part of this analysis, given its leading role in developing new doctrine and technology and the significance of current American-led military operations for new thinking on warfare. However, the book also reflects on alternative visions and questions whether approaches that fit one set of circumstances will necessarily suit others.

Why land?

Why bother with a book specifically on *land* warfare? There are in existence many excellent general histories of warfare: these have the value of placing land warfare into its broader historical and military context.[1] In considering the lessons of land warfare, history is the only evidence that we have, but military histories tend to conflate land warfare with warfare as a whole, missing those things that are unique about the land environment. If these works provide an invaluable aid to understanding broad developments in war and warfare, they do not focus on the fundamentals of land warfare itself. For example, *The Oxford Illustrated History of Modern Warfare* examines sea and air warfare in discrete chapters, whereas it examines land warfare implicitly as part of more general discussions on the development of warfare as a whole. Where books do focus on land warfare, they often consist of narratives of operations rather than a structured examination of the nature of land warfare. This is a strange omission given that there are many books that look at warfare at sea and in the air as a discrete topic. As this book demonstrates, whilst there are clearly many important points of commonality between warfare in its different environments, there are also important differences. Put simply, the fundamentals of land warfare are often different from those in other environments because of land itself: land, in the form of terrain, exerts a powerful influence on the methods required to fight over it. Success in land warfare matters: because human beings live on land, taking land or defending it successfully can have decisive political effects. However, effectiveness at land warfare cannot be achieved simply by transplanting techniques from the air or maritime dimensions: success requires developing and executing techniques rooted in the specific demands of fighting in the land environment.

One source for developing an understanding of the fundamentals of land warfare is the doctrinal publications of armies themselves. As a practitioner's guide to the subject, doctrine has the value of reflecting an understanding of the land environment from the bottom up; doctrine publications are rooted in an attempt to systemise lessons from the past, understand contemporary environments and look at the demands of the future. However, doctrinal publications can be narrow, impenetrable and often present as fact highly contested assumptions about history. *Understanding Land Warfare* bridges the gap between doctrine and general histories of warfare, by identifying, but also contextualising, the way that land forces have sought to manipulate the land environment to their own advantage.

Why warfare?

This is a book about warfare, rather than war. War, in a general sense, has been defined by the political scientist Hedley Bull, as 'organised violence carried on by political units against each other',[2] and by the influential Prussian strategist Carl von Clausewitz, as 'an act of force to compel our enemy to do our will'.[3] Studies of war encompass broader questions such as their origins and termination; their political, economic, social, ethical and technological context. Warfare comprises the conduct of war; it is concerned centrally with fighting.

The levels of war

Traditional military thinking recognised two levels at which war was fought: strategy and tactics. In general, the latter equated with battles and their conduct, and the former to those activities, such as manoeuvre, planning and preparation, which brought forces to the battlefield.[4] Tactics was often viewed as more important than strategy, in keeping with the belief that political victory in war was achieved through decisive military battle. Thus, the German Field Marshal Helmuth Graf von Moltke (1800–91) argued that 'The demands of strategy grow silent in the face of a tactical victory and adapt themselves to the newly created situation'. For von Moltke, strategy was thus merely 'a system of expedients' shaped by tactical outcomes and without any enduring principles.[5] The growing complexity of war and strategic thinking has led over time to a re-conceptualisation of war according to three levels (see Figure I.1).

The strategic level

The strategic level can be defined as (see Box I.1 for some alternatives):

> The level of war at which a nation, often as a member of a group of nations, determines national or multinational (alliance or coalition) strategic security objectives and guidance, then develops and uses national resources to achieve those objectives.[6]

The strategic level of war is sometimes subdivided into a higher level and a lower level:

- The highest level is the **grand or national strategic level**. This may be defined as 'the application of national resources to achieve national policy objectives (including alliance or coalition objectives)'.[7] Grand strategy embraces the definition of policy objectives, alliances, allocates resources and directs the 'national effort'. The purpose of grand strategy is to provide unity of purpose and coherence to all aspects of national policy. It is also at this level that policy is created. The term 'policy' provides a convenient short-hand for the political ends towards which national strategic means are directed.[8] As the Prussian officer Otto August Ruhle von Lilienstern (1780–1847) noted in his *Handbook for Officers* (1817–18), 'war as a whole always has a final political purpose, which means that war is undertaken … in order to realise the political purpose upon which the State has decided'.[9]
- The next level down is the **military strategic level**: 'the application of military resources to help grand strategic objectives.'[10] This is the military element of grand strategy and is concerned with developing resources, constraints, means, military objectives and end states that will support the grand strategic objectives sought.

The operational level

Beneath the strategic level is the **operational level** of war. The operational level is usually associated with a theatre of operations and the planning and execution of military campaigns through **operational art**; it is 'The level of war at which campaigns and major operations are planned, conducted, and sustained to achieve strategic objectives within theaters or other operational areas'.[11] The operational level provides the vital link between the tactical conduct of war and the third layer in the hierarchy, the military strategic level of war. As the noted strategist Professor Colin Gray notes, 'the operational level is that intermediate stage between tactics and strategy where campaigns come together (or not, as the case may be)'.[12]

The tactical level

The lowest level is the **tactical level**, which can be defined as: 'The level of war at which battles and engagements are planned and executed to achieve military objectives assigned to tactical units or task forces.'[13] At this level battles and military engagements are fought and it is the sequencing of tactical battles and engagements in time and space which gives meaning to the next level up, the operational level of war.[14] Overall, then, the tactical and operational levels provide the instruments of strategy; and strategy is the instrument of policy goals.[15] Military strategy is executed through operations and tactics.

Box I.1 British definitions of the levels of war

The strategic level:

> The strategic level of warfare is the level at which national resources are allocated to achieve the Government's policy goals (set against a backdrop of both national and international imperatives).

The operational level:

> The operational level of warfare is the level at which campaigns are planned, conducted and sustained, to accomplish strategic objectives and synchronize action, within theatres or areas of operation. It provides the 2-way bridge between the strategic and the tactical levels.

The tactical level:

> It is the level at which engagements are fought and direct contact is made with an enemy or opponent.

(JWP 0-01, *British Defence Doctrine*, 2-6–2-7)

As the title suggests, *Understanding Land Warfare* focuses its analysis at the **tactical** and **operational levels** of war. This focus is not intended to marginalise the importance of the context in which warfare is conducted: it is this broader context that shapes the purposes of warfare, the means used and the strategies pursued. However, this broader context is already well serviced with a range of excellent publications. A focus on warfare is also dictated by *Understanding Land Warfare*'s concern with examining the land environment. The nature of warfare on land is conditioned by the exigencies of the terrain over which adversaries fight. An army's ability to utilise terrain, whether for fire, manoeuvre, or protection, in the context of the means available and the nature of its opponent, is a central issue in determining success and failure on the battlefield. It can sometimes be easy to forget that warfare matters. As the historian John Keegan argued:

> it is not through what armies *are* but what they *do* that the lives of nations and of individuals are changed. In either case, the engine of change is the same: the infliction of human suffering through violence. And the right to inflict suffering must always be purchased by, or at the risk of, combat – ultimately of combat *corps a corps*.[16]

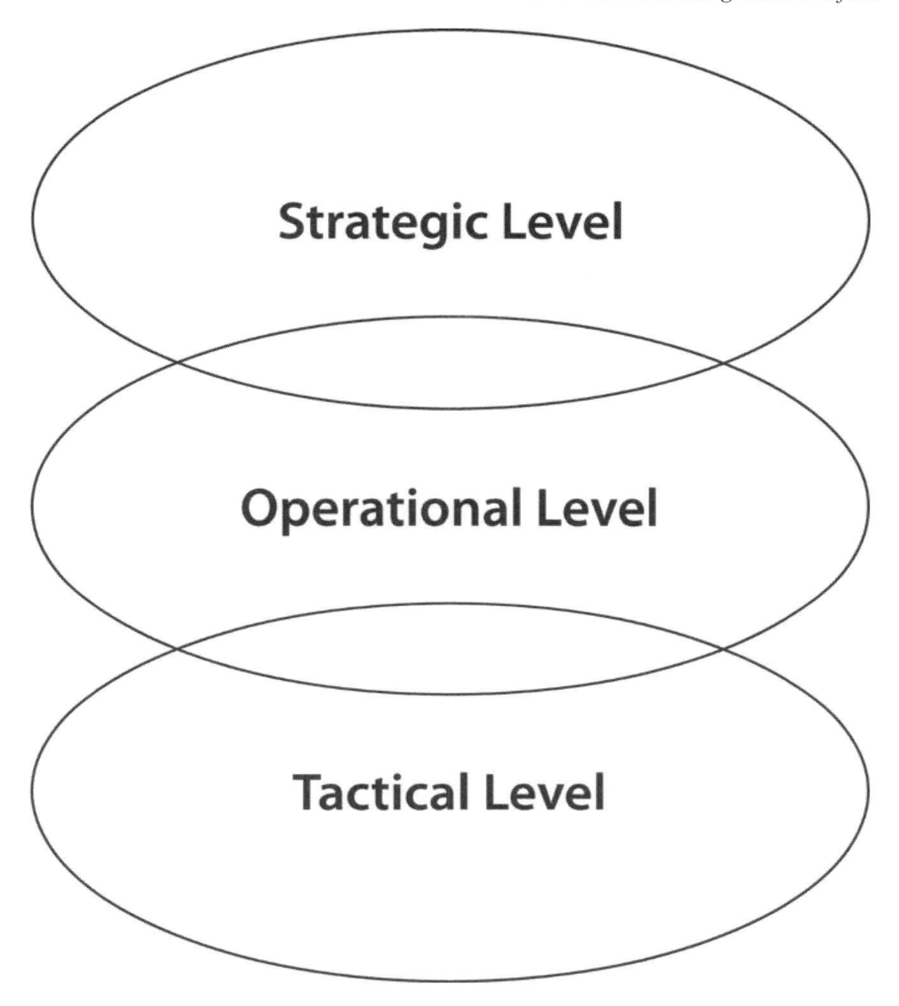

Figure I.1 The levels of war

The distinction between war and war*fare* has important ramifications for the scope of this book. This book does not provide a systematic analysis of war as a whole. This book investigates how it is that armies conduct military operations: it focuses on the theory and practice of conventional and unconventional operations.

What is modern land warfare?

Professor Colin Gray has commented that 'Just as punctuation alters meaning in literature, so does periodization by historians'.[17] For many, the word 'modern' might seem to imply 'recent', and so, from that perspective, this book might be one that should focus on warfare in the twenty-first century. However, the word 'modern' actually has a rather flexible meaning depending upon what one thinks that modernity entails. For some, the modern era began with the invention of the television; for some, it began with the Enlightenment of the late seventeenth and eighteenth centuries. The historian Theodore Ropp begins his *War in the Modern World* at the date 1415.[18] The focus of this book is land warfare since the beginning of the twentieth century. The choice of 1900 as a start date is, of course, simply one way of bounding the subject in hand, but actually, the use of the word 'modern' in the context of land warfare also has a rather specific meaning: it references the idea of the 'modern system' of land warfare. This idea, developed by the analyst Stephen Biddle in his seminal work *Military Power*, provides one of the central threads that runs through this book. The **modern system of warfare** describes a common approach adopted by armies through which they have sought to mitigate the growing effects of firepower: manipulating terrain is at the heart of this system. The modern system of warfare comprises, according to Biddle:

> a tightly interrelated complex of cover, concealment, dispersion, suppression, small-unit independent maneuver, and combined arms at the tactical level, and depth, reserves, and differential concentration at the operational level of war.[19]

Therefore, the term 'modern land warfare' also has a rather specific meaning: it is an examination of the tactical and operational level development of a specific system of land warfare. Because modern land warfare is focused on force employment – on the *methods* armies use – the technology deployed by an army does not determine whether an army is using the modern system: modern-looking armies with modern equipment can conduct themselves in decidedly 'unmodern' ways judged in terms of their tactical and operational level methods. There is no equivalent in naval or air warfare because their environments are more uniform and less subject to deliberate manipulation. With these points in mind, what is the scope of this book?

Contents

Understanding Land Warfare is divided into three parts:

Part I, 'The Development of Land Warfare', encompasses the first five chapters of this book. This part of the book sets out to introduce the key concepts associated with modern land warfare, and to chart how and why land warfare has developed since 1900.

Chapter 1, 'Land Warfare in Theory', explores the idea of 'land power' and explains why and how the land environment differs from the air and maritime environments. As part of this, the chapter examines the concepts through which debates on the conduct of land warfare have been expressed, from more traditional concerns such as fire, manoeuvre, attrition and suppression, through to jointery, combined arms and mission command. The purpose of this chapter is to give the reader an understanding of what these concepts mean, how they have been used, the difficulties associated with their implementation and the debates surrounding their validity as a means of understanding the basics of land warfare. This establishes a baseline of understanding for the subsequent chapters.

Chapter 2, 'The Development of Modern Land Warfare', looks at the way that **land warfare has changed** since 1900. This is an issue on which there is a profound lack of consensus. Focusing on the main schools of thought, Chapter 2 explores the different ways in which theorists have answered such questions as: Has land warfare developed in ways that are evolutionary or revolutionary? What have been the main agents of change: technology, new ideas, broader changes in society?

Chapters 3 and 4 examine, respectively, **modern tactics** and **operational art**. These chapters explore the historical conduct of land warfare since 1900, looking at the shift in the practical and theoretical conduct of military operations on land. These two chapters demonstrate the strong evolutionary underpinnings of land warfare today. Chapter 3 identifies the importance of the First World War as the catalyst for the emergence of modern land tactics. Chapter 4 highlights the significance of the period between the two world wars in the development of such important contemporary concepts as 'operational art' and 'manoeuvre warfare'.

Chapter 5 focuses on developing the idea of **variety in the forms** that land warfare can take. Whilst Chapters 3 and 4 chart the development of the 'modern system' of land warfare, this system has often been ignored by armies, or realised in different ways – but why is this? Answering this important question allows us to explore the importance of such factors as politics, economics and, in particular, the organisational culture of land forces themselves. There never has been a single kind of land warfare and there is unlikely ever to be so.

In Part II, 'What is Victory?', *Understanding Land Warfare* shifts its focus from conventional land warfare and focuses instead on forms of warfare that have variously been considered 'unconventional', 'low-intensity', or 'asymmetric': namely, counter-insurgency operations and peace operations. The unifying theme in Part II is the tension between the demands of military success on the one hand, and political success on the other.

Chapter 6, 'Counterinsurgency Operations', first examines the historical development of the theory and practice of counterinsurgency (COIN); it then goes on to investigate contemporary approaches to COIN. This chapter places a specific emphasis on explaining why it is that, despite well-developed theories for the successful conduct of these military operations, land forces have encountered **consistent difficulties** in executing them effectively.

Chapters 7 and 8 examine, respectively, 'Peace and Stability Operations' and the 'Challenges and Debates' associated with performing these tasks. Between them, these two chapters demonstrate simultaneously the **great flexibility of land forces**, which can undertake complex conflict prevention and conflict resolution tasks, but also **the tensions and difficulties** that can afflict these operations in practice. Peace and stability operations make demands on land forces that are not always compatible with the requirements of conventional warfare.

In Part III, 'Future Land Warfare', the final two chapters of this book examine some of the issues associated with preparing for land warfare to come. There is less consensus than one might suppose on what the challenges of the future will look like or, indeed, what kind of land forces might be required to meet them.

Chapter 9 looks at the **debates on the future of land warfare**. This discussion explores the main theories on the sources and implications of change: those who argue for a technologically focused Revolution in Military Affairs; others who see land warfare of the future conducted increasingly 'amongst the people'; and still others who see a hybrid future mixing conventional and unconventional warfare. Ultimately, this chapter shows that the future direction of land warfare is contested: we know much less than we think about what form land warfare might take in the future and what kind of land forces might be required to fight it.

The final chapter, Chapter 10, is entitled 'The Paradigm Army', and it uses the **US army as a case study** in the themes and issues developed by the preceding chapters of the book. It answers such questions as: How has the United States conceived of future land warfare? What doctrines, roles and structures has it created to realise this vision? How have these forces actually performed? Examining land warfare from a US perspective serves to reinforce the complex difficulties associated with conducting land operations successfully, but it also raises other interesting questions: How far can history provide a reliable guide to the future? How far are the dominant narratives on the likely shape of future land warfare a political or cultural expression of how we would *like* warfare to be? What do Iraq and Afghanistan really say about the sorts of land forces that we might need for the future?

Key themes

Understanding the complexities inherent in land warfare matters a great deal. The contemporary period has been an extraordinarily fertile one in terms of thinking on land warfare, especially in the United States. The experiences of the Gulf War, Kosovo, Iraq and Afghanistan have resulted in major debates on the sorts of future direction that land warfare might take. New technology and new doctrinal concepts have resulted in the articulation of a variety of ambitious and expansive developmental road maps for the US army. Major developments in US military doctrine such as those contained in the 2008 revision of Field Manual (FM) 3-0 *Operations*, and new manuals, such as FM 3-24 *Counterinsurgency* and FM 3-07 *Stability Operations*, have prompted debates about the tensions between more traditional land warfare roles and those of nation building; on the features of the future operating environment; on the meaning of 'balance' in land forces; on the ethos required to fight future wars; on the role of technology; and the significance of cultural awareness. At the same time, other states such as China, Russia and India have also been developing revised doctrines to cope with perceived changes in the context in which land warfare will take place.

However, fertile thinking is no guarantee of effective military performance. It is often the case that the failings of European armies in 1914 are blamed on military narrow-mindedness: a blinkered conservatism and lack of receptiveness to change. As this book shows, this was not the case: armies of the time did embrace new thinking. As land forces attempt to adapt at the start of the twenty-first century to conditions as complex and ambiguous as those of the early part of the twentieth, it would be unwise to assume that we have any better an understanding of the likely demands of future warfare on land. With this point in mind, collectively, the chapters in this book illuminate a number of recurring themes:

- First, *Understanding Land Warfare* demonstrates the **complexity** inherent in land warfare: the variety of ways in which land warfare is fought; the competing theoretical debates; the tensions and trade-offs.
- Second, it is clear that land warfare is **difficult to do well**: in distinguishing between the nature of land warfare and its character, this book identifies the problems in translating plausible-looking military theory into effective military practice.
- Third, in examining the interplay between change and continuity, this book argues for an **essentially evolutionary** explanation for the development of land warfare: less has changed between 1916 and 2016 than might be supposed.
- Fourth, despite the focus on the land environment, *Understanding Land Warfare* highlights the growing **importance of co-operation** – not just in terms of combined

arms within land forces but also in relation to inter-service interaction (joint warfare) and multi-national operations (combined warfare).
- Last, land warfare is marked by its **variety in form**: context shapes often radically different approaches to the conception and prosecution of land warfare. Debates on the future of land warfare too often reflect the projection forward of immediate Western experiences in Iraq and Afghanistan as a single narrative on the future of land warfare. In reality, land warfare, like war generally, is in part an expression of its political, economic and societal context. At any point in time there are likely to be many different ways in which land warfare is fought.

Thinking on land warfare is in a state of flux not seen since the period between the world wars. The experiences of Western forces in Iraq and Afghanistan have complicated already divergent post-Cold War debates on the relevance of past experience and the shape of future challenges. *Understanding Land Warfare* provides a systematic examination and critique of these ideas ranging from the broad debates about conventional versus unconventional futures (including a detailed evaluation of specific visions of the future) through to debates on the relevance of traditional conceptions of land warfare created by stability operations. *Understanding Land Warfare* is not intended as a replacement for a firm grounding in the history of land warfare, but it will, it is hoped, complement that history and aid in understanding the issues that stand at the heart of today's debates.

Notes

1 See, for example, Geoffrey Parker's edited *The Cambridge History of Warfare* (2005), and Jeremy Black's *War and the World: Military Power and the Fate of Continents* (1998).
2 Hedley Bull, *The Anarchical Society: A Study of Order in World Politics* (London: Macmillan, 1977), 184.
3 Carl von Clausewitz (edited and translated by Michael Howard and Peter Paret), *On War* (Princeton, NJ: Princeton University Press, 1976), 75.
4 Beatrice Heuser, *The Evolution of Strategy: Thinking War from Antiquity to the Present* (Cambridge: Cambridge University Press, 2010), 6.
5 Colin Gray, *Modern Strategy* (Oxford: Oxford University Press, 1999), 47.
6 Joint Publication (JP) 1-02, DOD Dictionary of Military and Associated Terms (8 November 2010).
7 Joint Warfare Publication (JWP) 0-01, *British Defence Doctrine*, 2nd edition, 1–8.
8 Gray, *Modern Strategy*, 20.
9 Heuser, *The Evolution of Strategy*, 11.
10 Ibid.
11 JP 1-02.
12 Quoted in Stuart Griffin, *Joint Operations: A Short History* (London: Training Specialist Services HQ, 2005), 12.
13 JP 1-02.
14 JWP 0-01, 1–8.
15 Gray, *Modern Strategy*, 20.
16 John Keegan, *The Face of Battle* (London: Jonathan Cape, 1976), 30.
17 Gray, *Modern Strategy*, 172.
18 Theodore Ropp, *War in the Modern World* (Durham, NC: Duke University Press, 1959).
19 Stephen Biddle, *Military Power: Explaining Victory and Defeat in Modern Battle* (Princeton, NJ: Princeton University Press, 2004), 28.

Suggested reading

Jeremy Black, *Rethinking Military History* (London: Routledge, 2004). An analysis of military history that provides an antidote to Western-biased and technology-centric accounts of the development of warfare.

Jeremy Black (ed.), *War Since 1900: History; Strategy; Weaponry* (Thames and Hudson, 2010). A lively and accessible introduction to the development of warfare since the beginning of the twentieth century.

Azar Gat, *A History of Military Thought: From the Enlightenment to the Cold War* (Oxford: Oxford University Press, 2001). A history of military thinking and strategic ideas that provides an important intellectual context for the practical development of land warfare.

Geoffrey Parker (ed.), *The Cambridge History of Warfare* (Cambridge: Cambridge University Press, 2005). This charts the development of Western warfare from the sixth century BC through to the future.

Charles Townshend (ed.), *The Oxford Illustrated History of Modern War* (Oxford: Oxford University Press, 1997). This book provides a collection of illuminating chapters that discuss developments in the practice and context of war from the eighteenth century onwards.

Part I

The development of land warfare

The first part of this book investigates the development of modern land warfare. Despite the many ways in which land, sea and air warfare are connected, they have many important differences. Our first step is to examine the character of warfare on land, its theories and concepts. This is the task of Chapter 1. Having outlined the theoretical basis of land warfare, the next four chapters examine the debates and controversies surrounding how we think about the development of land warfare; the development of the fundamentals of modern land warfare at the tactical and operational levels; and the factors that shape variable approaches to land warfare in the modern world.

1 Land warfare in theory

Key points

- Land warfare has many features that make it unique from warfare at sea or in the air.
- The root of these differences lies in the nature of land itself, in the form of ground or terrain.
- Land warfare is not wholly unique, however, and the effective prosecution of land warfare cannot be separated from such issues as strategy or effectiveness at joint operations.
- Land warfare cannot reliably be reduced to a simple set of principles; its conduct is conditioned by a range of difficult trade-offs.

Skill in the prosecution of land warfare matters. Whether one looks at the European armies of 1914–15, mired in tactical stalemate on the Western Front, or the Iraqi army in 1991 and 2003 locked in an unequal struggle against Coalition forces, problems in relative battlefield performance translate directly into high casualties, immense strategic difficulties and often decisive defeat. But why is land warfare so challenging? This first chapter provides the foundations for an answer to this question by examining the key concepts and ideas that underpin an understanding of land warfare. The recurrent theme throughout this discussion is **complexity**. It is a banal but fundamental observation that the key distinction between war on land and war in the other environments is **land itself**. Land, in terms of the physical ground upon which warfare is fought, has innate attributes that make it distinct from the sea, air, or electromagnetic dimensions; these attributes shape the character of those forces required to fight on it. The importance of terrain, its nature and variability make difficult and often changing demands on the forces required to fight over it.

The ideas presented in this chapter provide a basis of understanding for the debates presented in the remainder of this book. The chapter is divided into three parts. The first part examines the uniqueness of land warfare based upon the distinctiveness of the land environment and the consequences that this has had for the nature of warfare on land. The second part of this chapter examines the unity of land warfare: here the purpose is to identify the importance for the effective conduct of land warfare of thinking about how land warfare should be orchestrated at different levels and in relation to other instruments of power. Third, building on the previous two sections, the chapter explores some of the key concepts and principles that have developed as guides for the exercise of land warfare in the modern period. The ideas presented in this chapter are developed more directly in Chapters 3 and 4, which investigate how the themes explored here have influenced the development of a definably modern style of land warfare.

What are the characteristics of the land environment?

We focus first on examining what it is that is distinctive about land warfare. The effective prosecution of land warfare rests in part on understanding that land warfare cannot be approached in exactly the same way as war in the other environments. The root of the differences lies in the nature of land itself. Land embodies a number of **attributes** that shape the prosecution of warfare upon it. These attributes are what makes land warfare different from war on sea or in the air. Some of these attributes are, in a quite literal sense, easy to see; others are less obvious.

Box 1.1 Land forces and land power

For convenience, this book uses such terms as 'land forces' and 'armies' interchangeably. Strictly speaking, however, they are different. **Land forces** comprise ground-oriented military organisations including **armies**, marines, reserves, militias and so forth. So, all armies are land forces, but not all land forces are armies. Land power is the ability to exert influence on or from the land. Land forces constitute a crucial component of land power, but air and maritime forces can also make a vital contribution.

Variability

One of the most mundane observations regarding terrain is also one of the most important militarily: it is **highly variable**. Land is not a consistent medium. Unlike the sea or air, land embodies great variety in geography, vegetation and population density, different combinations of which can create difficult challenges for **land forces** (see Box 1.1). For example, mountains, jungles, urban areas and forests are problematic operating environments for vehicles; communication and visibility are difficult; movement is often channelled (see Box 1.2, for example). Deserts can pose fewer problems in relation to visibility and manoeuvre, the latter especially for tracked vehicles, but desert sand poses problems for the reliability of equipment and the stamina of personnel; navigation can also be a difficulty. Land's variability is magnified by the **effects of climate and weather**: excessive heat, snow, rain, or mud can transform the characteristics of even flat terrain, constraining manoeuvre, limiting visibility and/or depressing morale. Moreover, this diversity can be **heavily concentrated**: the same battlefield can exhibit widely differing terrain within a relatively limited distance; the impact of this terrain can vary depending on its location or the distance between features. The theme of variability can take many complex forms:

- First, conditions can be variable even **within a given class of terrain**: the 'desert warfare' of North Africa in the Second World War, for example, encompassed everything from open expanses of soft sand in the east through to the rocky ridges of Tunis with their cork and scrub woods.

- Second, the same terrain can change radically in its effects through **climatic conditions** or **human endeavour**. The open steppes of Russia that favoured the manoeuvre of German forces in the summer of 1941, were transformed later in the year first by heavy rain that created seas of mud, and then by snow and freezing temperatures. The Germans transformed the flat floodplains of Holland in 1944 by opening the sea dykes and flooding them.
- Third, geography and climate have **no uniform impact**: heavy cloud may have no effect on ground forces, but may reduce significantly the effectiveness of air mobility or close air support. Heavy woods or high hedgerows may create difficult conditions for vehicles to manoeuvre and fight, but they can increase the ability of infantry to move unseen. Many of the important effects of geography and climate are also **psychological** and depend to an extent on the training, morale and resilience of the forces concerned. Often, the challenges of combat can take second place to the misery of heat, cold, rain and/or isolation (see Box 1.3).
- Fourth, and related to the previous point, terrain and climate **do not necessarily affect both belligerents equally**. It is impossible for a given army to be perfectly adapted for all possible geographical and climatic conditions: thus, in any given context, one side is likely to be better prepared than the other – often the army fighting on home ground. Difficult terrain and climate can therefore constitute a relative advantage for an army. For example, in the early stages of Winter War against the Soviet Union (1939–40) the Finns were better prepared for fighting in both snow and forests than their opponents, inflicting heavy defeats on ponderous Soviet mechanised forces. Preparation for particular locales is complicated by the fact that military organisations must often fight simultaneously in many different conditions. In the Second World War, for example, the US army had to fight in such diverse environments as the jungles of New Guinea, the deserts of North Africa, the mountains of Italy, the hedgerows of Normandy, the forests of Germany and the coral atolls of the Pacific islands. One recurrent source of **achieving surprise** has been to use terrain that has been assumed by the enemy to be impassable: this was the stratagem used by the Germans in 1940 and 1944 in the Ardennes forest; by the Japanese in 1941 in their landward jungle assault against British forces in Malaya; and by the Iranian use in 1984–85 of the marshes of Hawr al Hawizah, where flooding prevented the Iraqis from bringing their tank superiority to bear.[1]

Box 1.2 Urban operations

In land warfare, urban areas are often crucial because they are significant political and economic centres that act as a hub for government services and transport infrastructure. But as the Germans found in Stalingrad in 1942 and the Russians in Grozny in 1994–95 and 1999–2000, urban operations are usually costly and time-consuming, especially for the attacker. Buildings obstruct observation, communication and navigation; they channel movement and obstruct manoeuvre; they make combined arms more difficult, and make indirect fire support challenging; psychologically, urban environments are demanding, with the prospect of unexpected, close quarter attacks ever present. A British officer in the Second World War called urban operations 'the most tiring and trying type of fighting [even] under the best conditions'; another argued that 'Street-fighting, like jungle-fighting, presents control problems ... it is really a science or skill of its own and should be practiced carefully beforehand.' Certain kinds of common techniques have developed to deal with urban operations: for example, squads of infantry with grenades and sub-machine guns move through houses on each side of a street, clearing them of defenders from the top down using explosives to blow a hole in party walls – so-called 'mouse-holing'.

(John Ellis, *The Sharp End*, 91)

Box 1.3 The psychological effects of climate

I knew for certain that the worst part of the war was not the shooting or the shelling – although they had been bad enough – but the weather, snow, sleet and rain, and the prolonged physical misery which accompanied them.

(US Lieutenant, Italy, 1944, in John Ellis, *The Sharp End*, 23)

Variability is one obvious characteristic of land as an environment, but there are other important themes to consider as well.

Political significance

Land has enormous **political significance**; people live on land and not on the sea, or in the air or in the electro-magnetic spectrum. Belligerents in war are therefore territorially defined and so it is land that, of all the environments, has the highest **symbolic and physical value.**[2] Because of this, in the end the highest order security interests of political actors tend to be associated inextricably with territory (see Box 1.4). Even where the principal focus of a war is in another environment, the political objective of war still remains the exercise of power over an adversary that is territorially constituted. Navies and air forces themselves rely on territorial bases for their continued functioning.

Box 1.4 The importance of land warfare

Since men live upon the land and not upon the sea, great issues between nations at war have always been decided – except in the rarest of cases – either by what your army can do against your enemy's territory and national life, or else by the fear of what the fleet makes it possible for your army to do.

(Sir Julian Corbett, *Some Principles of Maritime Strategy* (London: Conway Press, 1972), 14)

Therefore, whether it is the surrender of France to Germany in 1940, the taking of Saigon by North Vietnam in 1975, or the fall of Kabul to the Northern Alliance in 2001, the ability to take and hold an opponent's territory, or to threaten credibly to do so, remains a core metric of success in war.

Opacity

Land is **opaque**: whilst one can see, communicate or fire over land, it is difficult to do so **through** it. For example, in general, few units are ever likely to be able to see all of a battlefield all of the time. Even nominally flat terrain is, in reality, opaque: on the 'flat' North German Plain, for example, variations in elevation and vegetation mean that within 1,000 yards of a typical weapons position, around 65% of the terrain is invisible to the viewer.[3]

Resistance

Another feature is that land is a **resistant** medium: manoeuvre on land is **slower** relative to the air or sea environments and tends physically to be **much more wearing**. In land warfare, the problems of fighting the enemy can be dwarfed by the difficulties in moving and sustaining forces. For example, in its invasion of Russia in 1941 the German army was effectively 'de-modernised' by the enormous losses in vehicles and equipment caused by the extended lunge through the geography of western Russia.

Mutability

Last, land is a **mutable** medium: it can be rendered **less resistant** as a medium through the construction of roads, bridges, railways or canals, or it can be made more so by the destruction of the same. Land can be excavated, cleared, built in and built on to create defensive works or clear fields of fire that **render it more or less opaque**. In the Second World War, for example, the Japanese defenders of the 8-square mile island of Iwo Jima strengthened their defences with 800 pillboxes and 3 miles of tunnels.[4] This feature of land explains the continued relevance to armies of formations such as engineers and pioneers, as well as the pervasive influence of fortifications and entrenchments on the conduct of land warfare throughout history.[5]

Thus, in land warfare the ground itself is an omnipresent influence: **land shapes**, **channels**, **facilitates**, and **mitigates**. Land often defines the goals of war either in terms of political objectives or in terms of important ground that can be leveraged advantageously for further operations. The characteristics of land create **key terrain**: avenues of approach, obstacles, choke points, points of observation, or fields of fire. Indeed, land shapes the fundamental conduct of warfare on and over the ground: as one author notes, 'Fighting is mostly a matter of moving, hiding and shooting. The ability to do all three is conditioned by the lie of the land'.[6] We will return to these points many times over the course of this book, but the characteristics of land also have important consequences for the forces that fight on it.

The characteristics of land forces

The characteristics of the land environment have direct consequences for the nature of the forces required to fight on it.

Complexity

One consequence of the characteristics of the land environment is that land forces are **complex**. The need to control ground and to exploit the vagaries of terrain for the purposes of dispersal and camouflage has meant that whilst equipment 'platforms' such as tanks and armoured fighting vehicles have become progressively more significant since the First World War, land forces still tend to be made up of much larger numbers of disaggregated combat elements than navies or air forces: 'land power, of all forms of power, has perhaps the greatest density of such moving parts.'[7] This point is reflected in land forces in the continued importance of the role of infantry in land warfare. Whereas navies and air forces 'man the equipment', armies tend instead to 'equip the man'.[8]

Persistence

A second characteristic of land forces is that they are **persistent** (see Box 1.5): they can occupy terrain and remain there, sometimes indefinitely. Aircraft, on the other hand, can have only an ephemeral presence over land conditioned by their limited fuel and payloads, while ships, of course, are restricted to the sea.

Box 1.5 Persistence

> You may fly over a land forever; you may bomb it, atomize it, pulverize it, and wipe it clean of life – but if you desire to defend it, protect it, and keep it for civilization, you must do this on the ground, the way the Roman legions did, by putting your young men into the mud.
>
> (T.R. Fahrenheit, quoted in Milan Vego, *Joint Operational Warfare*, V-39)

Versatility

Land forces are also **versatile**: because land forces are personnel intensive, they are flexible. By focusing on changes to non-technological variables such as training, doctrine and military ethos, land forces can meet new military challenges by developing new techniques. For this reason, land forces have great relevance across the whole spectrum of warfare, from high-intensity conventional operations through to stability and security tasks (see Part II of this book).

Resilience

Land forces are **resilient**: disaggregation, the mitigating effects of terrain and adaptation in techniques can make it difficult to destroy land forces entirely. Moreover, with less relative reliance on capital-intensive platforms than navies or air forces, land forces can be regenerated more swiftly.

Objectives

Finally, the **military objectives** pursued by land forces tend to be physical in character. These objectives can encompass seizing important urban centres (not least the opponent's capital), taking economic centres, creating bridgeheads, avoiding the enemy, or destroying the opponent's army.

The strengths and vulnerabilities of land forces

The attributes of the land environment and the characteristics of land forces together shape a range of theoretical **advantages** and **disadvantages** for land power: these are related to one another and the degree to which the former may outweigh the latter is determined by context.[9]

Control

Crucially, land power has the ability to **exercise control** over land. Seapower and airpower can deny control through the application of firepower, thus preventing the enemy from taking ground, but they cannot physically take and hold territory. As General Norman Schwartzkopf noted, 'There is not a military commander in the entire world who would claim he had taken an objective by flying over it'.[10] Control of territory enables positive strategic options: control can be exploited for purposes beyond those simply of removing territory from the use of the enemy; control is therefore the foundation for imposing one's will on the enemy.

Nevertheless, land power is a **dependent** instrument: controlling land does not mean that one controls everything – it is a means not an end and controlling the wrong territory, or controlling it in the wrong way, or failing to exploit this control effectively might render control irrelevant or counter-productive. Therefore land power is dependent upon broader strategy for its effectiveness. Land power can also depend crucially upon the support of sea and air forces to exercise control: without seapower, for example, land forces may be unable to reach the necessary theatre of operation.[11]

Durability

Another advantage of land power is that it is **durable**: the control exerted by land forces can be temporally sustained, sometimes indefinitely, unlike the transitory effects of air- and seapower which may interact with land but must then retreat to their own environments. This provides opportunities for complex and sustained political and military interaction with the local population and/or enemy forces.

However, this durability can also translate into **vulnerability**: armies, unlike sea and air forces, have to occupy territory and this will result in prolonged contact with often hostile forces. This vulnerability can also extend to the lines of communication and logistics of land forces operating in hostile territory.[12] Moreover, durability also leads to **friction**: except where air mobility or sea or river transport is available, armies manoeuvre relatively slowly, often having to halt periodically to secure decisive geographic points and/or to consolidate.

Adaptability

Land power is also **adaptable**: land power is of central importance across the whole spectrum of war from high-intensity conventional operations through to complex political tasks associated with stability and security operations. In essence, land power gives political actors a wider set of strategic options than seapower or airpower.

However, land power is not infinitely adaptable – it still suffers from its **specificity**. Politics and geography create multiple contexts for the use of land power, but no land forces are equally capable in every situation. Limitations of time, resources, politics, training, experience, history and mindsets make land forces better suited for some contexts than others. So, the German army was very successful in 1940 against British and French armies in the terrain of Western Europe, but the challenging physical and psychological conditions of the Eastern Front proved much more problematic.[13]

Decisiveness

Cumulatively, the positive factors identified above create one overarching advantage: land power has the **power of decision**. As Admiral J.C. Wylie notes, 'The ultimate determinant in war is the man on the scene with the gun. This man is the final power in war. He is control. He determines who wins'.[14] In general, overall victory in war will usually require that a belligerent demonstrate an eventual superiority in land power. For this reason land power 'is *fortissimus inter pares*, the strongest among equals. It alone can achieve the greatest strategic and political effect'.[15]

On the other hand, land warfare also tends to embody **high costs**: because land power is exercised where people live, its exercise is tied inextricably both to higher levels of risk for the intervening forces and to the political and military consequences of the responsibilities associated with exercising control over human inhabitants. The exercise of land power can therefore be associated with profound controversies stemming from domestic and international scrutiny of the moral, ethical, legal, political and military conduct of land forces.[16]

Thus, the effective conduct of land warfare requires an understanding of the unique attributes of land itself. These attributes mean that land forces possess certain characteristics that can produce strengths and weaknesses, depending on the context. The effective use of land forces will require developing approaches that will minimise land power's weaknesses and maximise its strengths.

What connects land warfare to warfare in other environments?

However, at the same time as we need to understand land warfare's unique characteristics we also need to understand that the successful use of land forces requires an appreciation of those factors that should link warfare in its different environments. Trying to fight land warfare in isolation of these factors is a recipe for failure. As the contemporary strategist Colin Gray comments, 'To be good at fighting is important, but rarely sufficient'.[17]

Strategy and policy

Thinking back to the Introduction to this book, we identified that land warfare is fought at the tactical and operational levels. However, there is a **constant interaction between the various levels of war**: tactical events can have strategic implications; witness, for example, the profound political consequences of the abuse of detainees at Abu Ghraib prison in Iraq in 2003–04 when it was revealed by the media. The strategic level of war, in particular, can have a decisive effect on the lower levels of warfare in many ways:

- **Policy objectives**: the objective of land warfare is not simply to defeat the enemy's military capabilities, it is to **achieve policy goals** set at the strategic level of war – as US military doctrine asserts: 'The ultimate objective of military power is to achieve national policy objectives.'[18] However, what if these objectives are too ambitious or perhaps unattainable? The political purposes to which war is directed, as expressed in policy, are often undeclared, vague, poorly chosen and/or change over time, making it difficult to frame useful military objectives at the tactical and operational levels.

- **Strategy**: the connection between military means and political goals is provided by strategy: 'the bridge that relates military power to political purpose.'[19] If policy provides the 'ends', and military forces the 'means', then strategy provides the 'ways' that link the two. Strategy is reflected, amongst other things, in the **constraints, restraints** and **imperatives** placed on land operations, but what if the strategy chosen is confused, or simply wrong? Under such circumstances, there will be little meaningful link between tactical and operational land warfare and the objectives that it is supposed to achieve.
- **Long-term success**: the contemporary strategist Edward Luttwak comments: 'Because ultimate ends and means are both present only at the level of grand strategy, the outcome of military actions is determined only at that highest level: even a most successful conquest is only a provisional result.'[20] In other words, even great tactical and operational military successes may be undone if there is no strategic level thought given to how military victory can be translated into a long-term sustainable peace.

For these reasons, developments at the strategic level of war can have profound consequences for the tactical and operational level conduct of land operations. For example, strategic level failures in Vietnam undermined decisively the conduct of tactical operations, as one US officer noted:

> There wasn't anything different in Vietnam from the day I arrived to the day I left. We were fighting over the same terrain, in the same areas ... And the more you did this, you kept wondering, what the hell am I doing here? I'm not getting anywhere. It's the same thing over and over again.[21]

Equally, in the Second World War Germany was famed for the quality of its tactical and operational military performance, but it still lost the war, not least because of lamentable decision making at the strategic level. Ultimately, 'mistakes in operations and tactics can be corrected, but political and strategic mistakes live forever'.[22]

The difficulty of strategy and policy

Strategy and policy provides a crucial context that shapes the success or failure of land operations, but framing the right objectives and creating effective strategy is difficult for many reasons.

- War is a **human activity**: it is an art; it cannot be reduced to scientific principles that guarantee success.
- War is a **relational activity**: it is fought against an opponent, or opponents, who think, react and adapt.
- War is the **realm of friction**. Friction describes the accumulated influence of factors such as complexity, luck, fallible human judgement, enemy actions, weather, geography and so on which make war easier to plan than to execute effectively.
- War is **polymorphous**: the nature of war remains the same, but the character of every war is different. What works in one war may well not work in another because of

the operation of different political, economic, geographical or cultural factors.[23] Land warfare cannot be reduced to 'one-size-fits-all' templates.

There has emerged over time, by way of some compensation for these problems, a body of strategic theory that provides analysis and guidance in the face of the difficulties of strategy. This includes the works of such ancient writers as Sun Tzu, the Roman Vegetius, the Byzantine Emperor Maurice; officers such as Captain Basil Liddell Hart, Admiral J. C. Wylie, Raoul Castex and General Andre Beaufre; the American academics Edward Luttwak, Bernard Brodie and Thomas Schelling. Perhaps the dominant influence, certainly on Western militaries, has been the Prussian theorist **Carl von Clausewitz** (1780–1831) and his magnum opus, *On War* (see Box 1.6). However, in general, strategic theory is better at highlighting the nature of the problems of strategy than it is in providing concrete solutions. Strategy remains difficult to do well; as Clausewitz notes, 'Everything in strategy is very simple, but that does not mean that everything is very easy'.[24]

Box 1.6 Clausewitz's insights

The Prussian Carl von Clausewitz was a serving officer in the Napoleonic Wars and went on to direct the Prussian War Academy. His key work, *Vom Kreig* (On War), was published posthumously, and in an unfinished state, in 1832. More often quoted than actually read, Clausewitz's work was an attempt to develop a general theory of strategy. It is littered with illuminating ideas and concepts. These include:

- The importance of education: The proper use of theory for officers is to educate and to give insight, not to provide rigid templates for action.
- The complexity of war: Clausewitz notes that 'in war more than any other subject we must begin by looking at the nature of the whole'. War must be studied holistically so that its relationships and interdependencies can be understood. Clausewitz highlights, for example, the Trinitarian nature of war: war being composed of 'instrumental rationality'; 'primordial violence, hatred, and enmity'; and the 'play of chance and probability.'
- The importance of politics: War is an instrument of policy: this relationship is distinctive but also subordinate and so, whilst war has a distinctive 'grammar' it does not have its own logic. Politics permeates war, gives it meaning; errors in policy can be terminal for the likely success of a war.
- The distinction between war in theory and war in practice: Clausewitz highlights the central role of friction in warfare, noting that 'Action in war is like movement in a resistant element.' He also highlights the problems with relying on information and intelligence as a panacea.
- The human element in war: Clausewitz emphasises the importance of the moral factor in war. This focus on the importance of people extends beyond morale into the knowledge and attributes that contribute to success in war. Politicians, for example, need to have some understanding of military affairs; the best commander, the military genius, requires intuition and moral courage.
- Conceptual tools: Clausewitz identifies the importance of what he calls the centre of gravity of the enemy: 'the hub of all power and movement, on which everything depends': this centre of gravity can be military, political or economic and can be applied to all of the levels of war. As a tool it provides a focus of effort,

concentrating thinking on the enemy's key vulnerabilities. Clausewitz also highlights the significance of the culminating point: the point at which the power of the attack ceases to overmatch that of the defence; in doing so, Clausewitz highlights the paradoxical logic of strategy – the tendency for success to breed failure because it can encourage a belligerent to overreach itself.

(Gray, *Modern Strategy*, 92–99)

Joint warfare

The need to think about land warfare in a holistic way is a theme also evident in relation to the importance of joint warfare – operations that involve the participation of two or more of the armed services (see Box 1.7).[25] Joint warfare is not new, but it has become increasingly important since 1900 on the back of **technological developments** such as the introduction of aircraft, and the **growing complexity** of warfare that has driven functional jointery in areas such as logistics, communications, reconnaissance, close air support, air defence and so forth.[26] Military campaigns are unlikely ever only to require one of the services. As General Dwight D. Eisenhower commented in 1946:

> Separate ground, sea, and air warfare is gone forever. If we ever again should be involved in war, we will fight with all elements, with all services, as one single concentrated effort.
>
> (General Dwight D. Eisenhower, 17 April 1946)[27]

Box 1.7 Joint warfare definition

'Joint' connotes activities, operations, and organizations in which elements of two or more Military Departments participate. Joint matters relate to the integrated employment of military forces in joint operations, including matters relating to (1) national military strategy (NMS); (2) strategic planning and contingency planning; (3) command and control (C2) of joint operations; and (4) unified action with the US interagency and intergovernmental communities, nongovernmental organizations (NGOs), and multinational forces (MNFs) and organizations.

(Joint Publication 1, *Joint Doctrine for the Armed Forces of the United States* (25 March 2013), 1–2)

Indeed, the land environment is the least exclusive of the environments, especially in terms of the constant relevance of airpower to land operations. Joint warfare harnesses the principles of substitution and synergy:

- **Substitution**: the different services can cover one another's weaknesses. Airpower, for example, cannot hold ground. Land forces can, but air forces have much greater reach than land forces.
- **Synergy**: joint warfare can magnify the combat effectiveness of each of the services. For example, one way an enemy can reduce the effects of airpower is to adopt a more dispersed formation; this, however, will then make them more vulnerable to an attack by land forces.

Typically air and maritime forces can provide a wide range of additional capabilities in a campaign. Airpower, for example, can provide:

- Air control: protection from enemy air attack.
- Close Air Support (CAS): air attack on battlefield enemy targets.
- Interdiction: destroying or interfering with enemy communications infrastructure and movement.
- Mobility: tactically, often through the provision of helicopters, and strategically, through large transport aircraft.
- Intelligence, Surveillance and Reconnaissance (ISR): through over-flights of enemy positions.

Maritime forces also often provide crucial capabilities. Maritime air assets can provide the capabilities listed above, but naval forces can also give land warfare:

- Fire support: through the provision of naval gunfire and precision-guided munitions.
- Mobility and flexibility: through sea or river transport.
- Logistic support: through bulk transport.
- Air defence: using integral surveillance and anti-air capabilities.

The challenges of joint warfare

Despite the evident advantages of integrating the land, sea and air environments, joint warfare has **historically proven difficult to execute**. Ultimately, joint warfare requires a **strong understanding** of the distinctive attributes of each service and how these can be integrated to achieve a common purpose. In the United States, for example, joint structures are present at the military strategic level through the Joint Chiefs of Staff, and at the operational level through the creation of a single geographic commander (Central Command, Pacific Command, etc.) and theatre level commanders to take operational control of all forces within a campaign.[28] In theory, effective jointery can be assured through such initiatives as:

- Extensive **joint training**.
- Creating a specific **doctrine** for joint operations.
- Promoting exchanges between the services to create a **proper understanding** of the strengths, limitations and perspectives of other services.
- A **command structure** that facilitates unity of effort between the services, and service leaders with an ability to transcend a single-service perspective.

Historically, however, it has proven difficult to create this understanding, and even when the pressure of combat has created progressively more effective jointery, as happened amongst the Allies in the Second World War, in peacetime good joint practice often withers away. There are many reasons for this:

- **Organisational factors** relating to culture and politics. As one commentator has noted, 'jointness is as much a state of mind as a method of prosecuting war'.[29] However, creating this state of mind is difficult because each individual service has its own values, procedures and interests that create fundamental differences in outlook between armies, navies and air forces.

- **Resources.** Particularly in the financially constrained environments that can apply during peace, services often tend to prioritise core single-service capabilities and training at the expense of those for joint warfare since the latter are often viewed as a second order commitment.
- Certain **operational contexts** have also been problematic for joint operations. For example, contemporary operations in Afghanistan from 2001 and Iraq from 2003 onwards seemed to involve increasingly complex scenarios in which jointery had to be practised in complex multi-national and multi-agency environments.
- The need for **single-service expertise**: too rigid an insistence on the application of jointery can erode the single-service expertise and cohesion upon which effective joint operations depend. Distinct service cultures facilitate mastery of the demands of each separate environment: without this mastery, joint operations are built on foundations of sand.

In consequence, despite the advantages to a land campaign of effective joint action with air and maritime forces, joint warfare has often been conducted poorly, with the armed services failing to sustain the good practice learnt in earlier periods. Thus, the excellence achieved eventually by the United States in its Pacific campaign from 1941–45 did not survive the end of the Second World War. During the Korean War, the US air force, focused on its strategic bombing role, was unenthusiastic about providing close air support for ground forces and the US Marine Corps and army ran associated, but not joint, land campaigns. The Vietnam War from 1965–73 was even less joint than Korea, being run by a 'crazy-quilt command structure' and featuring separate air force and navy air campaigns, air–army tensions over control of helicopters and close air support assets, and a divided command in Washington.[30]

Psychology

Military effectiveness in any of the environments has a **physical component** (size of forces, quality of equipment and technology), and a **conceptual component** (quality of doctrine and military thinking); however, it also has a crucial **moral component**, which expresses psychological factors such as motivation and morale. Psychologically, warfare imposes immense burdens on its participants. The general pressures on the human mind exerted by warfare transcend the different environments. As the historian Michael Howard notes:

> After allowances have been made for historical differences, wars still resemble each other more than they resemble any other human activity. All are fought … in a special element of danger and fear and confusion. In all, large bodies of men are trying to impose their will on another by violence; and in all, events occur which are inconceivable in any other field of experience.[31]

Human psychology is central to war in practice. One US soldier recorded his feelings at the landings at Iwo Jima:

> Your mouth is a vacuum and speech is remote and any sound from it would be a turgid groan. Your mind looks at itself and it shrinks away and wonders whether or not to stand bravely or run and hide. Feelings, sense and physical motion are faint and far off, and all existence is a rushing wind in your ear.[32]

The pressures of combat can result in severe perceptual distortions (see Box 1.8). In land warfare, certain factors exercise a recurrent effect. For example, on land, belligerents may spend extended periods in close proximity to one another; the psychological pressure of extended combat can wear down troops relatively quickly (see Figure 1.1). Artillery fire can be a recurrent source of psychological battle casualties: one Private of the US 101st Airborne commented of his experiences in the Second World War that 'Being shelled is the real work of an infantry soldier, which no one talks about'.[33] US troops questioned in the Second World War identified a hierarchy of sources of threat (see Box 1.9).

Box 1.8 Perpetual distortions in combat

88% Diminished sound (auditory exclusion)
82% Tunnel vision
78% Automatic pilot
63% Slow motion time
63% Heightened visual clarity
61% Memory loss for parts of the event
60% Memory loss for some of your actions
50% Dissociation (detachment)
36% Intrusive distracting thoughts
19% Memory distortions
17% Intensified sounds
17% Fast motion time
11% Temporary paralysis

(Dave Grossman, 'Human factors in War: The Psychological and Physiology of Close Combat', in Michael Evans and Alan Ryan (eds) *The Human Face of Warfare: Killing, Fear and Chaos in Battle* (St Leonard's, NSW: Allen and Unwin, 2000), 16)

Box 1.9 US soldiers' attitudes to enemy weapons

Table 1.1

Weapon	*Percentage who rated it most:*	
	Frightening	*Dangerous*
88mm gun	48	62
Dive Bomber	20	17
Mortar	13	6
Horizonal Bomber	12	5
Light Machine Gun	7	4
Strafing	5	4
Land Mines	2	2
Rifles	0	0
Misc. (inc. Booby Traps, Tanks, Heavy Machine Guns)	4	2

(Ellis, *The Sharp End*, 90)

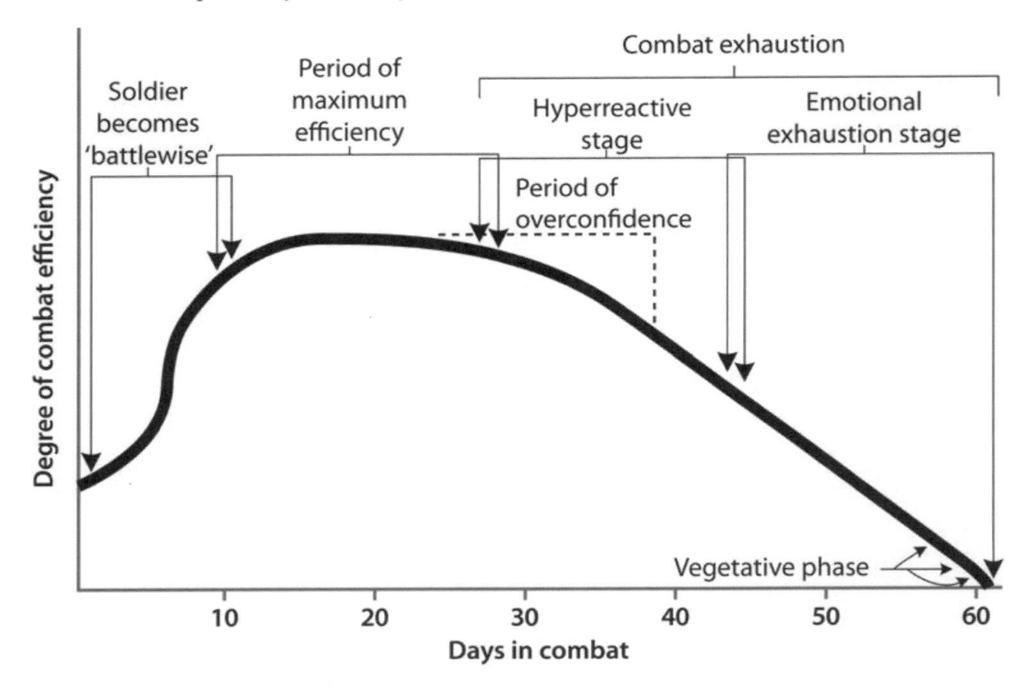

Figure 1.1 The effects of continuous combat
Source: (Dave Grossman, *On Killing: The Psychological Cost of Learning to Kill in War and Society* (New York, NY: Backbay, 2009), 44)

Land warfare is therefore no different from warfare in the other environments in that it remains a human endeavour in which those things that have a bearing on motivation and morale, such as training, group cohesion, leadership, physical conditions, and confidence in technology and doctrine, can have a decisive bearing on the performance of troops.

Doing land warfare well therefore requires not only an understanding of what makes warfare on land unique but also an understanding of the unity of warfare: that, whatever its differences, land warfare needs to be approached as an integral part of war as a whole. To be successful, land campaigns need to be integrated with other activity and subordinated to the needs of strategy and policy.

What are the key principles and concepts of land warfare?

Effectiveness in land warfare requires a grasp both of the uniqueness and the unity of land warfare. In the final part of this chapter we now turn to the more specific issues of **concepts and principles**. Which of these have proven themselves important to the successful conduct of modern land warfare? It is important, first, to consider the dynamics of land warfare – those factors that have most changed thinking and practice.

The dynamics of modern land warfare

The British soldier and strategist Basil Liddell Hart (1895–1970) gave a succinct summary in 1934 of some of the key changes that led in the twentieth century to the development of modern land warfare:

> The evolution of war since the mid eighteenth century has been marked by four main trends. First was the growth of size. From France under the Revolution and Napoleon, to America in the Civil War and Prussia under Moltke, the armies swelled to the millions of 1914–18. Second, came the growth of fire-power, beginning with the adoption of rifles and breech-loading weapons. This, imposed on size, conduced [sic] to a growing paralysis of warfare on land and sea. Third was the growth of industrialisation ... And fourth was the revolutionary growth of mobility, due in turn to the steam engine and the motor.[34]

The implications of the growing scale of war, firepower and mobility were indeed significant.

The scale of war

The **increasing size** of armies was the result of the emergence of 'industrialised people's war'. Industrialised war was the fruit of the emerging **industrial revolution** that gave states the material and bureaucratic capability to raise and sustain large armies. Industrialised *people's* war was the result of **political and social developments**, particularly ideologies like nationalism and a growth in literacy that harnessed the passions of society to the practice of war. The result was increasingly less limited wars in terms of the objectives pursued, the strategies used and the means employed.[35] By 1914, for example, the armies of Europe could muster collectively more than 20 million men. One consequence was the laying of the material conditions for the operational level of war. Armies were now so big that they had to be disaggregated: during the Napoleonic Wars, armies were separated into smaller army corps; 'separate to march, concentrate to fight' became the mantra. By 1914, states were able to deploy multiple armies, invalidating the Napoleonic paradigm of war built on the search for a single decisive battle. Armies were now so large that **no single battle could be decisive** and success would depend increasingly on co-ordinated tactical actions which might, cumulatively, bring victory. In the nineteenth century the Prussian/German army began to use terms such as *Kriegsoperationen* ('war operations') and *operativ* (loosely, 'operational') to describe the lower levels of strategy. Another consequence of the growing scale of war was the creation of armies so large that they could form a **continuous front**: with no flanks to turn, the pressing problem would become how this front could be broken and mobility restored.

Firepower

Allied to the growth in the size of armies was the **exponential growth** in their firepower. The nineteenth century had already seen major new developments: the muzzle-loading, smoothbore muskets and cannon of 1800 had by the end of the century been superseded by magazine-fed rifles, breech-loading rifled artillery and machine guns. The trend in firepower towards faster-loading, longer-ranged and more accurate weaponry continued: between 1900 and 1990 average artillery ranges increased by a factor of more than 20; small arms rate of fire increased between three or four times; the weapons payloads and unrefueled range of ground-attack aircraft increased by more than six times.[36] Increasingly, it was such long-range firepower as artillery and air attack that caused casualties to ground forces (see Box 1.10)

Moreover, the weapons technology of 1900 was further augmented by **new equipment**, including tanks, multiple-launch rocket systems and precision-guided munitions. The **growing lethality** of modern firepower posed an increasing challenge to the ability of modern armies to manoeuvre, resulting in the growing dispersal of forces in order to survive. Moreover, the extended ranges of new weapons technology opened up new possibilities for, and vulnerabilities to, attacks well beyond the front line, exposing reserves, headquarters and rear areas.

Box 1.10 Percentage of British battle wounds caused by different weapons during the Second World War

Table 1.2

Causal agent	*Percentage of wounds*
Mortar, grenade, aerial bomb, shell	75
Bullet, anti-tank shell	10
Landmine, booby trap	10
Blast, crush	2
Chemical (phosphorous)	2
Other	1

(Ellis, *The Sharp End*, 177)

Mobility

Strategic mobility had been revolutionised in the mid-nineteenth century with the application of railways for the movement of troops. However, railways were significant most at the mobilisation stages of war: they were largely irrelevant on the battlefield. **Tactical mobility** was revolutionised in the twentieth century by the military application of progressively more effective internal combustion engines. Thus, whilst in the First World War a soldier could move at some 6 km per hour on the attack and 25 km per day unopposed, a German tank in 1940 might be able to move at 40 km per hour on road and 18 km per hour across country. By 1991, a US main battle tank could reach speeds of 67 km per hour on roads, and travel more than 450 km before refuelling. Mobility was enhanced further by developments in airpower: airborne forces using parachute drops or gliders could utilise the speed and range of aircraft; by the 1960s helicopters allowed the movement of select infantry forces on the battlefield to be increased by a factor of four or five.[37]

We can add to Liddell Hart's list two other important themes that have shaped the development of land warfare from 1900 onwards: **logistics**, and **command and control**.

Logistics

The British General Sir Archibald Wavell (1883–1950) commented: 'It takes little skill or imagination to see *where* you would like your army to be and *when*; it takes much more knowledge and hard work to know where you can place your forces and whether you can maintain them there.'[38] Logistics has always been a central (if often overlooked) element in warfare, but developments in the twentieth century have provided both new challenges and new logistic opportunities. Modern logistics has been complicated by the **growing scale of war**, which has increased drastically the consumption of war materiel. By 1918, for example, the German army was expending 300 million rounds of ammunition a month: ammunition that had to be manufactured, moved to the front and then distributed.

Moreover, logistics had to cope with the **increasing variety** of items required to enable modern armies to function, as well as the greater mobility and dispersal of the forces that required those items. **Railways** provided an imperfect solution to the problems of modern logistics. As armies during the American Civil War discovered, railways could provide the means to move and supply large armies if those armies remained on, or close to, their railheads; however, once they moved from them, the problem became bridging the gap between the army and the railway line. Moreover, railways were inflexible: they were not always easily built and they proved vulnerable to enemy attack. It was the **internal combustion engine** that provided the means to link mobile armies to railways more effectively. Yet the internal combustion engine brought with it its own headaches, especially the need to provide large quantities of petrol, oil and lubricants (POL). In consequence, logistics remains a crucial source of friction in land warfare: in the 2003 Iraq War the longest single delay in advance towards Baghdad was imposed by logistics rather than Iraqi resistance.

Command and control

The exercise of command and control (C2) is essential to the ability of land forces to plan and execute military operations: good command and control is a crucial **force multiplier** that makes land forces more responsive, more efficient and more adaptable. However, such developments as the growing size of armies, their complexity, accelerating mobility and greater dispersal have all combined to make the exercise of command and control more difficult. **Technology** has provided some answers. Invented at the end of the nineteenth century, wireless radio proved a crucial development: as the technology matured, it allowed radio communication to be extended down to the sub-unit level. Since then, technology has also provided new means to acquire, process and present information including television, computers, mobile phones, remote sensors, aircraft, Unmanned Aerial Vehicles (UAVs), and data networking. These developments have been reflected in the expansion of the term C2 into C4ISR (Command, Control, Communications, Computers, Intelligence, Surveillance and Reconnaissance). **New organisations** and **procedures** for managing information have also been developed to address these complexities. Traditional approaches to command and control tended to be highly personalised. In 1870 the German Field Marshal Helmuth Graf von Moltke's headquarters as commander of the whole Prussian army consisted of only 70 men. Modern headquarters have become much larger, more complex and more specialised, a result of the need to create permanent staff systems, with specialised branches and formal procedures.

The sophistication of modern command and control arrangements often **carries costs**: very large quantities of data that must be interpreted and disseminated; vulnerability to attack; and the material outlay to staff and equip a headquarters organisation. Another problem posed by modern command and control is the tension it creates for the commander between remaining fixed in his or her headquarters with access to all of the information it can provide, or leaving the headquarters and 'going forward' to see what is transpiring on the ground. The former can isolate the commander from the realities of the battle; the latter can blinker the commander's perspective.[39]

Principles of war[40]

In a sense, one might think that the effects of scale, firepower, mobility, logistics, and command and control would exert a limited impact on the practice of land warfare given that most armies seem to believe that there exists a set of unchanging themes for the successful conduct of operations. These **principles of war** (actually better termed 'principles of warfare', since they apply to the military practice of war) display a significant similarity across military organisations (see Box 1.11).

Box 1.11 Other principles

There is a high degree of basic continuity on the principles of war between armies. Those in the text reflect US approaches; British and Australian principles are largely the same but add:

- Maintenance of Morale: The importance of sustaining the moral component of combat power through such techniques as effective leadership and fostering group cohesion.
- Flexibility: The need to promote the responsiveness of forces in order to adapt quickly to new challenges.
- Sustainability: The requirement to focus properly on those things required to maintain the combat power of land forces such as logistics and care of personnel.

The Indian army also adds two further principles:

- Administration: Ensuring that commanders have the requisite resources to perform their administrative functions.
- Intelligence: The acquisition and exploitation of information.

(British Defence Doctrine, JDP 0-01 (Developments, Concepts and Doctrine Centre, August 2001) 2-6; Indian Army Doctrine (Headquarters Army Training Command, October 2004), 23-24)

For strategists such as **Antoine-Henri Jomini** (1779–1869), war was science and success could be guaranteed, irrespective of the context, by the effective application of the right principles:

> There exists a small number of fundamental principles of war, which may not be deviated from without danger, and the application of which, on the contrary, has been in all times crowned with glory.
>
> (Antoine-Henri Jomini, *Precis de l'Art de la Guerre*, 1838)[41]

Thus, the dynamics of change outlined in the previous section should not have wrought fundamental differences. These principles of war encompass nine basic themes:[42]

- **The objective**: Military operations should be directed towards an **attainable and clearly defined objective**. This goal is not unalterable and may be subject to change during a conflict to meet evolving circumstances, but this objective is of overriding importance – it provides the purpose that should knit together activity at the different levels of war, and provides the measure of success for operations.
- **Offensive action**: An **offensive spirit** should pervade all operations, even defensive ones, at all the levels of war. Only the offensive can seize and maintain the initiative; only the offensive can create opportunities for exploitation and maintain the momentum of operations.
- **Mass**: Success requires the **massing of combat power** (which may not necessarily require numbers) at the decisive point. This may require economy of force elsewhere.
- **Economy of force**: Effective **prioritisation of effort** is required so that the minimum effort is expended on secondary tasks, allowing the maximum effort to be expended on first-order activities.
- **Manoeuvre**: Manoeuvre comprises **movement in relation to the enemy** to facilitate the maximum application of combat power, maximise one's own freedom of action relative to the enemy, and reduce one's vulnerability to enemy attack.
- **Unity of command**: The co-operation and co-ordination of forces towards the attainment of a **common goal** under the responsibility of a single commander. Only through this can unity of effort be achieved amongst the disparate elements that compose land forces.
- **Security**: Measures must be taken to **reduce the vulnerability of friendly forces** to enemy action. Security can include physical measures to protect troops, developing a better understanding of the enemy, risk management and so forth.
- **Surprise**: Inducing shock and confusion in an enemy by acting against them at a time, place, or through methods for which they are **unprepared**. Surprise can be a function of such themes as speed, secrecy, deception, manoeuvre and intelligence. The effects of surprise are temporary so they must be exploited.
- **Simplicity**: Orders, plans and concepts should be as **straightforward and uncomplicated** as possible in order to improve comprehension and to reduce as far as possible the chances of misunderstandings.

In reality, however, there are **no scientific principles that guarantee success** in land warfare. The difficulty is that, fundamentally, land warfare involves tactical and operational trade-offs, yet the value of the various trade-offs is context specific. To illustrate this point, it is useful to highlight some of the **key tensions** associated with the practice of land warfare. These tensions include those between: fire and manoeuvre; manoeuvre and attrition; destruction and disruption; centralisation and decentralisation; conformity and initiative; independent and combined arms; concentration and dispersal; and attack and defence.

Fire and manoeuvre

Firepower embodies the effects of lethal ranged weapons. The effects of firepower are complex and related: they include **destruction** (killing or wounding personnel and destroying or disabling materiel); **suppression** (preventing enemy forces from carrying out core tasks such as firing, moving, or observing); and **disruption** (the infliction of psychological shock).[43] The latter two effects can be powerful instruments that can render ineffective even well-protected enemy forces. Manoeuvre encompasses the movement of forces to gain an advantageous position. The relationship between fire and manoeuvre is at once antagonistic and synergistic. Firepower and manoeuvre can be antagonistic in two regards.

- First, manoeuvring forces tend to be **less effective at generating firepower** because movement can make fire less accurate and can make identifying targets more difficult.
- Second, in order to manoeuvre, troops must relinquish some of the benefits of cover, making them **more exposed to enemy firepower**.

Both of these observations contribute to the general point that, all other things being equal, **defence tends to be the stronger posture in land warfare**. However, **synergies** also derive from the interplay between manoeuvre and firepower:

- **Manoeuvre facilitates firepower** by placing forces in positions, especially on enemy flanks or rear, from which they can best maximise the effect of friendly firepower.
- Friendly **firepower facilitates manoeuvre** by destroying, suppressing and disrupting the enemy's ability to do the same to friendly forces.

In consequence, modern land warfare tends to be associated tactically with combinations of alternating fire and manoeuvre by sub-units, so that one part of a unit applies firepower against the enemy as another part moves.

Manoeuvre and attrition

Attrition describes the process by which units are worn down by human and material casualties. 'Attritional warfare' is often associated with such static warfare as the First World War. For this reason attrition is often seen as the opposite of warfare based on manoeuvre: attritional approaches are characterised as unimaginative and inefficient whereas such manoeuvres as General Schwartzkopf's left flank movement against the Iraqis in 1991 or General Douglas MacArthur's landings at Inchon in 1950 are seen as imaginative, bold and decisive. This perception of a dichotomy between attrition and manoeuvre has been reinforced by such developments as the internal combustion engine and wireless communications that have increased opportunities for operational level manoeuvre. In reality, **attrition and manoeuvre are related concepts**. Manoeuvre is a means, not an end: the purpose of manoeuvre is to place forces in an advantageous position relative to enemy: in effect to cause, or threaten to cause, favourable attrition rates through turning movements (passing round an enemy's flank); envelopments (passing round flanks and moving on the enemy rear); or encirclements (trapping enemy by blocking both flanks and rear, creating what the Germans termed a *Kesselschlacht*, or 'cauldron battle').

For this reason, manoeuvre and attrition are **synergistic**. Attrition gives manoeuvre meaning: on its own, high rates of mobility are unlikely to win campaigns – manoeuvre needs to be translated into attrition of the enemy if it is to generate military effect. Thus, the US campaign in the Pacific embodied extraordinary manoeuvre, but its success depended ultimately upon the wearing down of Japanese forces. Indeed, the strategist Colin Gray notes that whilst battles against Germany and Japan could be won through manoeuvre, defeating them in the war as a whole could only be done by a 'wearing out' fight. Moreover, attrition can be a vital precursor to manoeuvre in that it may often require attrition to wear down an opponent to the point at which breakthroughs (and subsequent mobile exploitation) are possible. For example, the extraordinary advance of General George S. Patton's 3rd Army in August 1944 during the breakout from Normandy was built on more than two months of bloody attrition in which Allied casualty rates were higher than those of the First World War.[44]

Destruction and disruption

Mirroring the distinctions between attrition and manoeuvre, a recurring division in military thinking exists between a focus on **destruction**, the physical elimination of enemy forces, and **disruption**, rendering enemy forces ineffective through psychological shock and dislocation. At the heart of the focus on the latter is the concept of **tempo**: 'the rhythm or rate of activity on operations, relative to the enemy.'[45] Advocates of disruption through tempo argue that enemy forces can be defeated through a higher speed relative to the enemy in decision making, execution of planning and operations, and speed of transition from one activity to another: in effect, the enemy army becomes fatally disrupted because its decision making is constantly out of date. For example, Liddell Hart saw the defeat of France by Germany in 1940 as the result of exactly this kind of high-tempo warfare:

> The pace of panzer warfare paralysed the French Staff. The orders they issued might have been effective but for being, repeatedly, twenty four hours late for the situation they were intended to meet.[46]

A focus on disruption as the route to success requires a corresponding emphasis on **exploitation**, which is the crucial route to sustaining a high tempo in operations. Exploitation embodies rapid follow-up offensive operations designed to maximise initial victories; consolidation, on the other hand, focuses on pausing to reconstitute forces before recommencing an attack. Disruption of the enemy is a temporary phenomenon because the effects of shock and demoralisation diminish over time. Achieving disruption-based success therefore requires exploitation as the means to sustain and magnify the psychological benefits of success. Maintaining the tempo of operations becomes crucial to **sustaining the initiative** in attack. Indeed, Richard Simpkin defines initiative as 'constantly creating new situations to be exploited'.[47]

In contemporary Western militaries, tempo is associated with manoeuvre and there is a preference for manoeuvre-based disruption over destruction as the route to success in land warfare – as Winston Churchill argued:

> battles are won by slaughter and manoeuvre. The greater the general, the more he contributes in manoeuvre, the less he demands in slaughter.[48]

However, maintaining the tempo required to create disruption can be challenging:

- Exploitation requires first the ability to **identify opportunities** – something that requires effective intelligence and intuitive commanders.
- Exploitation can only be carried out if the **means are available**, usually in the form of reserves, to develop them.

- Exploitation also carries with it **risk**: exploiting forces often advance quickly on narrow fronts making them vulnerable to counter-attacks, especially against their flanks. For example, the signal German success at Kharkov 1943 was built upon a counter-attack against overextended Russian breakthrough forces.

Centralisation and decentralisation

A fourth theme is the tension in land warfare between **centralisation** and **decentralisation**. It has already been noted that in war uncertainty is pervasive. Historically, land forces have attempted to manage uncertainty through two approaches: the first has been to **impose conformity** by centralising at the top; the second has been to encourage **greater local initiative** through decentralisation.[49]

Perhaps the most important manifestation of these choices is in the realm of command and control, which embodies the **exercise of authority** over, and direction of, subordinate forces.[50]

Centralised command

Centralised approaches to command and control **impose co-ordination and unity of effort**; commanders reduce the need for lower-echelon officers to manage uncertainty because the latter simply do as they are instructed, referring decisions up the chain of command. The problem with centralised approaches to command is that they trade certainty for time. Centralisation will slow decision making down because it reduces opportunities for local commanders to display initiative; instead, lower-echelon commanders will refer decisions upwards. Centralisation tends to place an emphasis on extensive formalised planning and on methodical execution: it is not conducive to maintaining the tempo of operations.

Decentralised command

The alternative approach is decentralised command, often termed '**mission command**' or *Aufstragstaktik*. The emphasis in mission command is on instilling in subordinates a clear understanding of what needs to be achieved (the commander's intent), but giving local commanders latitude over how this intent is realised. Devolved command increases the tempo of operations because many uncertainties are resolved by local commanders without waiting for orders from above. It also increases the relevance of command decisions since local commanders are likely to understand immediate circumstances better than more distant command echelons. Mission command facilitates opportunistic exploitation: it accepts more uncertainty at the higher level of command but increases the ability of local commanders to shape events.

Despite the potential advantages, decentralised approaches to command and control are **exceedingly demanding**. The complexity of operations can place great demands on the capabilities of more junior officers. Making mission command work requires training, effective communications, a supporting philosophy of command, and a military organisation that values displays of independence and initiative in even very junior leaders. Poorly executed, mission command can be a recipe for the progressive disintegration of unity of effort. Moreover, the exercise of decentralisation has been complicated by the pervasive importance of joint operations. Air activity, for example, needs to be heavily centralised at the higher levels of command to ensure the effective apportionment of effort to multiple tasks. Decentralisation can pose many problems for joint warfare, depending upon whether the services are merely de-conflicting their efforts, or co-operating with one another, or alternatively attempting an integrated effort.

Conformity and initiative

A related theme arises in relation to **military doctrine**. Doctrine can be defined as the 'guidance, mandatory or discretionary, on what is believed officially to be contemporary best military practice'.[51] Doctrine comprises:

> a common perspective from which to plan, train, and conduct military operations. **It represents what is taught, believed, and advocated as what is right** (i.e., what works best).[52]

As the British strategist Colonel J.F.C. Fuller (1878–1966) noted, doctrine is 'the central idea of an army'.[53] Doctrine performs crucial functions:

- it **interprets history** to derive relevant lessons for the present and future;
- it provides military practitioners with **intellectual guidance** on how to solve military challenges;
- it **provides cohesion** in thought and expression within a military organisation.[54]

Doctrine is reflected formally in written manuals, practical training and military education, and informally in the values of an organisation and its customary behaviour. Yet historically doctrine has also sometimes been a dangerous thing: doctrine is 'authoritative but requires judgement in application'.[55] Classically, four dangerous problems can manifest themselves:

- First, doctrine is based on an analysis of experience – the Soviet strategist A.A. Svechin (1878–1938) referred to it as 'the daughter of history':[56] yet the quality of historical analysis by militaries has often been poor, resulting in **selective learning** and the reinforcing of existing orthodoxies.
- Second, doctrine must simplify if it is to provide useful tools, but sometimes this simplification can become **over-simplification** and stereotyping.
- Third, the need to make doctrine relevant to future operations requires that judgements are made on what the future will look like. This carries the inherent **risk** that these judgements will be mistaken.
- Fourth, doctrine can become **politicised**: it can become symbolic of broader divisions within a military organisation based on age, experience, outlook, or background.[57]

The consequences of these problems can be, as France found in 1940, that the doctrinal foundations of land warfare espoused by an army can be coherent, well-understood and rooted firmly in an analysis of recent military experience, yet they can at the same time be utterly unsuited to war as it must actually be fought. French doctrine, based on 'methodical battle, firepower, centralisation and obedience', was an excellent doctrine for the challenges of 1918, but German operations in 1940 revealed that land warfare had moved on.[58]

Thus, the overarching danger is that doctrine ceases to be a common framework for effective thought and action in the context of the application of initiative, and becomes instead a mechanism for the rigid imposition of conformity (see Box 1.12). Achieving the right balance between initiative and conformity requires, amongst other things, a focus on professional military education and the development of a military culture that is dynamic, curious and open to new ideas.

Box 1.12 Doctrine

> In itself the danger of a doctrine is that it is apt to ossify into a dogma, and be seized upon by mental emasculates who lack virility of judgement, and who are only too grateful to rest assured that their actions, however inept, find justification in a book.
>
> (JFC Fuller, quoted in Hoiback, *Understanding Military Doctrine*, 13)

Independent and combined arms

Combined arms describes the process of using two or more combat arms (infantry, tanks, combat engineers, artillery and so on) in an **integrated** fashion. The key advantage of combined arms warfare is that, as Biddle notes, 'Combined arms integration reduces net vulnerability by teaming together weapon types of contrasting strengths and weaknesses'.[59] For example, artillery can project large volumes of firepower over great distances, but it is relatively immobile and inaccurate; infantry is more mobile and accurate but has less firepower. Used together, artillery can use its firepower to suppress defences enabling infantry to close sufficiently to bring its own advantages to bear.

Combined arms has a **long pedigree**: during the Napoleonic Wars, for example, the use of horse artillery to support cavalry was based on the principle that the most effective defence for infantry against cavalry (forming square) was also the most vulnerable to artillery fire. However, combined arms has become increasingly desirable in land warfare because overcoming the effects of modern firepower in the context of variable terrain requires a combination of mobility, protection and firepower best delivered by multi-arm integration. On the other hand, combined arms warfare is often **difficult to realise in practice**:

- First, combined arms warfare is **complex** and requires a high degree of skill, training, and command and control: troops need to understand other combat arms as well as their own; combined arms embodies combat arms with often different rates of mobility; combined arms techniques must vary according to circumstances and terrain.
- Second, the demands of combined arms create **tensions** in terms of the need to ensure sufficient combat arm specialisation: unless combat arms are effective at their core roles they will not be able to contribute to combined operations.[60]

Thus, whilst combined arms is in theory central to effectiveness in land warfare, in practice combat arms can often come to operate independently with only a modicum of de-confliction or co-operation.

Concentration and dispersal

It has already been noted that **mass** has long been established as a key principle of warfare. Clausewitz argued, for example, that the best strategy in war was 'always *to be very strong*; first in general, and then at the decisive point'.[61] **Differential concentration** (massing where the enemy is weaker) is an important means of achieving rapid breakthroughs which can then be exploited. The massing of forces can be facilitated by deception operations which may also magnify the effect of mass through surprise. Concentration also brings other benefits: forces that are concentrated are likely to be easier to command and control and it may also prove simpler to supply them. Despite these advantages, concentration carries potential costs.

- First, the modern battlefield carries with it a **counter-logic**: the increasing lethality of modern firepower can make mass extremely vulnerable: increasing the dispersal of forces may be a pre-requisite for their survival.
- Second, mass cannot compensate reliably for **differences in skill**. Thus, in the Ramadan/ Yom Kippur War of 1973 the Syrian attack against Israeli forces on the Golan Heights mustered a 6:1 superiority in numbers on a narrow frontage. Yet the Israeli superiority in tactical skill more than compensated for their inferiority in numbers, making the Syrian attempt at concentration an effective Israeli exercise in economy of effort.
- Third, dispersal can provide more flexibility in attack and defence by allowing **greater depth**. Deploying forces in successive echelons, or waves, provides attacking forces with successive forces for exploitation and provides defending forces with greater elasticity – uncommitted echelons in defence provide successive lines of defence and/or the means to counter-attack quickly.

Attack and defence

Finally, and as a function of the previous themes, there is a **tension between attack and defence** in land warfare. Victory is unlikely ever to be achieved without attack. Defensive operations may be a vital precursor for success, perhaps wearing down the enemy or giving opportunities to consolidate and mass friendly forces, but without an attack the initiative is ceded to the enemy and no positive outcomes can be achieved. However, whilst in the air and sea environments attack is often the stronger form of warfare, on land the reverse is true:

- When attacking, the **frictions** associated with command and control, co-ordination, integration, application of firepower, and supply are all worsened because attacking requires that forces manoeuvre in order to control ground and bring their firepower to bear.
- Moreover, defenders are able to **prepare their positions**: digging entrenchments; laying minefields; pre-registering artillery fire; placing weapons in positions of mutual support.
- These challenges have been worsened for attackers because **developments in firepower** have increased the accuracy and range at which attacking forces can be engaged by defenders.

In consequence, attacking at once embodies **opportunities** for positive outcomes but also the **risk** of serious defeat.

Conclusions

> [I]n war with its enormous friction even the mediocre is quite an achievement.
> (Helmuth von Moltke, quoted in van Creveld, *Command in War*, 13)

This first chapter has laid out a range of ideas and concepts that will be utilised in the remainder of this book. The character of land warfare has been shaped by a number of **distinct but interrelated challenges**:

- **The impact of land itself**: The nature of land itself shapes the character of land warfare. Land **matters politically**, but it is a complex medium – **variable**, **opaque**, **mutable** and **resistant**. If land warfare carries with it the possibility of decisive results, the nature of land creates a variety of strengths and weaknesses for land forces that can shape problems and opportunities for the exercise of land power, the operation of which will depend upon context.
- **Inherent strengths shape inherent challenges**: **Control**, **durability**, **adaptability** – these are traits that can make land forces eloquent tools of policy, but they can also create **risk**, **vulnerability** and **controversy**. There is nothing in its nature that dictates that land power should intrinsically be a successful, surgical tool of policy.
- **The impact of change**: Adding to the complexities posed by the nature of the land environment are those wrought by the processes of **dynamic change** inherent to the growing scale of war, and factors such as firepower, mobility, logistics, and command and control. Land forces have become **larger**, more **specialised**, increasingly **disaggregated**; their mobility and firepower has increased. These developments have not been an unmitigated benefit because war is relational: what benefits one army must disadvantage its adversary.
- **Trade-offs and complexity**: Nor can navigating a route through the consequences of these changes be achieved through the simple application of iron principles. Instead, land power is a more **complex phenomenon**. Responding to the implications of change since 1900 has been complicated by the wide array of **trade-offs** that exist in land warfare. These trade-offs do not represent binary choices: in practice armies operate on a spectrum conditioned, as Chapter 5 explores, by the qualities of the military forces concerned, the theatre geography, and the broader social, political and economic context. Nor can these themes be easily separated: land warfare is constituted by **synergistic combinations** of firepower, manoeuvre, attack, defence, centralisation and decentralisation, and so on.[62]
- **Unity and synergy**: All of these features need to be put into the context of the **unity of land power**. Separating the conduct of land warfare from strategy or joint operations is usually a sure route to failure, but each will place their own demands on land forces.

We turn now to consider the question 'How has land warfare developed since 1900?' Chapter 2 begins this process by looking at the debates on how and why land warfare has changed over time.

Notes

1 Patrick O'Sullivan, *Terrain and Tactics* (London: Greenwood, 1991), 27.
2 Colin S. Gray, *Modern Strategy* (Oxford: Oxford University Press, 1999), 213.
3 Stephen Biddle, *Military Power: Explaining Victory and Defeat in Modern Battle* (Princeton, NJ: Princeton University Press, 2004), 36.
4 John Ellis, *The Sharp End: The Fighting Man in World War II* (London: Aurum Press, 2009), 82.
5 See, for example, Anthony Saunders, *Trench Warfare, 1850–1950* (Barnsley: Pen and Sword, 2010).
6 O'Sullivan, *Terrain and Tactics*, 32.
7 Lukas Milevski, 'Fortissimus Inter Pares: The Utility of Landpower in Grand Strategy', *Parameters* (Summer 2012), 11.
8 Gray, *Modern Strategy*, 213.
9 Milevski, 'Fortissimus Inter Pares', 15.
10 David J. Lonsdale, *The Nature of War in the Information Age: Clausewitzian Future* (New York, NY: Frank Cass, 2005), 50.
11 William Johnson, *Redefining Land Power for the 21st Century* (Carlisle, PA: Strategic Studies Institute, 1991), 14.
12 Milan Vego, *Joint Operational Warfare: Theory and Practice* (Newport, RI: US Naval War College, 2009), II-26.
13 Gray, *Modern Strategy*, 165.
14 Admiral J.C. Wylie, *Military Strategy: A General Theory of Power Control* (Annapolis, MD: Naval Institute Press, 1989), 72.
15 Milevski, 'Fortissimus Inter Pares', 10.
16 Gray, *Modern Strategy*, 213.
17 Ibid., 362.
18 Johnson, *Redefining Land Power*, 20.
19 Beatrice Heuser, *The Evolution of Strategy: Thinking War from Antiquity to the Present* (Cambridge: Cambridge University Press, 2010), 490.
20 Edward N. Luttwak, *Strategy: The Logic of War and Peace* (Cambridge, MA: Cambridge University Press, 2001), 88.
21 Gray, *Modern Strategy*, 45.
22 Allan R. Millett and Williamson Murray, quoted in Williamson Murray, 'The Army's Advanced Strategic Art Programme', *Parameters* (Winter 2001), 31–39.
23 Colin Gray, 'Why Strategy is Difficult', *Joint Force Quarterly* Vol. 22 (Summer 1999), 6–12.
24 Carl von Clausewitz (edited and translated by Michael Howard and Peter Paret), *On War* (Princeton, NJ: Princeton University Press, 1976), 178.
25 Stuart Griffin, *Joint Operations: A Short History* (London: Training Specialist Services HQ, 2005), 10.
26 Roger A. Beaumont, *Joint Military Operations: A Short history* (Westport, CT: Greenwood, 1993), 186, 193.
27 Griffin, *Joint Operations*, 7.
28 Joint Publication (JP) 1, *Joint Doctrine for the Armed Forces of the United States* (25 March 2013), I-8.
29 Griffin, *Joint Operations*, 7.
30 Beaumont, *Joint Military Operations*, 149–51.
31 Michael Howard, 'The Use and Abuse of Military History', *RUSI Journal*, Vol. 138, No. 1 (1993), 13.
32 Ellis, *The Sharp End*, 104.
33 Ibid., 69.
34 Heuser, *The Evolution of Strategy*, 172.
35 See Stig Forster and Jorg Nagler (eds), *On the Road to Total War: The American Civil War and the German Wars of Unification, 1861–1871* (Cambridge: Cambridge University Press, 2002).
36 Biddle, *Military Power*, 30.
37 Ibid., 60–61.

38 Martin van Creveld, *Command in War* (Cambridge, MA: Harvard University Press, 1985), 231–32.
39 Ibid., 4.
40 Joint Publication 3–0, *Principles of Joint Warfare* (August 2011), A1–A5.
41 Quoted in J.D. Hittle (ed.), *Jomini and his Summary of the Art of War* (Harrisburg, PA: Military Service Publishing Company, 1958), 43.
42 Ibid., A1.
43 J.B.A. Bailey, *Field Artillery and Firepower* (Annapolis, MD: Naval Institute Press, 2004), 11.
44 See, for example, Max Hastings, *Overlord: D-Day and the Battle for Normandy 1944* (London: Pan Books, 1999), 289–90.
45 Charles Grant, 'The Use of History in the Development of Contemporary Doctrine', in John Gooch (ed.) *The Origins of Contemporary Doctrine*, The Occasional No. 30 (Camberley: Strategic and Combat Studies Institute, September 1997), 11.
46 Ibid.
47 Quoted in J.P. Kiszely, 'The Contribution of Originality to Military Success', in Brian Holden Reid (ed.) *The Science of War: Back to First Principles* (London: Routledge, 1993), 26.
48 Grant, 'The Use of History', 10.
49 Van Creveld, *Command in War*, 268.
50 JP 1, xvi.
51 Paul Latawski, *The Inherent Tensions in Military Doctrine* (Sandhurst Occasional Papers No. 5, 2011), 9.
52 JP 1, ix, emphasis in the original.
53 Quoted in Latawski, *The Inherent Tensions*, 8.
54 Ibid.
55 Ibid., 5.
56 Grant, 'The Use of History', 7.
57 See Harald Hoiback, *Understanding Military Doctrine: A Multidisciplinary Approach* (London: Routledge, 2013), 11–17.
58 Grant, 'The Use of History', 9.
59 Biddle, *Military Power*, 37.
60 Jonathan M. House, *Combined Arms Warfare in the Twentieth Century* (Lawrence, KS: University Press of Kansas, 2001), 4–6.
61 Clausewitz, *On War*, 204.
62 Gray, *Modern Strategy*, 160.

Suggested reading

Colin S. Gray, *The Strategy Bridge: Theory for Practice* (Oxford: Oxford University Press, 2011). A clear and insightful analysis of the nature, demands and complexity of strategy by one of key modern interpreters of Clausewitz.

Stuart Griffin, *Joint Operation: A Short History* (Training Specialist Services HQ, 2005). A concise overview of the history and challenges associated with joint operations: the book examines the topic through the lens of a variety of case studies.

Richard Holmes, *Acts of War: The Behaviour of Men in Battle* (London: Cassell, 2004). A marvellous evocation of the physical and psychological conditions of land battle from the perspective of the individual soldier.

Michael Howard, *Clausewitz: A Very Short Introduction* (Oxford: Oxford University Press, 2002). Provides a useful gateway into the study of Clausewitz and strategic theory.

Rob Johnson, Michael Whitby and John France, *How to Win on the Battlefield: The 25 Key Tactics of All Time* (London: Thames and Hudson, 2013). A light and accessible introduction to such key concepts in warfare as attack, envelopment, shock action and concentration.

Lukas Milevski, 'Fortissimus Inter Pares: The Utility of Landpower in Grand Strategy', *Parameters* (Summer 2012): 6–15. This article provides a concise summary of the advantages and disadvantages of land power.

Eitan Shamir, *Transforming Command: The Pursuit of Mission Command in the U.S., British, and Israeli Armies* (Stanford, CA: Stanford University Press, 2011). Shamir examines military change and the impact on approaches to command and control.

Hew Strachan, *European Armies and the Conduct of War* (London: Routledge, 1983). An illuminating analysis of the development of land warfare from the eighteenth through to the end of the twentieth centuries.

Martin van Creveld, *Supplying War* (Cambridge: Cambridge University Press, 1997). An innovative study of the relationship between military change, logistics and success in warfare.

2 The development of modern land warfare

Key points

- There is no single accepted narrative on the development of land warfare: writers differ on what they consider to be the key changes.
- One important point of contention is whether the structure of the development of land warfare can be characterised as one marked by irregularly paced evolution or a series of profound military revolutions.
- Another debate exists over the sources of change in land warfare: is technology crucial, for example, or should more emphasis be placed on conceptual innovations or the wider political, economic and social context?

Why and how has land warfare developed over time? The answer to this question might seem simple: it is, surely, driven by new technology. Longbows, pikes, gunpowder weapons, magazine rifles, machine guns, radios, tanks: each new development in military technology seems to drive forward new practices and new approaches in land warfare. This is reflected often in the language used to describe particular periods in military history: the 'horse and musket' period, for example, or the 'age of rifles', or 'war in the machine age'. In fact, though, the reality is much more complex than this: technology may often be one of the things that cause changes in the way that land warfare is fought, but there seem to be other critical factors at work. Indeed, even where technology is a crucial agent of change, its impact can be conditional on a range of other influences.

With these points in mind this chapter considers the answers to the question 'what drives changes in land warfare?' Picking up the themes of uncertainty and contention touched on in the Introduction to this book, this chapter identifies two key debates on this topic that lead to different answers to this question. First, there is debate on whether change in land warfare is driven by a process of rather **gradual evolution** over time, or whether instead change is caused by periodic occurrences of sudden, immense **revolutionary change**. Second, there is a debate on what drives the development of land warfare: is technology crucial, or are other factors, such as new ideas or broader political or socio-economic factors, the key agents? This chapter is structured according to those two debates. The first part explores the arguments between the revolutionary and evolutionary schools of thought; the second part reviews the contending theories on what drives change in land warfare.

This might all seem to be so much academic navel gazing, but a study of these debates shows that any given account of the development of land warfare will be based on assumptions, conscious or unconscious, that reflect where the author stands in these debates. Understanding, then, that there *is* a debate is an important conclusion: we know less objectively about the past than we think that we do. Just as important, the assumptions that we make about what has driven the development of land warfare in the past inevitably shape how we analyse land warfare today and in the future. We will return to this idea later on in this book.

Revolution or evolution?

We begin, then, by looking at the debate surrounding the structure of change in land warfare: has land warfare developed through the impact of a series of distinct military revolutions, or has it developed essentially by a long process of evolution?

Military revolutions and land warfare

Advocates of the 'change as revolution' perspective would see the history of land warfare as consisting of long periods of relative stasis in methods, interspersed by sudden, groundbreaking change that brings about periodic **paradigm shifts** (see Box 2.1): a complete change in the model of warfare, in which one type of warfare would be superseded by another. So, for example, the introduction of gunpowder could be conceived of as just such a revolution or paradigm shift in the sense that war before and after its introduction might be considered as very different things in terms of the technology, methods, military organisations and thinking.

Box 2.1 Paradigm shifts

The concept of 'paradigm shifts' was developed by the scientist Thomas Kuhn in his 1962 book *The Structure of Scientific Revolutions*. Kuhn argued that key developments in scientific thinking only usually took place in conditions in which one paradigm (world view or model of thinking) gave way to another. Until such a shift occurred, new evidence tended to be subsumed into old patterns of belief.

The idea that the history of warfare might be marked by distinct watersheds or paradigm shifts caused by rapid military change was first advanced in a developed form in 1956 by the historian Michael Roberts, who argued for the existence in history of what he termed '**military revolutions**'. Examining sixteenth- and seventeenth-century European warfare, Roberts argued that organisational adaptation amongst armies of the period, particularly the Swedes, led to changes in the conduct of warfare so rapid and fundamental that they constituted a **revolutionary break** with the past. Roberts argued that this revolution created a new way of war: the old model of land warfare was replaced by a new one that that was recognisably modern. Since then, many other commentators have supported this view, arguing that what drives the development of warfare forwards are periodic incidences of change so profound that they are revolutionary in their implications. Proponents of the idea of military revolutions argue that at various points in history, factors such as technology, economic change, political developments and organisational innovation have served to produce major discontinuities in warfare – sudden fundamental, structural shifts in the character of warfare.

Defining military revolutions

The concept of military revolutions thus expresses the idea of something more than just the emergence of important, but essentially evolutionary, changes within warfare. True military revolutions in land warfare are **game changing**: they undermine the value of existing equipment and ways of doing things.[1] When a military revolution takes hold, it 'fundamentally alters the character and conduct of military operations' and represents 'an observable breaking point between two recognisably different types of warfare'.[2] In general, supporters of the idea of military revolution see it as the harbinger of radical change, a 'paradigm shift' in war.

The policy impact

In addition to debates amongst historians and social scientists, the ripples from the military revolution debate extended into **military and policy-making circles** from the 1970s onwards. These analysts focused on the effect that new technologies, such as precision-guided munitions (PGMs), allied with new ways of using them, would have on the conduct of war. These debates reflected the assertions of historians like Roberts that warfare was marked by periods of rapid, transformational change. Further, advocates argued that ongoing developments in new technology along with new tactics was generating, or will generate, a revolution in warfare as significant as that advanced by Roberts for the sixteenth century.[3] As we will see later in this book, the 'military revolutions' perspective has had a profound impact on how many people have defined the dynamics of contemporary and future land warfare, and how, in consequence, many politicians and military commanders have sought to shape the organisation and practice of land forces.

In theory, the military revolutions perspective should make establishing the historical development of land warfare straightforward. All we would need to do is to identify the key paradigm shifts that have taken place in land warfare over time and establish what they involved and what effect they have had on how land warfare has been prosecuted. Sadly, the task is, in reality, much more difficult than this. The first problem that we face is that there is no consensus on whether such a thing as a military revolution does exist as a phenomenon. Instead, many commentators argue that, in effect, the development of land warfare should be conceptualised simply as a process of evolution, sometimes faster, sometimes slower.

The evolution argument

Those writers who favour this evolutionary perspective often focus, first, on the problems with the concept of a 'military revolution'. For critics, the concept of military revolutions is a vague and unhelpful one without much historical evidence.

Military revolutions: the definitional problem

The argument that the concept of military revolutions doesn't really hold water focuses in particular on four problems:

The historical evidence base

Even amongst those who believe that military revolutions have taken place, there is often disagreement about what those revolutions are. Arguing for greater rigour in the search for evidence for military revolutions, one writer has argued that 'History is the only reliable guide mankind has to the future'.[4] Using history, some commentators state the existence of military revolutions as a matter of fact. However, **history is not an objective commodity**. It is the result of a process of interpretation of usually incomplete evidence: the process of interpretation is influenced by social factors, culture and politics, intellectual fashion, bureaucratic politics and institutional self-interest. The contested nature of the historical evidence is reflected in the fact that there are few technologies, systems or events that command overwhelming agreement on their revolutionary character. So, longbows, for example, are identified by some as the cutting edge of a medieval infantry revolution; equally, their decisiveness is dismissed by others.[5]

Defining key terms

Other critics point to **definitional difficulties**. For example, use of the term 'revolution' in the military revolution concept implies change that occurs over a very short time span. Yet some commentators take a very extended view of the temporal dimensions of revolutionary change. Some, for example, associate military revolutions with change over periods of 20 or 30 years, arguing that: 'A military revolution need not take place quickly because its nature lies in its *impact* not necessarily the rate at which it takes hold.'[6] Thus, we refer to an *industrial* revolution even though this took place over a century or more. The problem with this view is that if the term 'revolution' is stretched to include change that occurs over an extended period of time, then it is difficult to distinguish revolution from evolution.

The scope of revolution

Another point of debate is **how *much* change** has to occur before something becomes 'revolutionary'. Does a military revolution have to alter the fundamentals of warfare across the board, as argued by some, or does it only have to result in significant changes within one element of the spectrum of war? How many militaries must adopt a new military technology or practice before it can be considered a revolution? Likewise, can a revolution be a revolution if other, older, forms of warfare continue to exist in parallel? Moreover, how 'military' does a revolution have to be in order for it to be considered as something that is conceptualised usefully as a military revolution? Since war is a social activity, don't all forms of political, economic, social and cultural change have military implications to some degree?

The impact on effectiveness

Critics also draw attention to the assumed relationship between **military revolutions and success**. The author Michael S. Neiberg comments that 'much of military history is written with the intent of explaining why the winners won and losers lost'.[7] Military success is often attributed to the leveraging by the victor of a new military revolution.

For many, military revolutions can be recognised by the scale of success achieved by participating states over non-participatory ones. Yet writers such as the strategist Colin Gray note that the sources of success or failure in war are broad: quality of strategy and policy are surely critical. This means that a military revolution is 'not a necessary, let alone a sufficient, condition for victory'.[8] Moreover, opponents learn and adapt, so that any marked advantages accruing to a belligerent through leveraging a military revolution are likely to erode over time unless victory is won quickly. Thus, even if they did exist, military revolutions might actually have a short shelf life in terms of creating overwhelming success.

The importance of continuity in military affairs

Aside from attacking the concept of military revolution itself, critics also point to a second general failing: that a focus on revolutionary change downplays the **role of continuity** in military affairs.

For advocates of a study of the evolutionary development of warfare, military change is characterised as a **continuous process**, sometimes less even and sometimes more so, but consisting generally of 'irregularly paced processes of innovating business as usual'.[9] In this vein, the historian Cyril Falls argues that: 'the student should not believe everything moves only when he sees the process at a glance, and stands still when he does not see it moving … the more scholarly the enquirer becomes, the more conscious is he of endless change.'[10] What apparently seem military revolutions may build to an important extent on diverse factors that may have been a long time in the making. Looking at the mechanics of technological development, R. Angus Buchanan argues that 'the image of the ratchet seems most appropriate, suggesting as it does the necessity for backward and forward linkages for every successful innovation'.[11]

This image of 'ratcheted change' where forward movement is rooted in the past, is something that advocates of military revolutions sometimes seem to miss. One advocate of the military revolution concept has argued that whilst evolutionary change is characterised by continuities that exist between generations, revolutionary change involves 'almost no continuity'.[12] Yet, if one looks at some of the proposed military revolutions, such as the Napoleonic revolution or *blitzkrieg*, other commentators argue for important points of continuity with the past, or indeed argue that revolution was built upon the coming together of **prior evolutionary developments**. The historian Geoffrey Parker, for example, has presented a powerful case that the growth in the *Trace Italienne*, a fortification style designed to meet the challenge of early artillery, featuring ramparts of earth, low-sloped, thickened walls, gun ports and flanking fire, prompted a military revolution, driving, for example, the growth in the size of armies. However, other commentators on military change have argued that other factors were just as important or that the key features of the *Trace Italienne* had been emerging from the fourteenth century through a process of evolutionary technological innovation.

Our first problem, then, in tracing the development of land warfare is that there is no consensus on how that development has been structured. For some, warfare on land has consisted of periods of relative stasis interrupted periodically by military revolutions that shift land warfare from one model of warfare to another. For others, though, it is continuity and evolution that are the critical factors that we should consider.

However, a second general difficulty that we have in analysing the development of land warfare lies in the problems of identifying what drives the changes that have taken place.

The drivers of change

Our problems in determining the development of land warfare are compounded by the fact that there is no consensus on what factors are most significant in driving that change forwards. This point is well illustrated if we return to the military revolutions perspective because, even amongst those writers who agree that military revolutions exist, there is a lack of agreement on what causes them.

Key terms

Discussing military revolutions isn't helped by the fact that there is a lack of a common vocabulary across the literature: many writers use the same terms to describe different things; many use different terms to describe the same thing. Whilst terms such as 'Military-Technical Revolution' (MTR), 'Revolution in Military Affairs' (RMA) and 'Military Revolution' crop up with frequency, some use the term RMA in the same way that others use Military Revolution; some use the term MTR in the same way that others use RMA. Some others use entirely different terms – 'sub-revolutions', for example.

Despite the varying views on military revolutions we can broadly divide the differing opinions into **three perspectives**, each providing rather different explanations for the causes of change. This general typology sorts out the ideas according to key themes rather than labels and focuses on the different weight accorded by each to factors such as:

- The importance of technology.
- The balance between internal and external factors in forcing change.
- The degree to which military revolutions are found or made.

For want of better terms, we can label the three approaches Military-Technical Revolutions, Revolutions in Military Affairs, and Political-Military Revolutions. In general, Military-Technical Revolution provides the narrowest explanation for military revolution, with the subsequent models adopting progressively broader explanations for the sources of change.

Military-Technical Revolutions

The term Military-Technical Revolution sees military paradigm shifts as a **predominantly technological phenomenon**. Radical military change is driven by technological innovation and consequent increases in the range, accuracy, mobility, complexity and interdependence of weapons systems (see Box 2.2). This view sees major changes in land warfare driven from below with new technology opening up new tactical and structural possibilities for armies, as well as shaping socio-political factors and opening up new strategic possibilities.

Box 2.2 Military-Technical Revolutions

Major-General Vladimir Sipchenko's six generations of warfare characterise this view, defining the development of warfare according to developments in weapons systems. Sipchenko characterises warfare as developing across six generations each being defined by key weapons systems:

- First Generation: infantry and cavalry forces without firearms
- Second Generation: gunpowder and smoothbore weapons
- Third Generation: rifled small arms and tube artillery
- Fourth Generation: automatic weapons, tanks and aircraft
- Fifth Generation: nuclear weapons
- Sixth Generation: data processing and precision weapons

(Randall G. Bowdish, 'The Revolution in Military Affairs: The Sixth Generation', *Military Review* (November–December 1995), 26)

Enthusiasts of this view argue that these kinds of technologically determined revolutions 'have been occurring since the dawn of history, they will continue to occur in the future, and they will continue to bestow a military advantage on the first nation to develop and use them'.[13] From an MTR perspective, revolutionary military change is a recurrent phenomenon caused by technological developments that have provided states with new opportunities to generate a military advantage over their adversaries. Much of the history of military transformation is popularly seen in this way: the history of the development of warfare is often seen as the history of changing technology. Viewed from an MTR perspective, the development of land warfare since 1900 is dominated by technological watersheds: machine guns, magazine rifles and quick-firing artillery produced deadlock; tanks restored mobile warfare; nuclear weapons marginalised large-scale, high-intensity conventional warfare; and precision-guided munitions and new sensor technology have created a new form of warfare that moves away from traditional verities such as the importance of mass.

Revolutions in Military Affairs

This perspective recognises that technology is rarely enough in itself to drive a military revolution: instead, **conceptual developments** are the crucial factor. In the early part of its introduction, most new technology does not fundamentally alter the way in which warfare is conducted. Often, new technology is used simply to serve existing objectives and is 'appliquéd' onto existing concepts. Sometimes, new technology may not be able to be turned into a practical military system at all: the technology may lack an appropriate operational concept, or the new system and concept may unacceptably challenge existing military culture or strategic precepts. Thus machine guns in most armies from the 1870s to 1880s, and French doctrine for the use of tanks in the inter-war period, provide examples where the potential impact of technology was much reduced by conservative methods of employment. In other words, the context in which technology is conceived and used must be conducive to change if radical military change is to occur.

So, the Revolution in Military Affairs perspective sees military revolutions developing from more than just emerging technologies: they require other components such as **operational innovation**, usually in the form of new doctrine, and **organisational adaptation**, to produce force structures able to leverage new technology or new methods of warfare. Revolutions in Military Affairs therefore involve 'the assembly of a complex mix of tactical, organizational, doctrinal, and technological innovations in order to implement a new conceptual approach to warfare';[14] they represent 'discontinuous increases in military capability and effectiveness resulting from combinations of new technology, doctrine and organization'.[15]

Box 2.3 Revolutions in Military Affairs

Lind highlights the interplay between technology and ideas. He identifies four generations of war in the modern period, each of which produced a new model of land warfare:

- First generation warfare featured the linear tactics of the smoothbore driven capabilities and limitations of muskets, as well as the social and conceptual implications of the French Revolution.
- Second generation warfare was driven by responses to rifled muskets, breech-loaders, barbed wire, machine gun and indirect fire. Linear tactics remained but with more emphasis on fire and movement. Telegraph, railways, and entrenchments led to the development of operational art.
- Third generation warfare was a response to increasing firepower and tactically saw the adoption of infiltration, dispersal, defence in depth, and operationally new techniques such as *blitzkrieg*.
- Fourth generation warfare is war of the future: non-state, irregular, decentralised.

(W.S. Lind, K. Nightengale, J.F. Schmitt, J.W. Sutton and G.I. Wilson, 'The Changing Face of War: Into the Fourth Generation', *Military Review* (October 1989), 3–4)

The Revolution in Military Affairs perspective tends to see the development of land warfare in terms of the introduction of **new techniques of war** which may often stem as much from using existing means in new ways as it does from new technology. These techniques include an 'artillery revolution' in the First World War, *blitzkrieg* in the Second World War, and manoeuvre warfare in the 1980s (these ideas are discussed in more detail in subsequent chapters); or organisational reforms such as the emergence of divisions and corps; or doctrinal innovation such as operational level theory, combined arms tactics, and indirect fire techniques.

RMAs and land warfare

By focusing on techniques and methods, a different story emerges on the development of land warfare – a story in which the changes in the prosecution of land warfare have been driven by successive military revolutions, such as:

- The **Infantry Revolution** of the fourteenth century, which saw the relative decline in the importance of cavalry and the emergence in importance of close formation

infantry based on the dissemination of new technology (longbows), new tactical innovations (close formation pike) and economic factors (the relative cheapness of foot soldiers).
- The **Artillery Revolution**, which compromised existing fortifications and was founded on technological improvements in the capabilities of cannon (range, speed of loading, mobility and size), and the growth in states (who could afford to exploit them).
- The sixteenth-century **Fortress Revolution**, which saw the emergence of new fortification techniques that helped negate some of the advantages of artillery.
- The **Gunpowder Revolution**, which saw combinations of new technology and new concepts lead to the emergence of the Swedish military system and linear tactics.
- The **Napoleonic Revolution**, built on the convergence of the developing Industrial Revolution and the political forces unleashed in the French Revolution, resulted in the increase in the size of armies, a shift to more mobile columnar tactics, and new forms of military organisation, especially the army corps.
- The **Land Warfare Revolution**, which stemmed from technology such as the railway, telegraph, and rifled weapons that enhanced strategic mobility, produced mass armies and redefined the relationship between military commanders and politicians.
- A **Mechanisation, Aviation and Information Revolution**, which emerged in the inter-war period, deriving from improvements in the internal combustion engine, aircraft, mechanisation and communications. This led to the emergence of new forms of organisation (such as mechanised divisions), and new concepts (such as *blitzkrieg*).
- Finally, a **Nuclear Revolution**, which occurred after 1945.[16]

It is still worth noting, though, that (as Box 2.3 illustrates) even within the RMA perspective authors differ on the sorts of RMAs that they identify.

Political-Military Revolutions

The Military-Technical Revolution perspective focuses on the transformative effect of new technology. The Revolution in Military Affairs perspective focuses on military methods, including how technology (new and old) is employed. What both of these have in common is their focus on the internal sources of the development of warfare: both views tend to see profound military change as something that originates from within military organisations themselves. Our third perspective, that of Political-Military Revolutions, sees military paradigm shifts as the result of the action of powerful **forces external to militaries**, such as **political, economic and social developments**: wider issues over which the military has no direct control. Advocates of this brand of military revolution have likened them to 'earthquakes': seismic shocks that originate outside of the military, with which armies must come to terms if they are to be effective. Reflecting this view, one author argues that 'Individuals or groups do not control military revolutions: they merely survive them'.[17]

The Political-Military Revolution school argues that perspectives on military change that focus on technology, doctrine and organisation, miss the complexity inherent in the nature of war. If warfare is 'violence articulated through strategy', then the nature of war is bound up with the complex interaction between a multiplicity of factors including politics, psychology, culture and belief.[18] The historian Charles Tilly, for example, highlights the **interdependent relationship** between the evolution of the modern state system and the prosecution of war. From the fifteenth century onwards, war became more complex and capital intensive. This helped to promote centralisation, bureaucratisation and monopolisation, especially in relation to finance and control over the means of coercion. This influenced the prosecution of warfare: supporting, for example, the shift from semi-private and mercenary forces to professional militaries, and the growing size of armies.[19] Likewise, the historian William McNeill identifies the important military implications of phenomena such as population growth, 'marketisation' and political-industrial revolutions.[20]

Military revolutions, then, are **driven by fundamental political, economic and societal changes**. These fundamental changes act as a major source of change in the character and conduct of warfare. Viewed from this perspective, the development of land warfare since 1900 needs to be understood as an expression of the wider shift from industrial to post-industrial societies. Major revolutions in the prosecution of land warfare have been the result of systemic change in politics and society. As such, these revolutions are marked, from a military perspective, by their unpredictability and essential uncontrollability – they have important potential military implications, but the military themselves have little ability to shape them. These major revolutions therefore change the capacity of states to create and project military power.

Political-Military Revolutions and land warfare

The Political-Military Revolutions perspective generates yet another way of constructing the development of land warfare. So, for the futurologists Alvin and Heidi Toffler, the development of warfare can be understood as comprising three generations, each of which has been driven by changes in the kinds of societies that wage it:

- **First Wave War** is Agrarian War: agriculture allowed societies to create surpluses to sustain war, and facilitated the creation of states. War in this form was limited, decentralised and personal.
- **Second Wave War** is Industrial War: this is based on mass production allied with mass participation as begun during the French Revolution, and marked the emergence of a new style of war. Unlimited, standardised, professionalised second wave warfare emerged in the nineteenth century and dominated the twentieth.
- **Third Wave War** is Post-Industrial War: knowledge-based economies have created knowledge-based warfare, based on information, communication, precision and the creation of networked systems.[21]

These forms of war **can exist in parallel**: for example, the Tofflers see the Gulf War of 1990–91 as a war between a Second Wave society (Iraq) and a Third Wave society (the United States), a fact that helps explain the enormity of US military success there. However, returning to a point made in the preceding discussion, even where commentators can agree on the origins of the changes in land warfare, this doesn't necessarily translate into an agreement on which developments constitute true military revolution (see Box 2.4).

Box 2.4 Alternative Political-Military Revolutions

In his book *The Scientific Way of Warfare: Order and Chaos on the Battlefields of Modernity*, Antoine Bousquet argues for a fundamental relationship between science and warfare: in particular, he argues that scientific worldviews have had a profound impact on military theory and practice. For Bousquet, therefore, the development of land warfare over time can best be explained in terms of four scientific metaphors:

- **Mechanistic Warfare** is associated with a time of basic mechanical constructs such as the clock. It is associated with a form of warfare based on close order drill and rigid tactical deployments such as those of the armies of the eighteenth century.
- **Thermodynamic Warfare** is associated with the development of engines and the application of ever growing flows of energy to warfare. Its military manifestations include mass mobilisation, motorisation and industrialisation in the period from the Napoleonic Wars through to the Second World War.
- **Cybernetic Warfare** was built on the harnessing of electro-magnetic forces encapsulated by the development of the computer. Developing after the Second World War, this is warfare characterised by decentralisation of control and ever greater automation.
- **Chaoplexic Warfare** emerged from the mid-1970s and was built on network technology and such scientific concepts as non-linearity. In a world increasingly chaotic and complex (hence 'chaoplexic') cybernetics and information offer future ways of warfare involving ever more decentralised and autonomous forces.

Conclusions

The Introduction to this book argued that one of the key themes inherent in an analysis of land warfare is uncertainty. In this chapter we have identified that this uncertainty applies even to as basic a question as 'Why and how has land warfare developed over time?' Any attempt to answer this question is plagued with a range of thorny issues:

- What balance should we strike between continuity and change? Has land warfare developed by a process of irregularly paced evolution, or can it be characterised as the move from one paradigm to another, each watershed marked by a stark military revolution?
- If military revolutions do exist then how many revolutions have there been? Some perspectives see a small number of fundamental changes; others see numerous military revolutions occurring.
- What drives change in land warfare? Does it just 'happen' as a result of broader political, economic and social change, or does it occur through the deliberate activities of innovating militaries?
- Does change derive from a perceived military challenge? Do new technology or new operational concepts emerge as a conscious response to threats, or do they occur anyway?
- What role does new technology play in the development of land warfare? Is it the decisive factor, is it an uncertain catalyst, or are ideas more important than materiel?

Why, though, given these uncertainties, would we muddy the waters so much by covering these issues at all? The importance of highlighting the debates on military change becomes immediately important when we consider the next two chapters of this book. Chapters 3 and 4 analyse the development of modern land warfare: Chapter 3 focuses on modern tactics, while Chapter 4 examines modern land warfare at the operational level. In terms of the debates outlined in this chapter, the analysis in Chapters 3 and 4 focuses on the importance of new ideas in the context of two definable watersheds: the First and Second World Wars. However, the essentially contested nature of the debate on the development of land warfare is also vital to understand because the way in which the history of land warfare has been interpreted by some analysts both inside and outside of the military has had a direct and significant impact on how armies have prepared themselves for war. Often, conclusions on the requirements of land warfare in the future have depended upon assumptions about what has been important in the past. We will return to the importance of this point in the third part of *Understanding Land Warfare*, but, as we will see, its implications have a resonance for almost all of the themes explored in this book.

Notes

1 Andrew F. Krepinevich, 'Cavalry to Computer: The Pattern of Military Revolutions', *The National Interest* No. 37 (Fall 1994), 30.
2 Colin S. Gray, *Strategy for Chaos: Revolutions in Military Affairs and the Evidence of History* (London: Frank Cass, 2002), 43; Benjamin S. Lambeth, 'The Technology Revolution in Air Warfare', *Survival*, Vol. 39, No. 1 (1997), 75.
3 See Antulio J. Echevarria and John M. Shaw, 'The New Military Revolution: Post-Industrial Change', *Parameters* (Winter 1992–93), 70–79.
4 Earl H. Tilford, Jr, *The Revolution in Military Affairs: Prospects and Cautions* (Carlisle, PA: Strategic Studies Institute, June 1995), 6.
5 Kelly DeVries, 'Catapults are Not Atomic Bombs: Towards a Redefinition of "Effectiveness" in Premodern Military Technology', *War in History*, Vol. 4, No. 4 (November 1997), 460–64.
6 Patrick M. Morgan, 'The Impact of the Revolution in Military Affairs', *The Journal of Strategic Studies*, Vol. 23, No. 11 (2000), 135.
7 Michael S. Neiberg, *Warfare in World History* (London: Routledge, 2001), 100.
8 Grey, *Strategy for Chaos*, 5.
9 Ibid., 53.
10 Cyril Falls, *A Hundred Years of War* (London: Gerald Duckworth, 1953), 12–13.
11 R.A. Buchanan, 'The Structure of Technological Revolution', *History of Technology* Vol. XVI (1994), 199–211, quoted in DeVries, 'Catapults are Not Atomic Bombs', 470.
12 Richard J. Dunn III, *From Gettysburg to the Gulf and Beyond: Coping with Revolutionary Technological Change in Land Warfare*, McNair Papers, No. 13 (Washington, DC: The Institute for National Strategic Studies, 1992), 3.
13 O. Hundley, *Past Revolutions, Future Richard Transformations: What Can the History of Revolutions in Military Affairs Tell Us about Transforming the U.S. Military?* (Monterrey, CA: RAND, 1999), (iii).
14 MacGregor Knox and Williamson Murray, 'Thinking About Revolutions in Warfare', in MacGregor Knox and Williamson Murray, *The Dynamics of Military Revolution, 1300–2050* (Cambridge: Cambridge University Press, 2003), 12.
15 Eric R. Sterner, 'You Say You Want a Revolution (in Military Affairs)?' *Comparative Strategy*, Vol. 18, No. 4 (October/December 1999), 299.
16 Krepinevich, 'Cavalry to Computer', 31–36.
17 Mark Grimsley, 'Surviving Military Revolution: The U.S. Civil War', in MacGregor Knox and Williamson Murray, *The Dynamics of Military Revolution, 1300–2050* (Cambridge: Cambridge University Press, 2003), 74.
18 Tilford, *The Revolution in Military Affairs*, 1.

19 Charles Tilly, *Coercion, Capital, and European States AD 990–1992* (Malden, MA: Blackwell, 2002), 67–95.
20 William H. McNeill, *The Pursuit of Power: Technology, Armed Forces, and Society Since 1000 AD* (Chicago, IL: University of Chicago Press, 1984).
21 Alvin Toffler and Heidi Toffler, *War and Anti-War: Survival at the Dawn of the 21st Century* (London: Little, Brown and Co., 1993).

Suggested reading

Jeremy Black and Donald M. MacRaild, *Studying History* (Basingstoke: Palgrave, 2000). Part I, in particular, is useful for illuminating the difficulties of using history as an objective tool.

Antoine J. Bousquet, *The Scientific Way of Warfare: Order and Chaos on the Battlefields of Modernity* (Chichester, NY: Columbia University Press, 2009). Another advocate of the RMA, Bousquet presents a contestable but nevertheless very interesting example of the Political-Military Revolution school.

Jeffrey R. Cooper, *Another View of the Revolution in Military Affairs* (Carlisle, PA: Strategic Studies Institute, 1994). Amongst other things, Cooper explores different types of RMA and highlights the uncertainties and debates surrounding the RMA concept.

Colin S. Gray, *Strategy for Chaos: Revolutions in Military Affairs and the Evidence of History* (London: Frank Cass, 2002). An incisive and comprehensive assessment of the nature and value of the military revolutions: it is particularly useful for its assessment of the different forms of RMA and the problems with the concept.

MacGregor Knox and Williamson Murray, *The Dynamics of Military Revolution, 1300–2050* (Cambridge: Cambridge University Press, 2003). A strong advocate of the value of the RMA concept; also presents an interesting composite perspective.

Clifford Rogers (ed.), *The Military Revolution Debate: Readings on the Military Transformation of Early Modern Europe* (Boulder, CO: Westview Press, 1995). A comprehensive overview of the competing debates on military revolutions from an historical perspective.

3 Modern tactics

Key points

- Modern tactics emerged during the First World War as a means of coping with the effects of the growing lethality of firepower.
- Modern tactics comprises a system that integrates decentralised command and control, fire and movement, dispersion, cover and concealment, suppression, and combined arms down to the sub-unit level.
- Developments since the First World War have not substantially changed the nature of the tactics of modern land warfare.

How do modern armies fight? The preceding chapter has outlined some of the key theoretical aspects of land warfare. The purpose of this chapter, and the one that follows, is to put this theory into the context of the historical development of modern land warfare since 1900. This chapter charts the development of **modern tactics**; it does this by examining the emergence of tactical concepts surrounding defence in depth and infiltration which form the core of modern tactical land warfare.

The hierarchy of the levels of war outlined in the Introduction to this book placed tactics at the bottom, below the operational level of war; however, this should not be taken to mean that tactics is the least important element in land warfare: this is not the case. Tactics is associated with the conduct of battles; whilst winning battles does not, in itself, guarantee victory in war, overall success is likely to be difficult to achieve if an army cannot outperform its opposition tactically. Tactical methods, however, have evolved over time. The tactics of modern land warfare emerged during the First World War; it was essentially innovation in tactics that restored manoeuvre to conventional combat on land. In terms of the themes outlined in Chapter 1, whereas traditional techniques tended to emphasise more such ideas as centralisation, concentration and conformity, the new approaches to tactics placed a much greater emphasis on such concepts as fire and manoeuvre, decentralisation, dispersal, initiative and, especially, combined arms.

This chapter begins by outlining the basic precepts of modern tactics. It then concerns itself with analysing how and why those tactics have evolved, focusing first on tactical methods prior to the First World War, and second, on the evolution that took place during the war itself. As the analysis demonstrates, there is a **strong evolutionary element** in the development of land warfare as we see it today. In terms of tactics, it was the First World War that acted as a catalyst for the development of the basic foundations for land warfare ever since; continuity and adaptation have been at the core of modern land warfare.

The fundamentals of modern tactics

The tactical level of war is:

> The level of war at which battles and engagements are planned and executed to achieve military objectives assigned to tactical units or task forces.[1]

Military tactics concern **the use of units to defeat the enemy in battle**; they comprise the lowest scale of military activity. Tactics constitute the currency of operational level military activity. Tactical success matters: as the strategist Colin Gray argues, 'Policy and strategy will propose, but ultimately it is tactics, which is to say people in combat, which must dispose'.[2] The British Field Marshal Sir William Slim (1891–1970) expressed this point when he asserted that there would come a moment in any war when:

> the General, however skilful and far-sighted he may have been, must hand over to his soldiers, to the men in the ranks and to their regimental officers, and leave them to complete what he had begun. The issue then rests with them, on their courage, their hardihood, their refusal to be beaten either by the cruel hazards of nature, or by the fierce strengths of their human enemy.[3]

As Chapter 1 has shown, one of the key agents of military change in the twentieth century was the growing lethality of modern firepower. One response historically to the challenge of firepower has been **technological innovation**: one of the dynamics in warfare is the constant competition between weapons platforms and the anti-weapons platforms technology designed to counter them. Tanks, for example, were one solution to the power of small arms and machine guns in the first half of the twentieth century, but these spurred the development of anti-tank weapons; similarly anti-aircraft weapons emerged to counter the developing capabilities of aircraft. However, the most significant method of countering the dramatic growth of firepower since the second half of the nineteenth century has been **tactical innovation**: developments in the methods and techniques used at the tactical level. The importance of tactical innovation is evident in the non-linear relationship between increases in firepower and casualty levels. Despite the growing range and lethality of firepower outlined in the previous chapter, relative casualty rates have, very broadly, declined. Average daily battle casualty rates in the nineteenth century were between 10% and 20% of an army. By the Second World War this had dropped to between 1% and 3%.[4]

Modern system tactics reflect the evolution over time of a number of themes that, collectively, have allowed armies to operate in the face of modern firepower. Tactically, effective application of these themes means that weapons technology does not reach in practice the lethality that in theory it might be able to deliver. So, what are the fundamentals of modern tactics?

The principles of modern tactics

Modern-system tactics comprise the elements of:

- Cover and concealment.
- Suppression.
- Dispersion.
- Decentralisation.
- Small unit manoeuvre.
- Integration.[5]

These themes constitute a system because they are mutually dependent upon one another. Tactics that attempt to apply only some of these themes are unlikely to work effectively against modern firepower. The first three of these themes constitute the means through which troops can mitigate the effects of firepower.

Cover and concealment

In order to reduce the effects of firepower, modern tactics emphasises **the use of terrain** for passive protection against fire. Terrain can be used to provide cover; that is, it can be used to provide physical protection against attack. For example, troops can place themselves behind walls or hills or within buildings and fortifications. Even if it does not provide much cover, such terrain as long grass or hedges can still conceal troops by hiding them from enemy observation, preventing the enemy from spotting them and bringing firepower to bear against them.

Suppression

Suppression is an active method of reducing enemy firepower. It comprises the use of heavy and continuous **friendly fire against enemy positions** to interfere with the enemy's ability to observe the battlefield, fire their weapons and co-operate effectively. The focus is not on killing or destroying the enemy but on using the threat of such outcomes to encourage their passivity.

Dispersion

Another key means of reducing the effects of firepower is to **widen the distances** between individual soldiers within a unit, and to increase the distances between units. This reduces exposure to such area effect weapons as artillery and mortars, and reduces exposure to the 'beaten zone' of such automatic weapons as machine guns.

The remaining three themes (decentralisation, small unit manoeuvre, and integration) provide the enabling conditions to utilise the previous themes effectively.

Decentralisation

To facilitate the use of cover, concealment and dispersion, modern tactics feature the **disaggregation** of larger military formations into small sub-units for the purposes of manoeuvre and combat. Typically, this sub-unit is a squad or section that is further divided into two fire teams (see Figure 3.1). This form of organisation requires the exercise of leadership and initiative by junior non-commissioned officers. Squads typically manoeuvre in a fluid fashion (see Figure 3.2).

Small unit manoeuvre

In addition to allowing the best use of the terrain, **manoeuvring in small sub-units** facilitates the use of suppressive fire at the tactical level through the means of fire and movement. Typically, one fire team acts as a 'fire base', laying down suppressive fire against an enemy position, whilst the other manoeuvres against the enemy, making best use of the terrain. This can also be done alternately by each fire team, allowing the squad to 'leap-frog' forwards. The principle of fire and movement is central to modern tactics.

Integration

Realising the previous themes requires effective integration. This integration is partly based on **combined arms**. Effective suppressive fire, for example, may require co-ordinating artillery, mortar and air units to support an infantry assault. This kind of integration requires training and organisational flexibility. However, integration is also required in those areas, such as command, control and communications, required to facilitate the co-ordination of often widely dispersed forces.

Collectively, these themes constitute the basis for the tactics of modern system land warfare, but why and how did this system develop? To answer these question we need to consider the tactical ferment that existed prior to the First World War.

Land warfare, 1900–14

The armies of the First World War began with much of the familiar paraphernalia of modern warfare: magazine rifles, machine guns, quick-firing artillery, and aircraft. They were supported by the political, economic, bureaucratic and social foundations of the Industrial Revolution: they were (with the immediate exception of the British Expeditionary Force) mass armies, mobilised by railways, controlled through telephone and telegraph, guided by professional military staff systems. Yet, the initial clashes between German and Allied forces seemed to demonstrate that an appalling gap existed between industrial and **technological modernism** and seemingly **archaic tactical systems** based on mass and linear tactics. The effects of this were evident to British troops at Mons in 1914: as they defended their position, one British soldier commented of the German attack:

> [They] advanced in companies of quite 150 men in files five deep … The first company were simply blasted to heaven by a volley at 700 yards and, in their insane formation, every bullet was bound to find two billets.[6]

The First World War was a conflict long expected by European militaries. Yet, in the face of radical increases in the lethality of modern firepower, the tactical systems of the armies of the time seemed almost wilfully inappropriate. The French infantry regulations of 1895, for example, finished its description of the desirable conduct of an infantry attack with the flourish: 'At a signal from the Colonel the drums beat, the bugles sound the advance and the entire line charges forward with cries of "*en avant, à la baionette*".'[7] Yet, the debates on military change in the period leading up to the First World War were, in fact, **extensive, far-reaching, and often well-informed**.

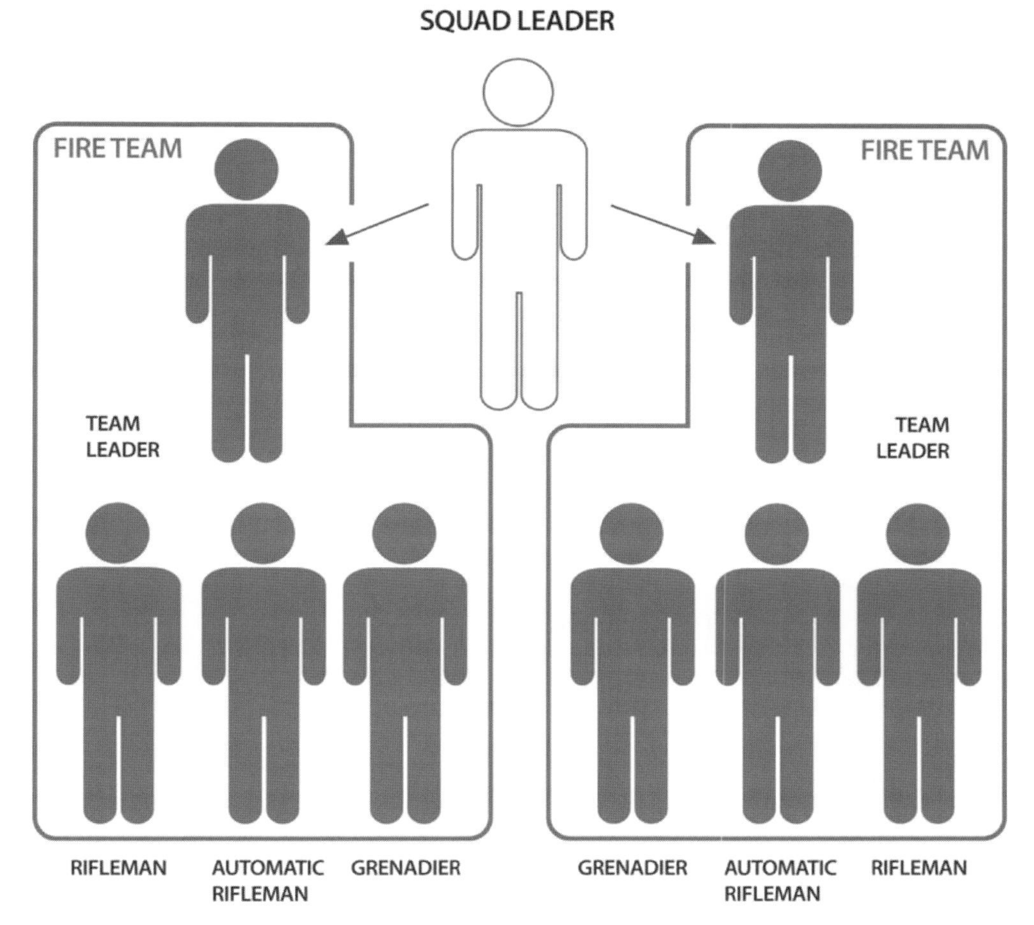

Figure 3.1 A modern infantry squad
Source: (Adapted from Field Manual 7–8, *The Infantry Platoon and Squad* (Headquarters of the Army, 22 April 1992))

Firepower versus the bayonet

As Chapter 1 has noted, the second half of the nineteenth century saw major changes in the technology of war: in particular, **battlefield firepower increased dramatically**. For armies of the period, there was ample evidence of the practical consequences of the increasing range, accuracy and rate of fire of infantry small arms, artillery and machine guns.

- In the **Boer War (1899–1902)** British troops in close formation had suffered badly at the hands of skilfully deployed Boer troops armed with Mauser rifles.
- In the two **Balkan Wars (1911–13)** all sides suffered heavy casualties as they attempted to execute tactical offensives against entrenched defenders buttressed by the support of machine guns.
- Perhaps the most salutary evidence of the implications of change came in the **Russo–Japanese War (1904–05)**. There, at the battle of Mukden, both sides

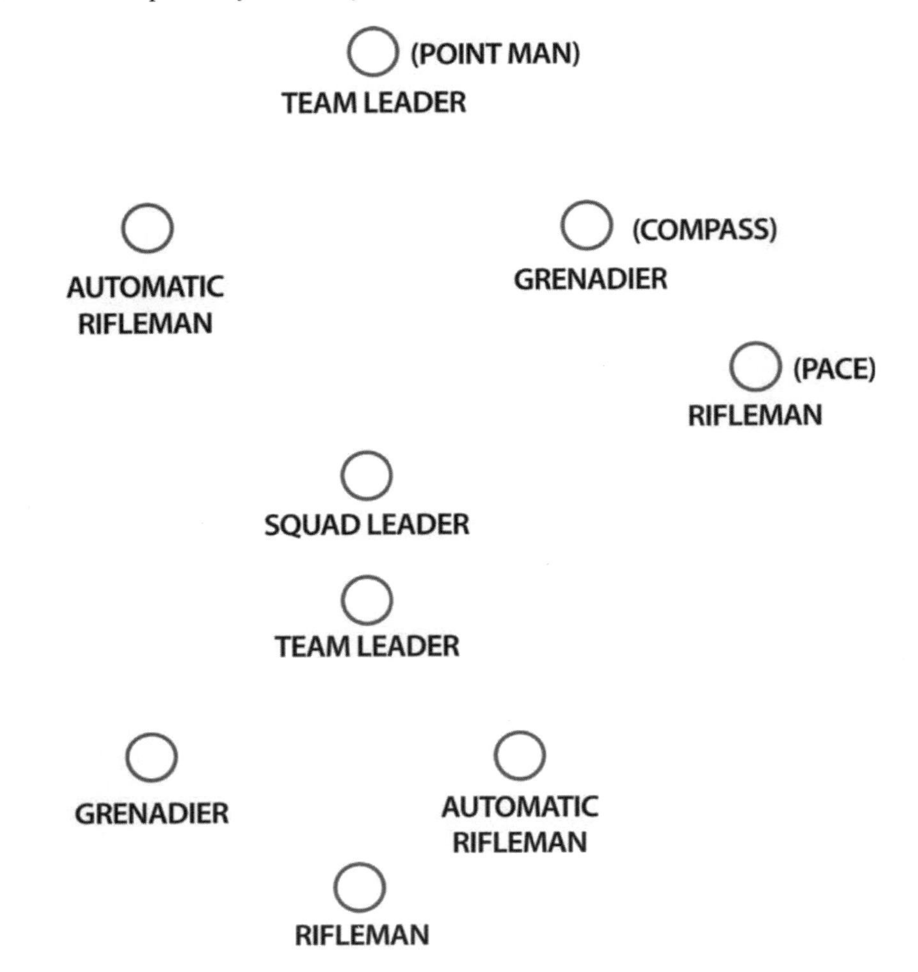

Figure 3.2 Disposition of a modern infantry squad
Source: (Adapted from Field Manual 7–8, *The Infantry Platoon and Squad* (Headquarters of the Army, 22 April 1992))

deployed more than 270,000 men, making it at that point the largest battle in modern military history. The casualties (90,000 Russian and 70,000 Japanese) were testament to the difficulties of offensive operations using tactical systems that continued to focus on mass rather than dispersion. The Russian army was defeated, but not destroyed, and the Japanese finished the war materially exhausted, having suffered very high casualties.

Yet, despite this evidence of the lethality of modern firepower, European armies began the First World War with tactical doctrines that continued to prioritise the **massing of troops** under the **close control** of commissioned officers (see Figure 3.3). Moreover, these tactics were put to work in the service of a 'Napoleonic paradigm' of warfare: a paradigm that saw the active pursuit of decisive battle as the fundamental route to success in war.[8]

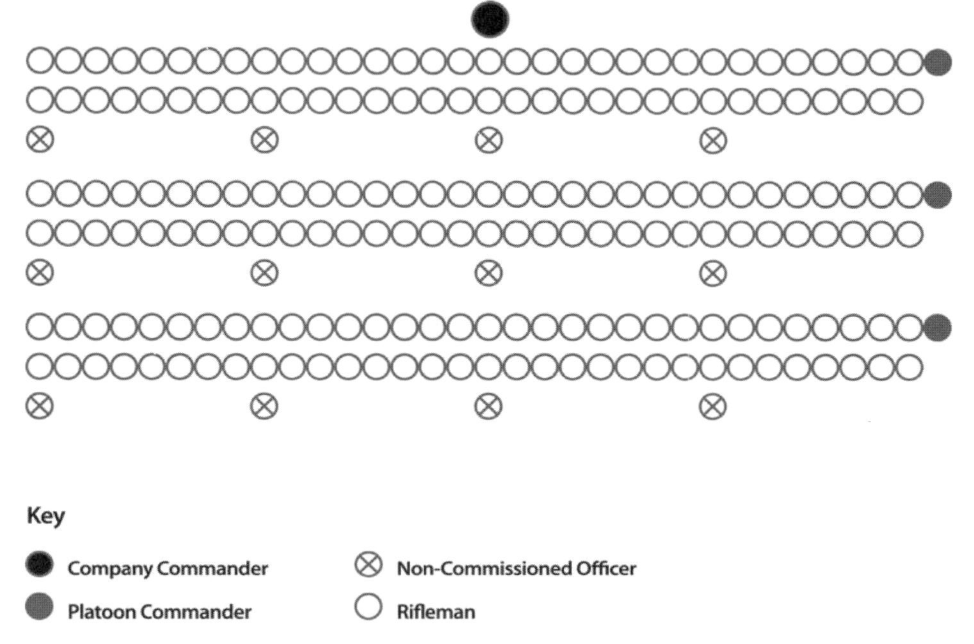

Figure 3.3 Column of platoons, 1914. Compare this with Figure 3.1.
Source: (Gudmundsson, *Storm Tactics*, 9)

In fact, though, this outcome was not the result of blind indifference to the implications of technological change on the battlefield. It was at the time widely recognised that modern battlefields were dangerous places. As **Captain A.D. Schenck** of the US army marvelled in 1895, 'Within a single generation … the art of war has changed more than it has ever changed since the advent of gunpowder'.[9] Nor were Schenck's views isolated ones: **General Victor Bernard Derrecagaix** (1833–1915) of the French army noted after the Franco–Prussian war that 'the rules of the military art have undergone changes which every day are affirming themselves more, and which put armies in a true period of transition'.[10] Military writers such as **G.F.R. Henderson** (1853–1904), **C.E. Callwell** (1859–1928) and **Colmar von der Goltz** (1843–1916), and civilians like **Henry Spenser Wilkinson** (1853–1937) and the writer **H.G. Wells** all contributed to a vibrant debate on technology and future warfare.

It was recognised that modern technology had widened the so-called 'zone of death' between opposing armies. However, there was a lack of consensus on how this problem should be resolved. Broadly, the debates on the implications of change fell into two camps.

- First, a **firepower school** argued that the implications of developments in firepower required radical changes in approach by armies: in particular, it would require tactical doctrines that focused on **dispersion**, **suppressive fire**, **decentralisation** and the exercise of **greater initiative** by ordinary soldiers.
- A second school, a **moral school** that espoused 'the cult of the offensive', argued that whilst modern firepower had certainly made offensive operations more difficult, this

problem was best addressed through a **focus on the moral dimensions** of warfare: heavy casualties were likely inevitable in modern warfare – victory could still be obtained if troops were **motivated enough** to cross the 'zone of death' and close with the enemy.

For example, even in the Imperial Russian army that had seemed so unprepared for the Russo–Japanese War, the debates on change had been ongoing since the 1850s:

- The **moral school** of thought was championed by **General Mikhail Ivanovich Dragomirov**. Drawing on the ideas of the eighteenth-century Russian hero General A.V. Suvarov, Dragomirov saw the solution to the problems of the late nineteenth-century battlefield as lying in military psychology: material factors could be overcome if sufficient **élan and esprit de corps** were generated in the troops. Thus, a focus on *vospitanie* (indoctrination) and *obrazavanie* (training) would give troops the courage and capacity to continue offensive operations despite the problems of the modern battlefield.
- However, other Russian officers argued that a return to the past was unacceptable and that the new conditions of warfare required radical changes to modern armies. **General Aleksei Nikolaevich Kuropatkin** and **Colonel Alexandr A. Neznamov**, for example, noted amongst other things that future battlefields would become **increasingly lethal**, and that modern conditions would require greater **dispersion** and **less centralisation** in command systems.[11]

Resolving the debate

By 1914, the moral school was in the ascendant, a result that with hindsight seems inexplicable given the demonstrated increases in firepower. But there were many conditions that shaped this outcome:

The demands of strategy

First, and a point developed more fully in Chapter 5, debates on military change have a broader context. One of these contexts is **strategic need**. There were those before the First World War that argued that the 'cult of the offensive' would lead to disaster. The French socialist Jean Jaurès argued that the Napoleonic paradigm was an historical fallacy and that technology had now made defence, not attack, the stronger form of war. Similarly, the Polish financier **Jan Bloch** (or Jean de Bloch) argued in his *Modern Weapons and Modern War* that increases in firepower would lead inevitably to the triumph of the defence, a situation that would lead future wars to be marked by entrenchment and protraction. However, many others regarded a **short war as a necessity**: either because the mass mobilisation of manpower and industry required for modern warfare could not be sustained economically or politically for an extended period of time or because large early successes would be a prerequisite for success in the long term.[12]

Cultural determinants

Another of these broader contexts relates to **values**. Some of these beliefs were transnational in nature: the power generally, for example, of the notion of **Social Darwinism** in the period before the First World War encouraged a focus on offensive doctrines as the natural expression of a strong society whilst defensive doctrines were characterised as those of the weak. It also encouraged observers to discount some of the evidence from other wars because the difficulties experienced were taken to be racially based reflections of a lack of competence: thus some of the lessons of the Russo–Japanese War could be dismissed because the participants were dismissed as actual or quasi-Asiatics.

Domestic politics

At a national level, military assessments were influenced by **prevailing political assumptions**: in Russia, for example, the power of Dragomirov's views, with its echoes of the hero Suvarov, were powerfully reinforced by nationalism and the desire of the so-called 'national school' for a military system rooted in specifically Russian experience. At the level of military organisations themselves, further assumptions shaped the limits of acceptable military adaptation. In an era of conscript armies, for example, it was simply assumed that ordinary soldiers would be incapable of complex tactical evolutions. A US army officer observer of German manoeuvres in 1893 commented:

> They evidently intend to handle their infantry in close lines in the next war. The average German private is not a person to be turned loose in a skirmish line and left to a certain degree to his own devices … They prefer to lose men than to lose control of the officers over them.[13]

Assessing the lessons of combat

Another problem was that the **relevance of battlefield experience** of the past to current conditions was **not always easy to discern**. For example, there were many European military observers present during the American Civil War. Features of the war such as the dominant role played by entrenchments, the cost and difficulty associated with battlefield offensives, the importance of skirmishers, and the role of cavalry as mounted infantry rather than a battlefield shock force were all noted and debated. However, it was difficult for armies to determine which of the lessons of the war might be of general applicability and which were specific to the US context. European armies tended to conclude that the character of the American Civil War was dictated by peculiarly American conditions: the reliance on militia armies; an assumed lack of military professionalism rooted in inexperience and an individualist political culture; and American geography.

What were perceived as swift, decisive conflicts such as the **Danish War of 1864**, the **Austro–Prussian War of 1866**, and the **Franco–Prussian War of 1870–71**, seemed to demonstrate that aggressive offensive operations could still work even in the face of rifled (and in the case of the 1870–71 war, breech-loading) small arms. Whilst the Russo–Japanese War should perhaps have been a more direct and immediate exemplar for the future, its lessons too were opaque. The Japanese had, after all, won, even if they had taken heavy casualties, and they been trained by French and German military advisers, so their victory was taken as a vindication of the existing European tactical systems.[14]

Practical difficulties

Importantly, however, there were **clear practical problems** with the solutions espoused by the firepower school. Technology had developed in some areas much faster than others. Firepower had increased dramatically at the tactical level, but **tactical communications** had not. It was widely recognised at the time that dispersal was an obvious way of mitigating the increases in battlefield firepower. Equally, it was accepted that suppressing enemy defences through the use of artillery barrages and small-arms fire was another desirable way of enabling attacking forces to close with the enemy. However, without wireless communication, dispersion would place the bulk of troops beyond the command and control of their officers: attacks would therefore lose forward momentum, dissolving into an exchange of fire that would allow the defender to bring his artillery to bear on the attacking troops; suppression, too, was difficult to rely on as a decisive element, since without wireless communications it would prove extremely **difficult to co-ordinate** an infantry attack with indirect artillery fire. On the basis of lessons from the Boer War, for example, the German army was by 1902 experimenting with so-called 'Boer Tactics' – replacing close platoon columns with dispersed skirmish lines. However, skirmish tactics spread a battalion over a frontage of 3,000 metres, making it impossible for officers to maintain command and control over the formation. By 1903, therefore, the German army had reverted to its previous close-order methods.[15]

Military doctrine in 1914

Thus, in 1914, the belligerents continued to pursue a general concept of war that focused on rapidity, manoeuvre and concentration, directed towards bringing on a decisive battle. Tactically, however, the doctrines employed by the armies of 1914 were not blindly conservative; they were, instead, **hybrid** constructs mixing a strong element of the moral school with aspects of the firepower approach.

The influence of the moral school

Of European armies, it was perhaps the French who were the most enthusiastic adherents of what became characterised as the 'cult of the offensive'. Fervent supporters such as **Colonel Louis Grandmaison** (1861–1915) inculcated in the French military the notion of the 'offensive *a l'outrance*': it was the attack that would seize the initiative from the enemy and break them in the contest of wills, a notion articulated by the French military in the belief that 'The will to conquer is the final condition of victory'.[16] The French 1914 infantry regulations reflected this outlook, stating that 'The bayonet is the supreme weapon of the infantryman'.[17] However, the moral school percolated through the other European militaries as well. Lieutenant General Sir Ian Hamilton, a British observer during the Russo–Japanese War, dismissed 'all that trash' written by Bloch and argued instead that 'War is essentially the triumph … of one will over another weaker will'.[18] The German Drill Regulations of 1906 echoed these views: 'Infantry must cherish its inherent desire to take the offensive; its actions must be guided by one thought, viz, forward upon the enemy, cost what it may.'[19]

The influence of the firepower school

However, this focus on the moral school did not exclude new technology or, indeed, many of the facets of the firepower school. Even arch exponents of the cult of the offensive were clear that modern armies needed to embrace change. The French **Colonel** (later Marshal) **Ferdinand Foch** (1851–1929), for example, was an enthusiastic exponent of the importance of the moral aspects of war; however, he saw this as founded upon, and not at the expense of, integrating modern technology – providing troops with the full panoply of modern weapons technology was a necessary prerequisite for building their morale. Dragomirov was even more explicit in embracing some aspects of the firepower school: he recognised that modern battlefield firepower required **dispersion**, **flexibility** and **concealment**, and that this would in turn require **small unit manoeuvre** and **decentralised command** structures that would require capable, educated junior officers. The German Field Regulations of 1908 were clear in regard to the importance of these latter points:

> It is no less important to educate the soldier to think and act for himself. His self-reliance and sense of honor will then induce him to do his duty even when he is no longer under the eye of his commanding officer.[20]

In this vein, **entrenchments** and **machine guns**, which would form such an important part of the fabric of the battlefield in the First World War, were also in a fashion embraced. Machine guns were deployed at a ratio generally of two per battalion, and were allocated roles such as protecting flanks or providing fire support from the rear. Entrenchments, and attacks on entrenchments, were a routine part of military exercises in the period immediately preceding 1914.

The complexity of doctrine in 1914

This theoretically hybrid approach was **reflected doctrinally**. German doctrine, for example, saw the bayonet charge as the only way to seal victory; firepower, however, would be necessary in order to establish the conditions under which the bayonet charge could be delivered successfully. Thus, German troops would enter the battlefield in close order: if artillery fire could suppress the defenders then they would stay in close order and deliver the charge; if the defenders were not suppressed, however, the formations would be broken down into smaller groups and would advance by rushes to a point 400–500 metres from the enemy, where they would establish a firing line. This firing line would endeavour then to suppress the enemy in order that a final charge in close order could be launched.[21]

However, these doctrines were always likely to be **difficult to employ effectively** because they were **untested** in actual combat and they were based, therefore, on specific, and unproven, **assumptions** regarding the character of future land warfare. Entrenchments, for example, were viewed as a likely temporary feature in any combat because the expectation was that any future war would be marked by **fluid encounter battles**: troops either would not have time to dig in extensively, or these static defences would be quickly flanked. In this dynamic war of manoeuvre, machine guns would simply be too heavy to keep pace with the infantry; they would, moreover, expend too much ammunition or simply jam. In a mobile war, it was expected that artillery would need **small, mobile field guns** because they would be used in a direct fire role against troops in the open. For this reason, application of hybrid approaches was poor: field exercises often paid only lip-service to them and in 1914 the **approaches used varied** from rigid mass through to more dispersed formations, depending upon the views of local commanders.

Not all of the potential lessons of the period were erroneous. On the basis of their analysis of the Russo–Japanese War, for example, the German army invested in howitzers and heavier artillery in order to deal with the potential problems of the well-entrenched defence. However, in general, armies in 1914 focused on close order tactics and offensive action in the context of a war of manoeuvre; this focus left armies ill prepared for the challenges of the First World War.

Adaptation and the First World War

> In effect the Germans invented the modern battlefield.
>
> (Murray, 'The West at War', in Parker (ed.) *The Cambridge History of Warfare*, 295)

The military clashes of 1914 exposed the inadequacies of pre-war military thought. The German plan drawn up by Alfred Graf von Schleiffen, Chief of the General Staff from 1891–1905, and developed by his successor, Helmuth von Moltke the Younger, was founded on assumptions of a mobile war of rapid manoeuvre: aiming to create a gigantic envelopment of the French forces, 1.3 million German troops were committed to what was hoped would be a 39-day campaign. Victory against the French would allow German forces to be transferred to the east on day 40 in order to fight the Russians. However, the Schleiffen plan fell apart under the friction of real war, and attempts to restore mobility through turning movements ended after the 'race to the sea' led by November 1914 to the **creation of a continuous front** of defences on both sides. With no flanks to turn, assaults had to be launched frontally, but here, the **defensive power** of machine guns and magazine rifles were compounded by the poor performance of light field artillery in indirect fire against entrenchments.

Adaptation, 1914–15

However, adaptation to the real conditions of the war took place relatively quickly, a development notable given the problems posed in training, equipment and command by the large, conscript armies that existed:

- First, mass infantry formations were replaced by **skirmish lines** in order to reduce casualties.
- Second, more emphasis was placed on **technology**: hand grenades, for example, were widely distributed to all troops. Flamethrowers were in use by January 1915. In April 1915, gas was first deployed.
- Third, lavish investment was made in **heavy artillery**. Massive investment in the number and size of guns was accompanied by efforts to improve their effectiveness in an indirect fire role, such as deploying forward observers, using map-based firing, aerial observation and the use of timetables, flares and signal equipment to try to improve artillery–infantry co-operation.

On the basis of these developments, three general offensive concepts emerged during the first part of the war:

- The first focused on **restoring mobility** through a three-stage battle: artillery preparation; infantry breakthrough; and then exploitation by cavalry that would collapse the enemy's front. This was, for example, the general structure of the British plan on the Somme in 1916.

- A second concept, '**bite and hold**' was advocated by those, such as General Sir Henry Rawlinson, commander of the British 4th Army, who believed that breakthroughs could not be practicably achieved: artillery would crush the enemy defences, infantry would then occupy and consolidate in a series of limited operations.
- A third concept was reflected in that used by the Germans at Verdun in 1917. There, breakthrough was rejected, and the purpose of the massive use of artillery was simply to bleed the French army to death through massive **attrition**.[22]

In all cases, artillery was the key: the mantra of attack became increasingly 'the artillery conquers, the infantry occupies'.[23] On the Somme in 1916, for example, the British employed 1,437 artillery pieces, firing 1.5 million shells over a period of seven days.[24] At Verdun the Germans deployed nearly 10,000 artillery pieces and trench mortars and fired over four months some 24 million shells.[25] By 1918, 25% of the British Expeditionary Force's manpower was in the artillery.

The difficulty was that whilst these techniques could usually ensure that attacks would take the enemy's front-line trenches, the results tended to be strictly limited. The reason for the limited nature of success lay in a number of related difficulties.

- First, the use of massive artillery preparation created **self-defeating dynamics**. The preparatory artillery bombardment relinquished surprise and indicated to the enemy the coming point of attack. Artillery bombardments created a moonscape out of the terrain over which advancing infantry would have to pass: this broke up infantry skirmish lines and made command and control more difficult.
- Second, the reliance on artillery created **structural pauses** in an attack: once the infantry had reached the limit of the range of their supporting artillery they would have to halt and wait for the guns to be moved up again.
- Third, **command, control and communication** were problematic. The reliance on field telephones connected by landlines meant that higher command essentially lost immediate control of assaulting troops once they left their trenches; it made it difficult to co-ordinate artillery and infantry forces in the attack, and difficult also to time effectively the commitment of reserves. Couriers provided a slow and unreliable means of communication; detailed timetables, or systems of flares and so on, were equally susceptible to the friction of war.
- Fourth, defending forces had ample opportunity to prepare **counter-attacks** against the disordered attacking forces.[26]

Cumulatively, these factors created the condition of '**war on a tether**': assaulting troops could advance as far as the range of their artillery support: beyond this, the attack would quickly culminate.

Yet, despite the cost and apparent sterility of tactical land warfare during this period, a **gradual process of evolution** was taking place. This process of evolution would result in the development of recognisably modern offensive and defensive land warfare tactics that would have at their heart a set of common principles: **decentralisation**; **small unit manoeuvre**; **combined arms**; **dispersion**; **suppression**; and **cover and concealment**. The academic Stephen Biddle argues that by 1918 a process of 'convergent evolution under harsh wartime selection pressures' had produced 'a stable and essentially trans-national body of ideas on the methods needed to operate effectively in the face of radically lethal modern weapons'.[27] These ideas focused on:

- Reducing exposure to hostile fire.
- Enabling friendly movement.
- Slowing the enemy.

These techniques first magnified, but would then eventually break, trench deadlock and would go on to define the standards for successful tactical land operations in the post-1918 period. These developments reflected 'nothing less than a complete and rapid reorganisation of tactics'.[28]

Modern tactical defence

It was the German army that was at the forefront of the development of the new tactics, not least because of the more open nature of their military organisation (see Chapter 5). Because the German army in 1915 and 1916 was largely on the defensive, modern system defensive tactics emerged first in the German army. Some changes were technical in nature. For example, realising the defensive utility of machine guns, the ratio of machine gun companies to ordinary infantry companies increased steadily in armies from an average of 1:12 in 1914, to 1:4 by 1916, and by 1917, often 1:2.[29] However, the most significant development was the gradual evolution of techniques for **elastic defence in depth**, a development that reflected a step change from the linear character of orthodox defence. Defence in depth emerged as a response to the destructive power of attacking artillery and the recognition that trying to hold front-line positions in numbers exposed defending troops to the full power of the attacker's artillery. In what the Germans termed *die Materialschlacht* ('the battle of materiel'), exemplified by the Somme in 1916, forward defence merely increased the losses of the defenders by exposing them to the full effect of artillery attack.

Evolving over time, the new principles for defence were enshrined formally by the German army with the publication in December 1916 of *The Principles of Command in the Defensive Battle in Position Warfare*. Defence in depth embodied three key characteristics:

- **Echeloned defence**: The first of these was an 'echeloned defence', creating a deep defensive zone consisting of **multiple defensive lines** (see Figure 3.4). The forward defences would consist of an outpost zone, perhaps 500 or 1,000 metres deep; this would be lightly manned to reduce casualties from artillery, and was essentially merely a defensive 'trip-wire'. The main battle zone might be more than 2 km deep and consist of perhaps three lines of trenches. Where possible, defences would be augmented with strong points sited for all-round defence to allow defending troops to continue to fight even when outflanked.[30] Echeloned defence harnessed the 'entropic effect' of depth: attackers would have to travel further, they would make slower progress through multiple defensive lines, and they would be further enervated by each successive defensive line. This entropic effect would slow the attacker's breakthrough and give the defenders more time: time to bring up reserves; time to choose the moment and place for a counter-attack.

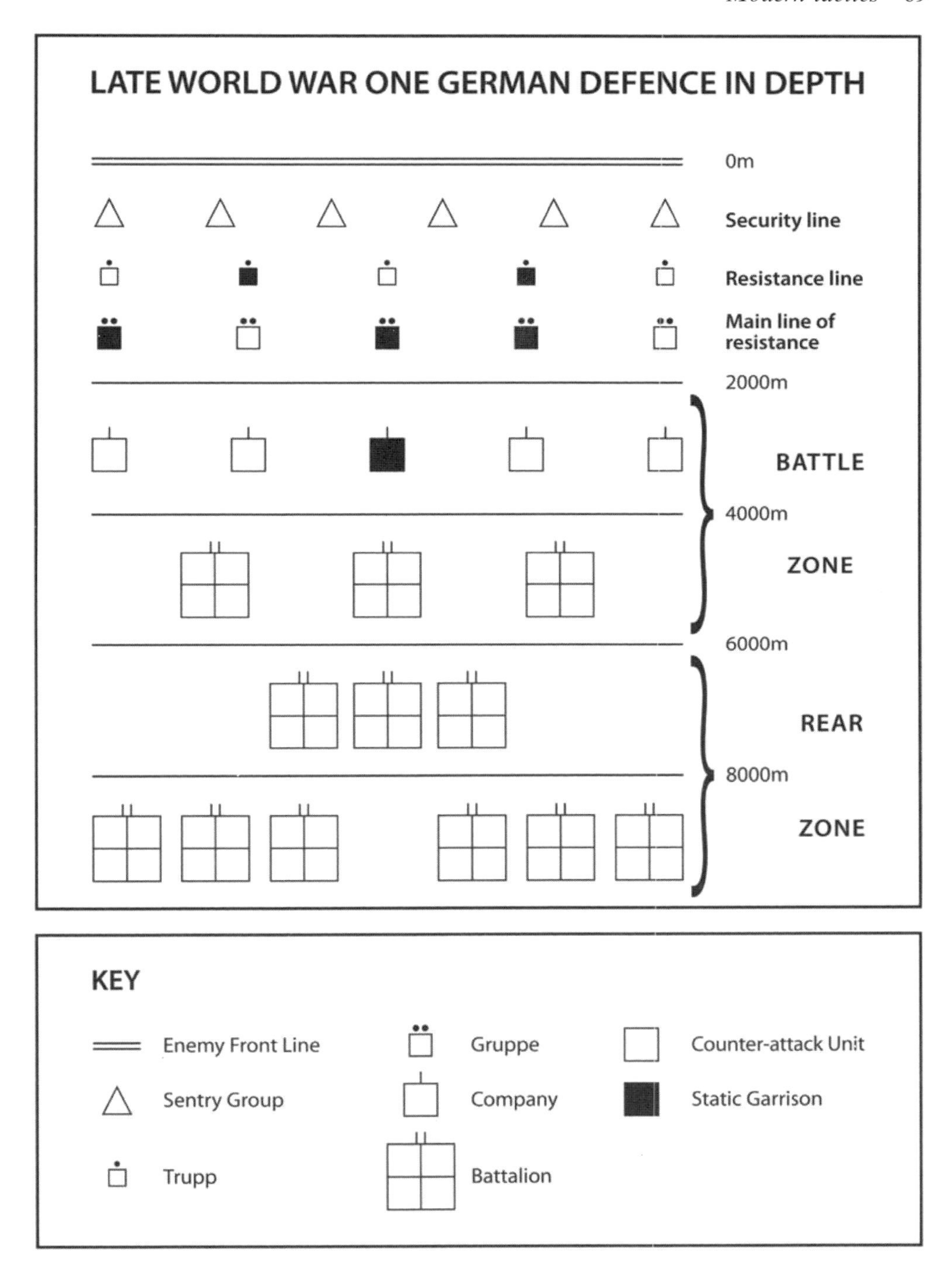

Figure 3.4 Defence in depth

Source: (Martin Samuels, *Doctrine and Dogma: German and British Infantry Tactics in the First World War* (London: Greenwood Press, 1992), 79))

- **Aggression**: Indeed, **counter-attack** was the second key principle of elastic defence in depth. Having given ground initially, aggressive counter-attacks would be launched as quickly as possible to leverage the effects of enemy disorder.[31] For this reason, the majority of the defending forces were positioned in reserve, reducing their exposure to artillery attack and providing plentiful forces to re-take ground lost in the attacker's initial assault.
- **Decentralised command**: The preceding two developments seem straightforward with hindsight, but they required an enormous practical advance in command and control associated with the third key principle: **decentralisation**. Sector commanders were given control over all of the forces in their area, so that authority for counter-attacks did not have to be sought from above. Thus, by 1917–18 German battalion commanders in the front battle zones could order remaining battalions in their zone to counter-attack; forward divisional commanders could commit reserve divisions into the counter-attack.

Modern tactical offense

Paradoxically, however, just as defence in depth strengthened the power of the defence in the short term, it also provided an additional incentive for the evolution of offensive counter-tactics. These methods of attack were demonstrated in **Operation Michael**, the German offensive launched on 21 March 1918 in order to take advantage of the forces released by Germany's victory over Russia in 1917. In contrast to the limited success of previous offensives of the war, Operation Michael broke through British lines on an 80-km front and tumbled British forces back for a week. Moreover, 1918 as a whole saw a series of offensive operations by both sides that resulted in the exchange of some 13,000 square miles of territory.

This apparently sudden empowerment of the attack was **not due to the development of new technology or the massing of greater material superiority**:

- **Gas** had become a routine element in attacks and was used especially in mixed shell barrages of gas and high explosive. However, practical difficulties, such as the variability of weather, and counter-measures, such as gasmasks, robbed it of any decisive effect.
- **Tanks** were first used en masse at Cambrai in 1917, but these weapons were slow, technically unreliable, and vulnerable to counter-measures that were quickly developed such as anti-tank obstacles and armour-piercing rounds. In August 1918, for example, more than 80% of the tanks employed to attack Bapaume were lost due to technical failure or enemy counter-measures.[32]
- **Aircraft** played an increasingly important role in 1918, especially for reconnaissance and ground attack, but payloads were limited, and the technology and mechanisms for ground-air co-operation were underdeveloped. Indeed, in Operation Michael, Germany deployed only nine tanks and operated without control of the air.
- Nor was success simply a function of even **greater mass**: Germany succeeded in Operation Michael with a local superiority in men and guns far lower than that enjoyed by the failed British Somme offensive of 1916.[33]

The tactics of infiltration

The roots of Operation Michael's success lay not in technology and numbers, but in **new methods** that developed in an evolutionary manner from 1915 onwards, spurred by a combination of ad hoc local initiatives in response to the conditions of the battlefield, specific lessons drawn from battles like Verdun, and the conclusions reached through the activities of experimental assault detachments. The Germans were at the forefront of these developments, but the competitive dynamic of war ensured that the techniques were passed, though with varying degrees of rigour and completeness, to other belligerents. There was a strong relationship between the precepts developed for defence in depth and the new offensive techniques: the catalyst for applying the principles of the former to the challenge of the latter receiving renewed emphasis by the German victory in the east in 1917 and the time-limited opportunity this gave them to use the troops for a knock-out blow against the Allies before US intervention could have a decisive effect. In particular, the new offensive approaches took the principle of **decentralisation** and applied it to create what was often termed **infiltration or stormtroop tactics** (*Stosstrupptaktik*). The crucial tenets of this tactical approach encompassed decentralisation in a variety of forms.

Manoeuvre

The first form was **decentralised manoeuvre**. Whereas in 1914 the smallest unit of independent manoeuvre was generally the company of 250 or so men that would manoeuvre in close proximity to other companies, infiltration tactics focused on the *Gruppe*, or squad, of around eight men. The section would manoeuvre in the attack without focusing on maintaining immediate contact to either flank with the rest of the platoon or company: instead, the section would advance flexibly, mitigating defensive firepower by making use of dispersal, cover and concealment. Unity of effort in the company would be maintained by a focus on a common objective.

Firepower

The second form of decentralisation related to **firepower**, in order to unshackle attacking infantry from a wholesale reliance on heavy artillery. Whereas infantry battalions in 1914 were overwhelmingly rifle-armed, as the war progressed infantry units were equipped increasingly with much more firepower, including mortars, grenade launchers, automatic rifles and light machine guns. The organisational consequences of this were to drive combined arms co-operation down to much lower levels. At the squad level, the basic *Gruppe* would comprise a light machine gun and riflemen; the light machine gun would provide the suppressive firepower that would enable the rest of the section to manoeuvre against enemy defenders. In parallel, infantry battalions, which had been in 1914 almost purely rifle battalions, were augmented by special weapons platoons or companies, creating battalion level combined arms battlegroups. By 1918, a battalion assault **battlegroup** might consist of three or four infantry companies, a trench mortar company, a battery of light artillery, a flamethrower section, a signal detachment and a pioneer section.[34]

Command and control

The third form of decentralisation related to **command and control**. With units now manoeuvring as sections, a robust capacity for small unit leadership was crucial. As the German General Erich von Ludendorff (1865–1937) noted, 'The position of the NCO as group leader thus became more important. Tactics became more and more individualized'.[35] Moreover, the new tactics also required the exercise of high levels of initiative: as German doctrine argued, 'Each attack offers opportunities for self-designated activity and mission-oriented action, even down to the level of the individual soldier'.[36]

Embracing these forms of decentralisation allowed the Germans to systematise a new approach in offensive tactics: **infiltration tactics**. Infiltration tactics were focused primarily on:

- A **psychological rather than physical approach** to land warfare.
- An emphasis on **disruption** rather than destruction.
- A focus on **greater depth in attack** in order to facilitate rapid **exploitation**.

What this meant in practice was that attacks would be launched by successive echelons of troops. The first echelon would push through enemy positions, bypassing enemy centres of resistance, continuing to advance as quickly as possible without reference to maintaining a continuous front or protecting flanks. The second echelon would then be left with the task of mopping up. By tasking the lead echelon with advancing, rather than eliminating the enemy, these lead forces could sustain the tempo of the attack, seizing and retaining the initiative, spreading chaos and dislocation in the defenders' rearward defensive zone. The impact of infiltration tactics was augmented by developments in **supporting artillery**: the use of a scientific system of gunnery to allow mathematic pre-registering of artillery targets, the so-called 'Pulkowski Method'; short, intense, 'hurricane barrages' rather than extended saturation of the target; a mix of explosive and gas shells to force defenders into gasmasks; the use of 'creeping barrages', or *Feuerwalze*, to support advancing troops; a focus not on destroying enemy entrenchments but attacking instead command posts, headquarters, bridges, communications centres. The purpose of this artillery support was not to obliterate enemy defences, but instead to create surprise, shock and **suppression**, the better to facilitate infiltration. Pioneered on the Eastern Front by the German Colonel Georg Bruchmüller, these tactics aimed, in Bruchmüller's words 'to break the morale of the enemy, pin him to his position'.[37]

The foundations of German success

The basis for the successful implementation of these new tactics lay in the formal creation of **new doctrine**, **force structures** and **training regimes**.

- **Codification** of new techniques at battalion level and below was enshrined in the publication in May 1916 of *Instructions for the Employment of an Assault Battalion*, and significantly revised editions of general infantry doctrine, *The Training Manual for Foot Troops*, produced in December 1916 and January 1918. At regimental level and above, the new methods were set out in January 1918 in *The Attack in Position Warfare*.[38]
- **Structurally**, the new tactics were reflected in the creation of specialist assault troops to lead attacks. Developing from experimental assault companies formed in 1915, by 1918 specialist attack divisions were created that received the best personnel and equipment. These would comprise the first echelon of any assault, ensuring that the lead infiltration troops were the most aggressive and capable.

- **Rigorous training** in the new techniques was achieved by rotating units through specially established training schools, often utilising the specialist assault battalions as a training cadre. By early 1918 the German army had put up to 70 divisions through this training programme.

The techniques utilised by the Germans in March 1918 were adapted by the Allies, and built, in any case, on the ideas developed by individuals such as Captain Andre Laffargue and ad hoc initiatives developed during the battle of Verdun. The consequence was a rash of successes for offensive actions in the last portion of the war: including battles on the Marne in July 1918, Amiens in August and the '100 Days' offensive from September to November 1918. The new offensive techniques were still extremely costly – British losses from August to November 1918 were greater than those of the Passchendaele offensives of 1917 – but infiltration tactics had nevertheless restored some mobility to the battlefield.

Modern tactics

The period 1915–18 therefore saw the birth of the modern tactical system of land warfare: tactics based upon combined arms down to the lowest infantry sub-unit; tight-knit fire and manoeuvre; decentralisation; dispersion; suppression; and cover and concealment. Tactics changed dramatically between 1900 and 1918, but developments after 1918 were **evolutionary**. The tactical principles developed by 1918 would be made more complex in the future by new technology such as anti-tank guns and mines, but would not be altered in their fundamentals.

Indeed, technological developments after 1918 would make the modern system even more important. The theoretical lethality of modern firepower continued to increase. At lower levels firepower was augmented by the increasing presence of machine guns, assault rifles, rifle-grenades and man-portable anti-tank weapons in infantry formations. New technology continued more generally to increase the firepower available on battlefields: the developing capabilities of tanks, aircraft and artillery; better sensor technology (such as night-vision equipment); better target acquisition (such as laser designation systems). However, the modern system of tactics continued to be important for two reasons:

- First, **terrain and dispersion** continue to reduce the effectiveness of firepower. For as long as this is the case, tactical employment of forces remains a crucial means of mitigating the effects of firepower. Where armies can use cover, concealment and dispersion better than their opponents, they have a significant advantage.
- Second, **counter-technologies** often provide potentially effective antidotes to modern firepower, if such counter-technologies can be integrated effectively and used to suppress the enemy. All weapons platforms are vulnerable to something. For example, the developing capabilities of tanks have prompted the deployment of more effective infantry anti-tank weapons utilising wire- and laser-guided munitions.

The internal combustion engine has complicated modern system tactics by creating mechanised and airborne forces with much greater mobility than ordinary infantry, and introducing many more weapons systems. However, the principles for the effective use of such forces have remained the same. Modern **battlegroups**, for example, might be created at the company, battalion and brigade levels, integrating infantry, armoured and mechanised forces, anti-tank, reconnaissance, anti-air and engineering forces (see Figure 3.5). Modern battlegroups may be multi-national, and might even embody forces for non-military reconstruction tasks. However, the basic principles guiding the employment of these forces continue to be focused on the modern system of tactics.

Conclusions

At the tactical level, modern land warfare was forged on the battlefields of the First World War. From 1915–18, principles, many of which already existed, were systematised and forged into a tactical approach to land warfare that is still at the heart of battle today. To cope with the dramatic increase in firepower created by modern technology, armies embraced:

- Decentralisation.
- Combined arms.
- Fire and movement.
- Dispersal.
- Cover and concealment.
- Small unit manoeuvre.

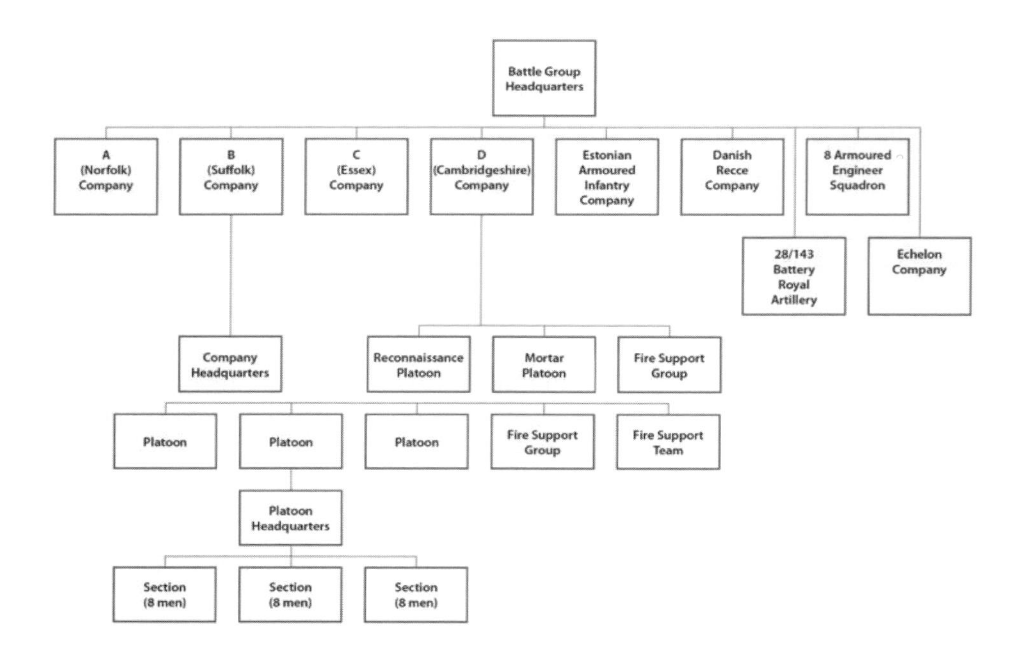

Figure 3.5 British battlegroup, Afghanistan, 2012
Source: (ukarmedforcescommentary.blogspot.co.uk/2012/07/the-infantry-of-army-2020.html)

These principles were applied to both offensive and defensive tactics. This modern system of land warfare at the tactical level created conditions in which offensive breakthroughs were possible, even against defence conducted in depth, because the modern system of tactics enabled land forces to mitigate the effects of modern firepower.

However, developments at the tactical level of war could not, in themselves, deliver decisive military success. In the period from 1918 onwards, military theorists began to consider ways in which new technology, particularly the internal combustion engine and wireless communications, could be harnessed to translate battlefield successes into a broader framework of victory. This leads us to the next chapter, which considers the second key component of the modern system of land warfare: operational art and the operational level of war.

Notes

1 Joint Publication 1–02, *Department of Defense Dictionary of Military and Associated Terms* (November 2010), 277.
2 Colin S. Gray, *Modern Strategy* (Oxford: Oxford University Press, 1999), 26.
3 Quoted in Gray, *Modern Strategy*, 39.
4 Hirsch Goodman and W. Seth Carus, *The Future Battlefield and the Arab–Israeli Conflict* (New Brunswick, NJ: Transaction Publishers, 1990), 165–66.
5 Stephen Biddle, *Military Power: Explaining Victory and Defeat in Modern Battle* (Princeton, NJ: Princeton University Press, 2004), 3.
6 Michael Glover, *Warfare from Waterloo to Mons* (London: Book Club Associates, 1980), 248.
7 Jay Luvaas, *The Military Legacy of the Civil War: The European Inheritance* (Lawrence, KS: University of Kansas Press, 1988), 167.
8 Beatrice Heuser, *The Evolution of Strategy: Thinking War from Antiquity to the Present* (Cambridge: Cambridge University Press, 2010), 137.
9 Brian McAllister Linn, *The Echo of Battle: The Army's Way of War* (Cambridge, MA: Harvard University Press, 2007), 41.
10 Heuser, *The Evolution of Strategy*, 173.
11 John W. Steinberg, *All the Tsar's Men: Russia's General Staff and the Fate of the Empire, 1898–1914* (Baltimore, MD: Johns Hopkins University Press, 2010), 41–43, 215.
12 Colin McInnes, *Men, Machines and the Emergence of Modern Warfare, 1914–1945* (Camberley: Strategic and Combat Studies Institute, 1992), 4.
13 Bruce I. Gudmundsson, *Stormtroop Tactics: Innovation in the German Army 1914–1918* (Westport, CT: Praeger, 1995), 1.
14 Jeremy Black, *War: Past, Present and Future* (New York, NY: St Martin's, 2000), 212–13.
15 Gudmundsson, *Stormtroop Tactics*, 21.
16 McInnes, *Men, Machines and the Emergence of Modern Warfare*, 8.
17 Griffiths, *Forward into Battle: Fighting Tactics from Waterloo to the Near Future* (Chichester: Anthony Bird Publications, 1981), 87.
18 McInnes, *Men, Machines and the Emergence of Modern Warfare*, 8.
19 Gudmundsson, *Stormtroop Tactics*, 55.
20 Ibid., 17.
21 Ibid., 21–22.
22 McInnes, *Men, Machines and the Emergence of Modern Warfare*, 21.
23 Timothy T. Lupfer, *The Dynamics of Doctrine: The Changes in German Tactical Doctrine During the First World War* (Lawrence, KS: Combat Studies Institute, July 1981), 4.
24 Williamson A. Murray, 'The West at War', in Geoffrey Parker (ed.) *The Cambridge History of Warfare* (Cambridge: Cambridge University Press, 2005), 294.
25 Biddle, *Military Power*, 85; McInnes, *Men, Machines and the Emergence of Modern Warfare*, 6.
26 Jonathan M. House, *Combined Arms Warfare in the Twentieth Century* (Lawrence, KS: University Press of Kansas, 2001), 35–36.
27 Biddle, *Military Power*, 28.

28 Griffiths, *Forward into Battle*, 95–96.
29 Gudmundsson, *Stormtroop Tactics*, 94–95.
30 Lupfer, *The Dynamics of Doctrine*, 12.
31 Ibid., 16–19.
32 House, *Combined Arms Warfare*, 48.
33 Biddle, *Military Power*, 89.
34 Gudmundsson, *Stormtroop Tactics*, 98–100.
35 Ibid., 94.
36 Ibid., 139.
37 Lupfer, *The Dynamics of Doctrine*, 45.
38 Ibid., 41.

Suggested reading

Stephen Bull and Gordon L. Rottman, *Second World War Infantry Tactics* (Oxford: Osprey, 2008). Illustrates the strong continuity that existed between First and Second World War infantry tactics, despite developments in technology.

Robert M. Citino, *Quest for Decisive Victory: From Stalemate to Blitzkreig in Europe, 1898–1940* (Lawrence, KS: Kansas University Press, 2002). The first five chapters are especially useful: the Russo–Japanese and Balkan wars are important, but often forgotten, case studies in the emerging challenges of modern firepower.

John A. English and Bruce I. Gudmundsson, *On Infantry* (Westport, CT: Praeger, 1994). A classic examination of the development of infantry doctrine and organisation.

Field Manual 3-21.8, *The Infantry Rifle Platoon and Squad* (Headquarters Department of the Army, March 2007). armypubs.army.mil/doctrine/DR_pubs/dr_a/pdf/fm3_21x8.pdf. Worth comparing with Bull and Rottman: again, the continuities are evident in the tactical aspects of modern land warfare.

Bruce I. Gudmundsson, *Stormtroop Tactics: Innovation in the German Army 1914–1918* (Westport, CT: Praeger, 1995). An authoritative analysis of the development of German infiltration tactics.

Michael Howard, 'Men Against Fire: The Cult of the Offensive in 1914', Peter Paret (ed.) *Makers of Modern Strategy: From Machiavelli to the Nuclear Age* (Princeton, NJ: Princeton University Press, 1986). Examines the impact of the moral school of tactics on infantry doctrines in 1914.

Gary Sheffield, *Forgotten Victory: The First World War – Myths and Realities* (London: Headline, 2001). Provides an engaging illustration of how far tactics had developed by 1918.

4 Modern operational art and the operational level of war

Key points

- Modern land warfare comprises not just activity at the tactical level, but activity at the operational level of war through the conduct of operational art.
- The key material foundation for the emergence of operational art was the expansion of the battlefield.
- The key conceptual foundations for the theories of operational art were laid by the Soviet Union and Germany in the inter-war period.
- Despite the importance of the operational level of war and operational art, different armies have often realised these concepts in different ways.

How does an army achieve military success in conventional land warfare? The obvious answer to this question might be to win battles; but in the period from the First World War, there has been a growing body of thought that has argued that focusing narrowly on winning a battle in modern land warfare actually is a recipe for failure. Instead, so the argument goes, army commanders need to raise their eyes and focus instead on winning campaigns. Of course, winning campaigns requires winning battles, but thinking about campaigns, and the relationship between campaigns and strategic goals, begins to pose for the commander such questions as: What battles should I fight? When should they be fought? In what sequence? From thinking such as this the second key component of modern land warfare has emerged: **operational art**, and also its associated idea of an **operational level of war**.

This chapter charts the emergence of operational level thinking in modern armies. First, we discuss some contemporary definitions and concepts in order to develop a general appreciation of the nature and demands of operational art and its relationship to the operational level. Then the chapter explores why and how these ideas have become so important. If the First World War was the key source of modern tactical land warfare, then it is the period from 1918 to 1941 that saw operational art become an important feature of the modern system of war. Developments after the Second World War saw the formalising in Western armies of a distinct operational level of war. Taken together, modern system tactics and modern system operational level thinking comprise the heart of modern land warfare.

What is operational art?

Our first task is to develop an appreciation of what operational art is, and its relationship to the operational level of war. We need to note, though, that these definitions and relationships are **subject to debate**: so what we are developing here is a foundation of understanding that we will later open up to critique.

Defining operational art

In a general sense, the term 'military operation' simply describes carrying out a military action or executing a military mission.[1] So, troops may be 'carrying out an operation', or they may be 'on operations', or they might be 'operationally effective'. However, the term **operational art** has a different and very specific meaning. The essence of operational art as a concept is the moving of military thinking beyond a focus on a single army in a single battle, to the co-ordination of the activities of **multiple forces in a wider theatre** of operations; it comprises **synchronising** such activities as attack and manoeuvre, breakthrough and exploitation across wide fronts and to great depth.

Theatres, campaigns and the operational level

Contemporary definitions of operational art are often derived from the concept of the **operational level of war**. As Chapter 1 has identified, the operational level of war is an intermediate level between tactics and strategy. It is associated with military campaigns and military theatres. Military **theatres** (often termed theatres of operation or theatres of war) are large geographically defined areas of land, water and air. An individual battlefield would comprise only a tiny part of a theatre.[2] Theatres can be geographically contained (as with the Allies' Italian theatre from 1943–45), or very large (as with the United States' Pacific theatre from 1941–45). Where the boundaries are drawn to distinguish one theatre from another will depend upon the specific physical, political and military context and may change over time. A **campaign** is a series of linked tactical military operations that take place within a theatre directed towards a common purpose.[3] Thus, we can talk of Napoleon's campaign in Russia in 1812, or the Normandy campaign of 1944, each of which comprised many different, but related, manoeuvres and battles.

Bearing these points in mind, the **operational level of war** can be defined generally as 'that intermediate stage between tactics and strategy where campaigns come together'.[4] **Operational art** is what a commander does at this level. It is:

> the orchestration of all military activities involved in converting strategic objectives into tactical actions with a view to achieving a decisive result.[5]

Box 4.1 contains some alternative definitions but these alternatives still reflect the same basic elements. Thus, the purpose of the operational level of war is to ensure **coherence** between the objectives and imperatives that flow from the **strategic level** and what is being done at the **tactical level** by myriad military units.

Box 4.1 Definitions

The operational level of war

The operational level links the tactical employment of forces to national and military strategic objectives. **The focus at this level is on the design, planning, and execution of operations.**

(Joint Publication (JP) 3-0, *Joint Operations*, 11 August 2011, I-13)

The operational level of warfare is the level at which campaigns are planned, conducted and sustained, to accomplish strategic objectives and synchronise action, within theatres or areas of operation. It provides the 2-way bridge between the strategic and the tactical levels.

(Joint Doctrine Publication 0-01, *British Defence Doctrine*, November 2011, 2–9)

Operational art

Operational art is the use of creative thinking by commanders and staffs to design strategies, campaigns, and major operations and organize and employ military forces.

(JP 3-0, II-3)

Operational Art is the orchestration of a campaign, in concert with other agencies, to convert strategic objectives into tactical activity in order to achieve a desired outcome. It embraces a commander's ability to take a complex and often unstructured problem and provide sufficient clarity and logic (some of which is intuitive) to enable detailed planning and practical orders.

(Joint Doctrine Publication 01, *Campaigning*, December 2008, 3–6)

For example, in the Gulf War of 1991 General Norman Schwartzkopf was the operational level commander: he commanded the coalition forces in the Kuwaiti Theatre of Operations. It was his responsibility to plan and execute the campaign to achieve the strategic goals given to him by the strategic and military strategic levels. Schwartzkopf was responsible for planning and executing a campaign that would ensure synchronisation and harmony of effort amongst all the different tactical activities within his theatre of operations. Schwartzkopf's plan was to envelop the Iraqis from his left flank. Without this operational level of command, the fighting in Kuwait would have dissolved into an uncoordinated series of parallel activities, as subordinate commanders on land, sea and air, and from different countries, fought their own battles without reference to one another or to a unifying purpose (see Figure 4.1).

The operational level is defined by two themes: **function** and **context**. The key **function** of the operational level of war is to link tactical level activity to strategic level intent. The operational level commander is an interface between higher level political and military guidance coming from superiors outside of the theatre of operations and the tactical activity that goes on within a campaign (see Figure 4.2). Put simply, politicians and superior military commanders tell the operational level commander what he or she must achieve within a military theatre, and the latter is then responsible for crafting a military campaign that will achieve this on the ground. As contemporary

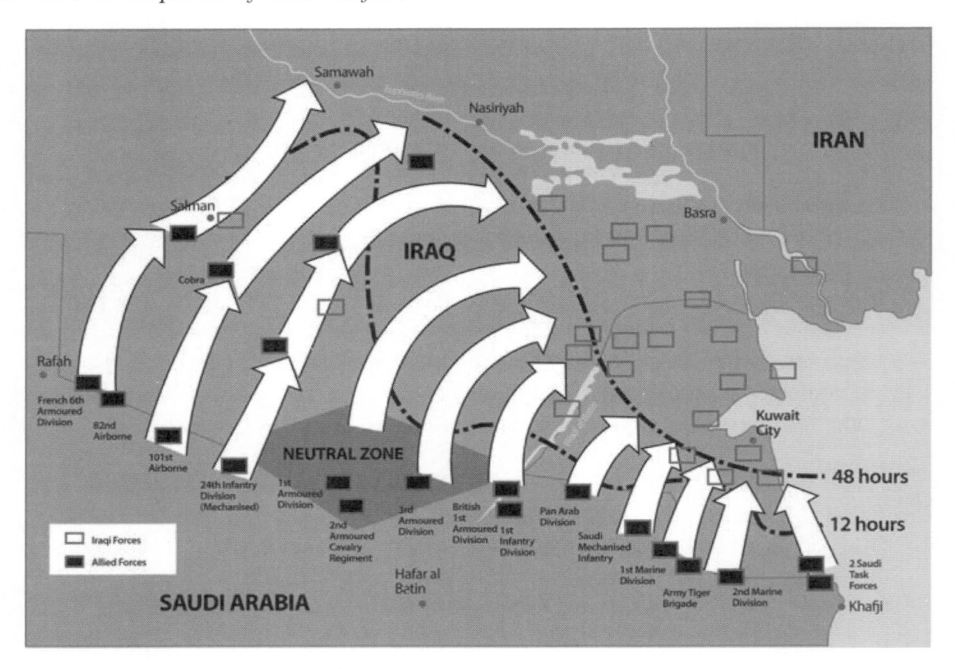

Figure 4.1 The Kuwaiti Theatre of Operations (KTO), 1991
Success depended upon the theatre-level synchronisation of coalition forces towards a common purpose.

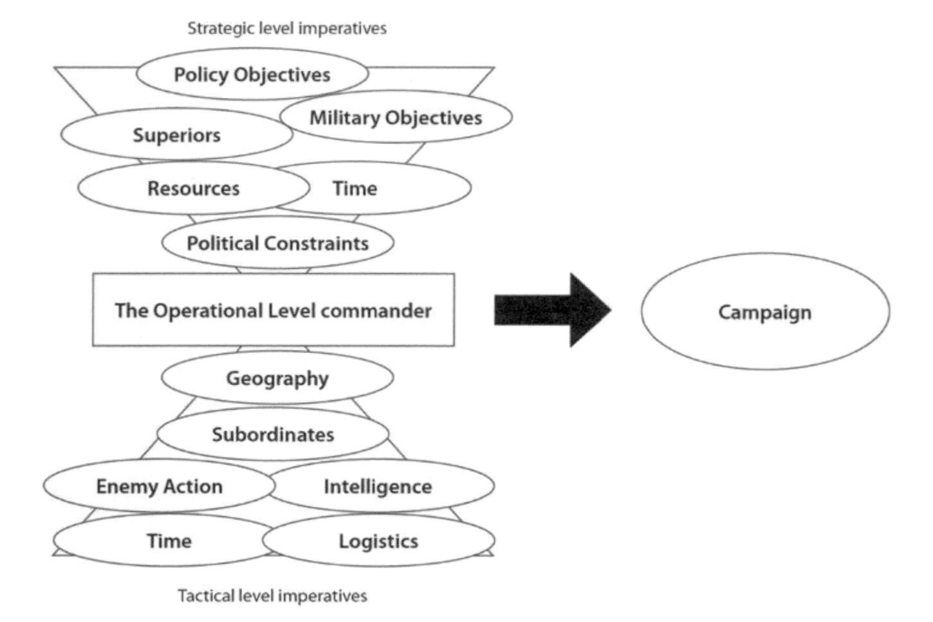

Figure 4.2 The operational level as an interface
The operational level commander must reconcile competing influences and imperatives from above and below in order to craft an effective campaign.

doctrine notes, the operational level of war provides 'the crucial link between the setting of military objectives and the employment of military forces at the tactical level'.[6] However, every **context** is different, and so the operational level commander must perform this function according to the dictates of local circumstances.

Key variables

Whilst each context is different, skill at operational art will usually involve a range of common themes. First, strategic **purpose** and the operational **environment** will exert a powerful influence:

- **Objectives**: an operational level commander will have tactical and strategic goals. He or she must seek to reconcile them.
- **The theatre of operations**: operational art is executed in a physical environment. The geography of the theatre of operations provides challenges and opportunities that the commander needs to mitigate and exploit.

Second, there will be common **means** available to realise operational art:

- **Co-ordination**: to achieve strategic goals, the operational level commander must impose unity of purpose on his subordinate tactical level commanders, sequencing when and where tactical activity takes place to create a single coherent campaign. Unity of purpose should also extend to other national contingents where operations are multi-national in nature and other agencies where non-military contingents are present.
- **Manoeuvre**: sequencing is not just about fighting battles, but the co-ordinated movement of friendly forces into positions of advantage. Manoeuvre can sometimes render battle irrelevant – for example, cutting an opponent's lines of communication can force him to withdraw without combat. Manoeuvre can impose crippling psychological blows against an enemy, as well as being critical to sustaining the tempo of an operation.
- **Jointery**: few theatres will involve only one service. Effective operational art is likely to require skill in joint warfare and the utilising of non-land assets.

Third, operational art will also be subject to the influence of key **enablers**:

Intelligence: good intelligence will allow the identification of enemy centres of gravity (see Chapter 1). Intelligence is central to identifying proper objectives, identifying enemy vulnerabilities and creating an effective campaign plan.
Deception: this can facilitate breakthrough and manoeuvre by allowing concentration of friendly forces at the critical point and encouraging enemy dispersion.
Reserves: these can provide the means to respond to new conditions, both positive and negative.
Logistics: effective operational art must be framed by what it is logistically possible to sustain.
Concept of war fighting: a capacity for effective operational art is likely to be enhanced by theoretical/conceptual underpinnings that shape the capacity to act operationally. Critically, this is likely to involve a decentralised approach to command and control.

The development of operational art

Operational art is therefore associated with the need to think beyond a single battlefield: to knit together activities across often very large geographical distances so as to construct a single integrated campaign. Why did this function become so important? Why were modern tactics regarded increasingly as an insufficient tool?

Historical origins

Operational art is a child of the period from 1918–41, but the foundations for operational art were laid much earlier than this, and were rooted in **material** and **conceptual** developments.

The 'strategy of the single point'[7]

Arguably, operational art had always existed to some extent in that it had always been relevant for commanders to consider how their tactical actions should be co-ordinated to achieve their broader goals. However, tactics, in the form of a focus on winning battles, was more important. Traditionally, states had struggled to raise and sustain more than one main army at a time. Consequently, the focus of land warfare tended to be tactical: commanders sought a decisive battle that would destroy the enemy's main army and end the war (see Figure 4.3).

This condition changed. As Chapter 1 has noted, one of the key developments in warfare since the eighteenth century was the **growth in its scale**. This had material and

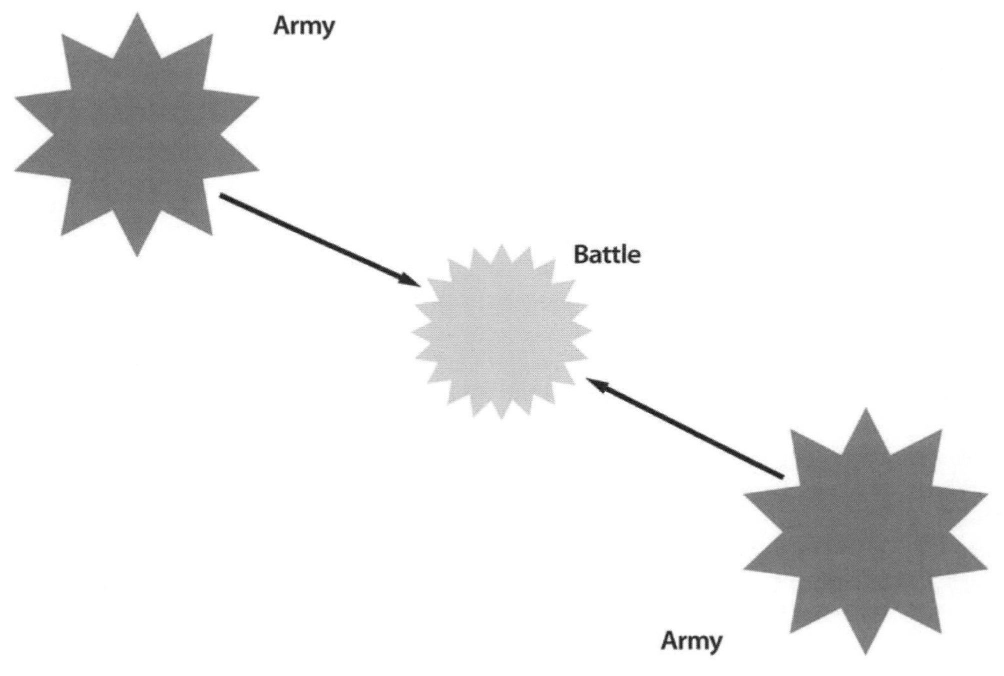

Figure 4.3 Traditional campaigning

conceptual roots. The fruits of the Industrial Revolution combined with the political consequences of the French Revolution, which introduced the concept of mass conscription, provided the **material resources** for states to employ larger and larger armies. The growth in the size of armies provided the material basis for the emergence of operational art as a vital component of modern warfare.

However, it required **conceptual change**, development in the way that officers thought, before operational art became a potentially decisive factor. One of the crucial developments was Napoleon Bonaparte's creation of the army corps (*corps d'armée*). Napoleon's *Grande Armée* was simply too large for it to be concentrated into a single force: it could not be supplied or moved effectively. Napoleon therefore subdivided his army into **all-arms army corps**, each strong enough to hold off a much larger force for a period of time. By doing this, Napoleon opened up new possibilities for fixing a larger enemy force with one portion of his army and then using the remainder to manoeuvre in the wider theatre of operations against the enemy's flank or rear.[8]

By the time of the Franco–Prussian War (1870–71) material developments, especially the railway and electric telegraph, allowed states to **deploy and manoeuvre several armies simultaneously**, each composed of multiple army corps. This disaggregation of armies had two effects.

- First, success required **co-ordinating** the movement of a number of often **widely separated** military forces.
- Second, military success could now be influenced heavily by **manoeuvre beyond the battlefield**. A commander's skill in synchronising forces within a wider theatre of operations now became increasingly important. Reflecting this, terms such as 'grand tactics' emerged to describe the co-ordinated handling of separate forces off the battlefield.

However, there was still an important point of commonality with the past: the purpose during the nineteenth century was still to concentrate the separate elements of an army against the enemy's forces and to fight a **single decisive battle**. Hence this approach is sometimes referred to as the 'strategy of the single point' (see Figure 4.4).

Linear strategy

By 1914, the forces of industrialisation, especially in the form of railways, and nationalism had allowed European powers to field armies that totalled millions. States could now sustain **multiple armies** spread out across the theatres of operations, creating what has been termed 'linear strategy' (see Figure 4.5). There were several important consequences of the emergence of this 'linear strategy'. One was the **expansion of the battlefield**: battles were now much larger, and spread over extended fronts. Moreover, with multiple armies, wars now embodied **multiple battles**, often fought simultaneously. This created the need to develop new structures to **command and control** these battles. No longer was an army commander also the political ruler (as in the case of Napoleon) or answerable directly to him or her. Just as armies were now composed of several army corps, so armies were brought together for command and control purposes into 'army groups', establishing an intermediate level of military command. Clearly, under these conditions, no single battle could be decisive; a state's army could not be destroyed in a single battle. Instead, victory in land warfare could only come through

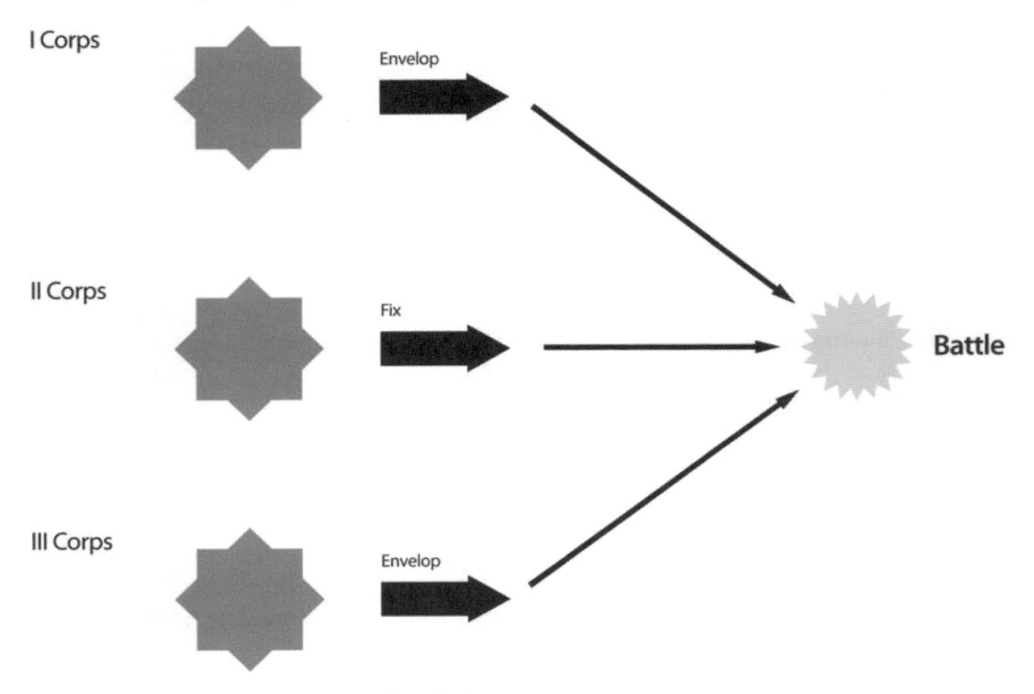

Figure 4.4 'The strategy of the single point'

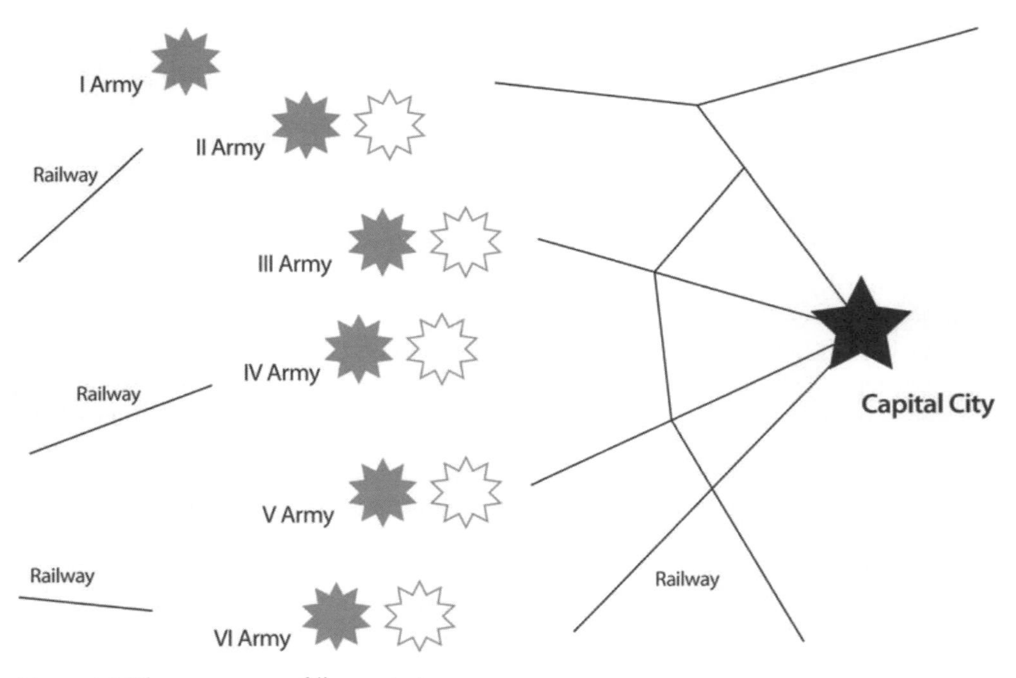

Figure 4.5 The emergence of linear strategy

synchronising the activities of multiple armies in a single theatre; success would come through a ***process***, an accumulation of tactical actions over time, rather than a single tactical event.

1918 and the inadequacy of tactics

The implications of the emergence of linear strategy became evident in the First World War. As the preceding chapter has explained, modern system tactical warfare had restored the potency of offensive operations on the battlefield. Yet the potential of such operations was undermined because it proved difficult to **translate local tactical successes into strategic effect**. The solution to this problem was the development of an intermediate level of thinking between tactics and strategy that would impose coherence on the former in pursuit of the goals of the latter.

The limitations of modern system tactics

The need for a system of thought above that of tactics was amply demonstrated by the ultimate failure of Germany's **Operation Michael** (discussed in the previous chapter), an operation that the Germans finally cancelled on 5 April 1918. Operation Michael was very successful relative to the offensives that had preceded it, but ultimately, the Michael offensive was only a tactical success: it failed to create a larger breakthrough. The key problem was that the infiltrating forces had **neither the tempo nor endurance** to prevent the Allies from using their reserves to establish a new defensive line.[9] The cumulative effects of attrition, physical exhaustion and a rate of advance tied to the movement of infantry meant that the Spring Offensive culminated quickly, despite large gains.

Other problems lay in the realms of command and control. First, it proved technologically impossible for the German High Command to **maintain control** over mobile operations for more than five or six days. Second, there was no **organising concept** behind Operational Michael that would link its various parts towards a common goal: asked about the operational objective of the offensive, Field Marshal Erich von Ludendorff replied: 'I object to the word "operation". We will punch a hole in [their line]. For the rest, we shall see.'[10]

Depth and technology

During 1918, it was recognised that the solution to this problem might be to attack the enemy's defences to a **much greater depth**, undermining their ability to bring fresh troops forward. For example, later campaigns featured a greater emphasis on the use of aircraft to try and **isolate the battlefield** and **disrupt enemy reserves**. Ultimately, however, what the campaigns of 1918 lacked was some means of moving and co-ordinating attacking forces at a higher tempo, so that they could not only break through an opponent's front line, but also exploit quickly over long distances to overthrow completely the whole depth of the enemy's defences.[11]

After the First World War, **new technology** would become available that would make operational art an increasingly important part of modern land warfare. In the inter-war period, the coming of age of two key technologies, **wireless radio communications** and the **internal combustion engine**, provided the technological means to support a new concept of warfare in which the whole theatre of operations would be orchestrated as a single battle. Efficient engines provided the foundation for mechanised forces and more capable aircraft. Wireless allowed the activities of these mobile forces to be co-ordinated over wide distances.

The inter-war period and operational art

The period from 1918 onwards saw operational art emerge as a foundation of modern system land warfare. Nevertheless, it is worth identifying that the concepts of operational art that developed varied by country, shaped by different strategic, political, economic and social contexts. Three themes in particular can help us to understand the nature of the differences in national approaches:

- Some countries **codified** operational art, introducing it into written doctrine; others did not, allowing it to remain an organic and **ad hoc** approach to warfare.
- Some countries focused their concept of operational art on the **destruction** of enemy forces; others, however, focused on psychological **disruption** and **shock**.
- Some countries concentrated their thinking on **operational art**; others focused more formally on the **operational level** of war.

Britain, France and the United States

Many theorists were seized by the possibilities presented by technology. In Britain, the staff officer **J.F.C. Fuller** outlined in May 1918 his own vision of how the Western Front could be collapsed through the mass use of tanks. His 'Plan 1919' envisaged a force of 5,000 Allied tanks used in combination with aircraft to effect a breakthrough of the German front line. Fuller further envisaged the use of these forces for subsequent deep exploitation: having broken through the enemy line, tank forces would then push on far into the depth of the enemy defences, overrunning enemy command, control, communications and logistics, destroying their morale, cohesion and their capacity to respond. **Captain Basil Liddell Hart** expounded similar views, but saw the necessity of a fully mobile but more balanced force encompassing mechanised infantry and capable of advances of up to 100 miles a day.

In France, **Colonel Joseph Doumenc** argued for the creation of large mechanised infantry formations, and **Colonel Charles de Gaulle**, in his 'The Army of the Future' (*Vers l'Armée de Métier*) of 1934, argued for the establishment of a fully professional mechanised force of 100,000 men organised for offensive operations. In the United States, cavalry officers such as **Adna R. Chaffee** were instrumental in the development of forward-looking principles for the conduct of mechanised warfare.[12]

One of the key unifying elements in these approaches was the **focus on disruption rather than destruction**: on using deep exploitation to paralyse and shock the enemy in order to win without the grim attrition that had occurred in the First World War. Liddell Hart, for example, argued that:

> The key to success … lies in rapidity of leverage, progressively extended deeper – in demoralising the opposition by creating successive flank threats quicker than the enemy can meet them, so that his resistance as a whole or in parts is loosened by the fear of being cut off.[13]

However, these theorists had only a **limited impact** on their own armies: economic, political and organisational factors conspired to ensure that their views were marginalised. Nor did they articulate a specific theoretical concept of operational art. It was in the Soviet Union and in Germany that these ideas received more serious application.

The Soviet systematisation of operational art

It was the Soviet Union that first systematised the concept of 'operational art'.[14] Soviet thinking was spurred on by many factors:

- Marxism's modernist **worldview**.
- Russian **military experience** in the First World War, the Russian Civil War and the Polish–Russian War (see Box 4.2).
- The activities of a cadre of exceptionally talented **ex-Tsarist officers** including Mikhail N. Tukhachevsky and Alexander A. Svechin.

Box 4.2 Soviet experience

The Western Front was not the whole of the First World War: in the east, especially, the size of the theatre created lower force-to-space ratios that allowed what the Germans termed *Gummikrieg* (rubber war): see-sawing patterns of advance and retreat exemplified by Russia's 1916 Brusilov offensive against Austria-Hungary. In the Russian Civil War (1917–23) and the Russo–Polish War (1919–21), mobile operations played a prominent role. For example, the Red Army's First Cavalry Army under Marshal Semyon Budennyi conducted deep raiding operations against the Whites, utilising mobile combined arms forces: cavalry, armoured cars, mounted infantry, cart-mounted machine guns, air reconnaissance and even an armoured train. Similarly, the White General Konstantin Mamontov's IV Don Cavalry Corps utilised air reconnaissance to find gaps in the Red Army's lines before utilising the mobility of his force to raid deep into Red territory, creating chaos in their rear areas. In Poland, where Tukhachevsky had commanded, the Poles had survived early tactical defeats because of a Russian focus on key geographical objectives rather than the Polish army. These experiences provided fertile ground for the emergence of operational concepts founded on depth, manoeuvre and combined arms.

The Soviet answer to the problem of how to convert tactical breakthroughs into rapid, decisive victories had a number of key features. Surprise, deception and secrecy were identified as foundation elements of success, but Soviet approaches placed particular emphasis on three other factors:

Deep operations

The principle of **deep operations** recognised the fact that the broad front operations of the First World War were indecisive. What were required instead were **combined arms attacks** launched **in parallel** across a number of narrow fronts that would be combined with attacks against the **depth of the enemy**. Combined arms attacks would achieve a tactical breakthrough; then mechanised forces would develop this breakthrough into operational success with a rapid push deep into the enemy defences, pursuing the enemy, breaking their reserves and establishing favourable conditions for the next operation. Depth in attack would also be realised through joint operations: Tukhachevsky envisaged airpower as a key means of deep attack and even experimented with integrating airborne drops (*vozdushnyi desant*) into attacks. Whilst the German Spring Offensive achieved an advance to a final depth of between 6 and 22 miles, Soviet theorists envisaged operations that might achieve a penetration of 150 miles into the depth of the enemy (see Figure 4.6).[15]

Successive operations

The concept of **successive operations** was built on the conclusion that a single operation against an enemy would not defeat them decisively. Instead, a succession of rapid, high-intensity operations would be required in order to prevent the enemy from withdrawing and forming a new defensive line (see Box 4.3). The principle of successive operations would require the capacity to sustain an uninterrupted series of offensive operations, all co-ordinated towards achieving a common purpose.

Box 4.3 Successive operations

> [I]t is impossible, with the extended fronts of modern times, to destroy the enemy's army at a single blow ... a series of destructive operations conducted on logical principles and linked together by an uninterrupted pursuit may take the place of the decisive battle that was the form of engagement in the armies of the past.
>
> (Soviet General Mikhail Tukhachevsky, 1923, quoted in Kelly and Brennan, *Alien*, 44)

Scale

Deep and successive attacks required the co-ordination not just of armies, but of **groups of armies**. The co-ordination of armies and army groups introduced an intermediate scale of war that was clearly well beyond the management of a single battle and which instead encompassed the co-ordination in time and space of large forces towards a common purpose within a theatre of operations.

The consequence of Soviet thinking was their identification in the 1920s of a specific concept of **operational art** (*Operativnoye Iskusstvo*): the theory and practice of preparing for and conducting combined and independent operations by large units (armies and groups of armies). Svechin made the first use of the term in 1922. Strategy pursued the objectives defined by policy; tactics was the realm of immediate military problems. For Svechin:

> operational art governs tactical creativity and links together tactical actions into a campaign to achieve the strategic goal.[16]

The path to the strategic goal might require a number of different operations, each composed of many tactical actions, and each of which might have different

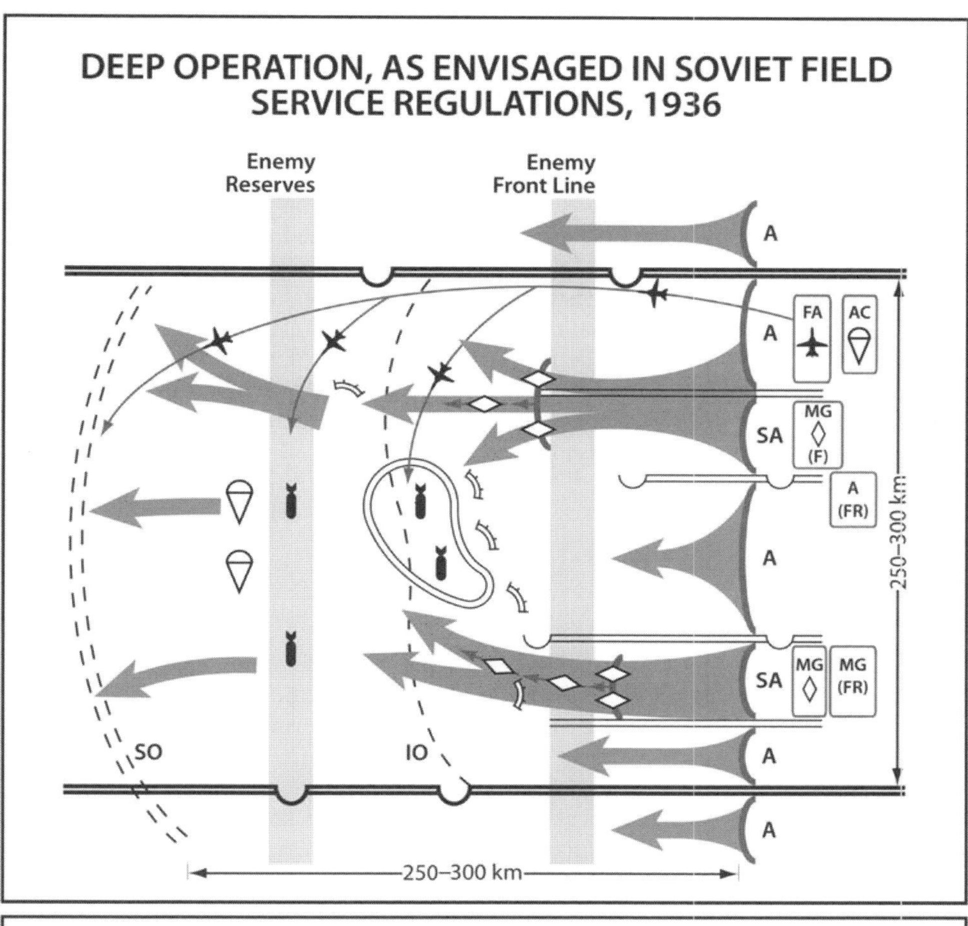

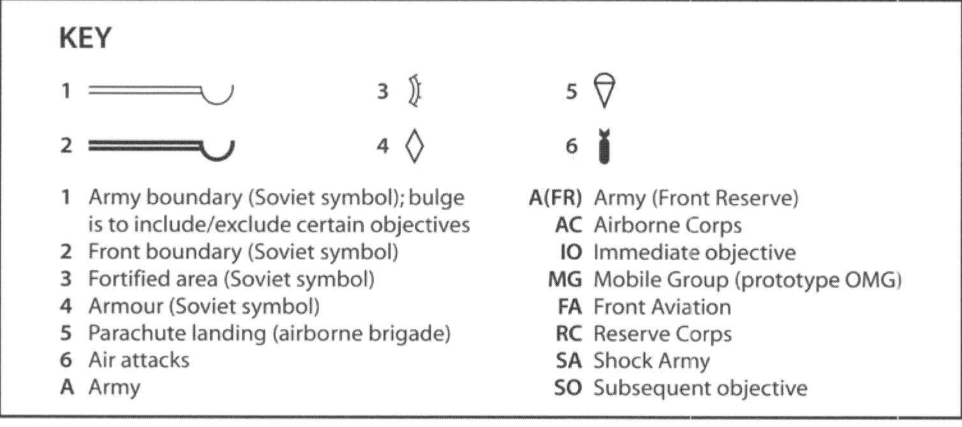

Figure 4.6 Deep operations
Source: (Christopher Bellamy, *The Evolution of Modern Land Warfare: Theory and Practice* (London: Routledge, 1990), 90)

intermediate objectives. These operations might be widely separated geographically or chronologically. As Svechin noted:

> tactics make the steps from which operational leaps are assembled; strategy points out the path.[17]

Whereas theorists such as Fuller focused their thinking on inducing paralysis in the enemy, Soviet thinking focused on **encirclement and destruction**. Tukhachevsky argued that the aim of an operation was 'Not the destruction of some hypothetical, abstract nervous system of the army, but destruction of the real organism'.[18]

By 1936 the Soviets had a formal concept of operational art that was reflected in the training and force structures of the Soviet army, but this positive basis for advanced thinking on the character of modern land warfare was badly affected by Stalin's purge of Red Army officers in 1937–38. The effect of this was evident in the poor showing of the Soviet army in the war against Finland in 1939–40. Yet the Soviet potential was demonstrated in the east in August 1939: less affected by Stalin's purges, Soviet forces in Mongolia under Georgi Zhukov inflicted a stinging defeat on the Japanese at Kalkhin Gol.

German blitzkrieg

German approaches to operational art have gained wide fame under the label of *blitzkrieg* ('lightning war').[19] Wireless radio and mechanisation were embraced wholeheartedly by Germany. As early as the mid-1920s, the German army began to fit its vehicles with radio mounts, and great effort was put into the development of an effective range of radio sets to equip vehicles and headquarters. Mechanisation, expressed most visibly in the increasing capabilities of the tank, promised to increase dramatically the speed at which forces could manoeuvre, attack and exploit success.

Material developments

The key organisational expression of this approach was the creation of **mobile forces**, especially the panzer division, which would have the ability to break through an opponent and then to engage in rapid exploitation far into the enemy depth. The first three panzer divisions were created in 1935 and built on earlier experimentation with the Reichswehr's motorised transport elements. Tanks were an important component of this organisation, but other factors were recognised as being just as significant:

- The panzer division was a **combined arms** formation, including in its order of battle reconnaissance units to obtain intelligence and to screen its movement; mechanised infantry and artillery to reduce centres of enemy resistance and to hold ground; and combat engineers to facilitate its mobility.
- Diversity in composition gave the panzer division **flexibility** in combat, allowing it to form combined arms battlegroups to suit the needs of the moment.
- **Joint operations** were also recognised as being a central element of operational level mobile warfare. German joint doctrine actually preceded the creation of the Luftwaffe in 1933, and the importance of air support was reflected in the creation of specific assets for close air support, including the JU-87 (Stuka) dive bomber.

Mechanisation produced rapidity; combined and joint organisation created flexibility: a British General noted in 1941 that his panzer division adversary 'always has the right tool available to deal with whatever opposition appeared before them'.[20]

Conceptual foundations

The concepts that would guide the employment of these forces were articulated by **General Hans von Seeckt**, Army Chief of Staff from 1920–26. Looking to the demands of future warfare, von Seeckt argued that future war would be a war of **manoeuvre**, and would require the **co-ordinated employment** of formations of corps level and above. Success would require the application of:

- Combined arms.
- Surprise.
- Rapidity.
- Concentration of force.
- Depth.
- Simplicity.[21]

German doctrine extolled, in particular, **decentralised command and control**, expressed through the notion of *Auftragstaktik* ('mission command'), a style of command that focused on the setting of objectives whilst leaving the methods of attaining those objectives to the initiative of the commander. These ideas were reflected in the Reichswehr's first field manual, 'Leadership and Combat of Combined Arms' (*Fuehrung und Gefecht der verbundenen Waffen*), issued in 1921, and new regulations issued in 1933 entitled 'Troop Leadership' (*Truppenfuehrung*).[22]

In practice, *blitzkrieg* techniques as used in the early part of the Second World War had a number of features:

- The objective overall was to exert pressure against the enemy's front, whilst simultaneously seeking to **envelop** his forces and cut off his lines of retreat.
- **Surprise** was a central means through which the maximum number of forces could be concentrated at narrow points in the enemy defences and through which the psychological effect of the attack could be magnified.
- **Combined and joint attack** would create multiple breakthroughs at these points, breakthroughs that would then be exploited by mechanised forces.
- **Mechanised forces** would be employed independently and on a large scale. By 1940, all of Germany's mobile divisions were concentrated into four panzer groups, each of two to four panzer and motorised corps, on the basis that 'The panzer groups, like armies, would be capable of independent long-range tasks, driving wedge-like blows that would tear open the enemy's front and help the infantry armies forward to their operational objectives'.
- These breakthrough forces would then move through the enemy's depth, **bypassing centres of resistance** and focusing instead on acting as giant pincers to encircle wide swaths of the enemy.
- **Aircraft** would provide mobile firepower, helping to support the ground attack and ranging far into the enemy defences to add depth to the attack and to magnify the dislocation of the enemy caused by the swift advance of the breakthrough forces.

Thus, in theory at least, the mobility of the combined arms panzer division harnessed to doctrines of manoeuvre, decentralisation and depth, solved the problems of 1918: the exploitation of breakthroughs was no longer tied to the speed of advance of the infantryman on foot; future war would be *Bewegungskrieg* – the 'war of movement'.[23]

The ad hoc nature of blitzkrieg

However, German approaches were far from constituting a coherent, consensual body of doctrine: one thing that *blitzkrieg* did not embody was systematisation. The term *blitzkrieg*, for example, was never for the German army an official doctrinal term. Indeed, it was rarely used at all by the German army. When he first heard of it in late 1941, Hitler commented that it was 'a very stupid word'.[24] As the historian Heinz Frieser notes, the term *blitzkrieg* was simply a synonym for 'modern operational level war of manoeuvre'. German operational art was not a formal doctrine, it was not published; it was not taught explicitly, nor was it practised on pre-war exercises. Instead, *blitzkreig* emerged from the deliberate German focus on *freie operationen* ('free operations'): on achieving coherence in military operations not through the rigid conformity imposed by detailed doctrine but instead by **educating** individuals to think in the right way – encouraging **creativity and imagination** through rigorous processes of officer **selection, education and training** combined with war games and a General Staff system. The German army had no explicit concept of operational art, yet in practice, the synthesis of aggression, enterprise, large-scale manoeuvre and advanced combined arms tactics would conform to the modern-day understanding of operational art.[25]

For these reasons, Germany's early successes in operational art were not the result of a coherent and rigorously inculcated doctrine: they were the result, amongst other things, of the application of the tactical concepts developed by 1918, including a focus on surprise, infiltration, manoeuvre, combined arms, successive echelons and shock, harnessed to technology, and then applied on a larger scale. In executing *blitzkrieg* in practice, ad hoc approaches driven by the initiative of commanders on the ground were crucial.

The Second World War

Even in the German army many continued to resist the ideas of radicals such as **Colonel Heinz Guderian**, author of *Achtung Panzer!* (1937), an enthusiastic advocate of the aggressive use of panzer forces. Indeed, the campaign in Poland in 1939 was actually quite conservative: German armour was not used independently on an operational scale and continued to be shackled to the infantry divisions; rates of advance were unremarkable, at an average of 11 miles per day;[26] and the campaign demonstrated weaknesses in German combined arms and defensive capabilities.

***The triumph of* blitzkrieg?**

It was only in the 1940 campaign against France that the operational level use of panzer divisions began with the creation of panzer group Kleist: eight divisions strong, five of which were panzer divisions and three of which were motorised. It was at this **army group level** that one could see operational art as a sequence of tactical actions directed towards a common purpose. Even radicals, though, were surprised by the early success of operational level mobile warfare. The 1940 campaign was expected by Germany to be longer; indeed, Hitler referred to the stunning German success against France as 'a miracle, an absolute miracle'.[27]

The limits of blitzkrieg

In France, the campaign that opened on 10 May 1940 lasted 46 days, but in reality the war was won by Germany within ten: within three days of the war beginning, Guderian's XIX Panzer corps had created a 60-mile gap in the Allied centre, and a week later German forces had reached the Channel.

The tempo of German exploitation had created chaos in Allied command, control and cohesion (see Box 4.4). Whilst Allied forces continued to present on occasion some tough tactical resistance to the German advance, no coherent operational or strategic response was able to emerge on the part of Allies under the pressure of exploitation by German mobile forces. German implementation at an operational level of the principles of surprise, concentration, speed, depth, combined arms, and decentralised command and control crushed Allied forces which had a numerical and, in many cases, technical superiority. When an armistice was signed on the 25 June the Allies had been defeated utterly, losing half their forces at a cost of only 27,000 German dead.[28] The British Prime Minister, Winston Churchill, was shocked, commenting that:

> I did not comprehend the violence of the revolution effected since the last war by the incursion of a mass of fast-moving heavy armour. I knew about it, but it had not altered my inward convictions as it should have done.[29]

Box 4.4 Attack order, Panzer Group Kleist, France, 1940

> Success springs from speed. The important thing is to forget about your right and left flank and to push quickly into the enemy's depth and to take the defenders again and again by surprise.
>
> (Attack order, Panzer Group Kleist, France 1940, in Frieser, *The Blitzkrieg Legend*, 100)

The 1940 campaign validated the progressive ideas of those such as Guderian, and in 1940 the operational scale employment of armour formally appeared in German regulations.

In June 1941, Germany launched Operation Barbarossa, and invaded the Soviet Union. By September 1941, German operational mobile warfare had achieved a series of staggering offensive successes, lunging deep into Russia, achieving breakthroughs, exploitation and encirclements that inflicted on the Soviets losses of 4 million casualties and 3.5 million men taken prisoner.[30] Yet, Germany's war in the east ground to a halt and, by 1942, land warfare seemed to exhibit by degrees an increasingly positional character: despite operational art, protraction and attrition seemed to have reasserted themselves in the conduct of battle.

The limits of operational art

Why did attrition reassert itself, and did this render operational art irrelevant in land warfare? Two factors were key: the demanding nature of operational level mobile warfare; and enemy adaptation.

Operational level mobile warfare

The failure to replicate later in the war the spectacular successes of 1940–41 was in part inherent in the **demands of deep, mobile operations**, which required:

- A high degree of **mechanisation**.
- The **logistics framework** to support rapid, successive advances.
- The capability to manage the inevitable **attrition** of the exploitation forces.

Germany had none of these, being in many respects only a 'semi-modern' army.

- **Mechanisation**: In 1940 in the west, only ten of the 141 divisions were panzer or motorised divisions; in 1941 against Russia, it was still only 17 out of 152.[31] Indeed, the Germans used in the Second World War nearly twice as many horses as they had used in the First World War (2.7 million versus 1.4 million).[32] For this reason, the forces available to the German army for breakthrough and operational level exploitation were relatively small.
- **Logistics**: Mechanised forces solved many tactical problems, but they generated new logistical ones. In 1914, an infantry division required 100 tons of supplies daily; in 1940, a panzer division required 300 tons.[33] Thus, even the relatively small German mobile forces generated huge new logistical requirements. In Russia, the Fourth Panzer Army alone required 16,000 tons of supply to provide one load of ammunition; one day of petrol, oil and lubricants required 6,000 tons of supply; and one day of rations 600 tons. To deliver all of this required 46 trains.[34] However, the German army was equipped with only a tenth of the vehicles required to supply its 900-mile advance. As early as mid-July 1941, Barbarossa began to degenerate into increasingly ad hoc offensive surges, followed by enforced pauses to resupply and wait for non-mechanised forces to catch up.
- **Attrition**: Moreover, even in victory, losses were high due to Soviet counter-attacks and logistic problems. By the beginning of November 1941, the panzer divisions had only a third of their motor vehicles and around 35% of tanks still operational.[35] By the end of November 1941 personnel losses amounted to 750,000 casualties, including 200,000 dead; the army's reserves were already exhausted and the army in Russia was 340,000 men under-strength.[36] These problems meant that operational level offensives could not be maintained indefinitely and the consequent pauses provided opportunities for the enemy to regroup and counter-attack.

Enemy adaptation

The second difficulty in replicating the early German success in the Second World War was the **relational dynamic of war**. Germany's early successes were to an extent situational:

- The **Polish** army was badly equipped and deployed; it had, moreover, to cover its eastern borders against the Soviets.
- The **French** forces suffered poor morale, a lack of mobile reserves, and an inadequate armoured doctrine that 'penny packeted' much of the armour, and consigned much of the rest to hastily raised and unbalanced formations.
- The **Soviet** army in 1941 was still reeling from the purges that had removed a swath of the officer corps. The effects of this were compounded by the subsequent expansion and reorganisation of the Red Army, in circumstances of inadequate equipment and inexperienced officers, which left the Soviets with an excellent doctrine that was beyond the capabilities of the army to execute. In addition, the Soviet leadership worsened the situation, first by forcing the Red Army to defend too far forwards, and then in forcing commanders into questionable offensive and defensive operations.[37]

Thus, against an opposition that was better, an opposition that was given, for example, the time to learn and adapt, operational level mobile warfare was always likely to prove less decisive.

For this reason, operational level mobile warfare was confronted progressively by the rational operational level application of the tactical principles of **defence in depth**. If *blitzkrieg* was focused on extending the depth of attack, so the obvious counter was to extend the depth of the defence. Indeed, the Russian deployment in June 1941 was an attempt to apply this principle, but it was poorly executed thanks to the Red Army's general disarray. By the middle of the war, the defence had reasserted itself thanks to the competent execution of these techniques. Operational defence in depth relied on the same principles as its tactical forebear:

- **Successive lines of defences** consisting of **mutually supporting positions** sited for all-round defence.
- The positioning in **reserve** of extensive forces held in readiness for the **counter-attack** once the attackers had exhausted themselves.

There was new technology available for the defence: minefields, for example, and anti-tank guns both helped to neutralise the effects of enemy armour. However, the real difference was the **scale** on which these principles were executed.

In **Operation Citadel**, the German attempt at a double envelopment of the Kursk salient in July 1943, the Soviet defences consisted of eight echeloned defence zones extending for more than 70 miles in depth and embodied more than 3,000 miles of trenches. Each zone was further subdivided: for example, the first defence zone was 3 miles deep and employed five parallel lines of deep entrenchments. Protecting the front of the zone were nearly a million mines. The main defence zone consisted of mutually supporting, well-camouflaged positions integrating anti-tank guns, artillery, infantry, machine guns and dug-in tanks.[38]

Indeed, mechanisation, which had provided the means for *blitzkrieg*, could also be harnessed to the defence. The Germans, in particular, became adept at 'manoeuvre on interior lines', using mobile reserves to counter-attack enemy breakthrough forces. At Kharkov in February and March 1943, Field Marshal Erich von Manstein conducted a brilliant counter-blow against an offensive by Soviet forces. At Kursk, Russian mobile forces were also harnessed to the defence. Behind the static defence zones, and positioned for counter-attack, was the Steppe Front consisting of six Soviet armies with more than half a million troops, 1,600 tanks and 9,000 guns and mortars.[39]

The evolving character of operational art

For these reasons, the character of operational art as it developed in the Second World War was marked by a greater degree of **continuity** with the past than might be supposed. Once defensive solutions had been found to operational level mobile warfare, 'the succeeding modern operational and strategic history of war on land was unremarkable'. In the end, attrition became a vital prerequisite for Allied victory.

The continued relevance of operational art

This did not make operational thinking redundant: far from it. In the period from 1942 onwards there were still examples of extraordinary offensive successes. On 19 November 1942, for example, the Soviets launched **Operation Uranus**: combined arms attacks opened up gaps in the German front; cavalry and mechanised exploitation in depth created pincers that snapped shut on the 21 November, encircling the German 6th Army with 290,000 men. In Normandy, from the 25–31 July 1944, the US army's **Operation Cobra** achieved a breakthrough that allowed General George C. Patton's 3rd Army to exploit to distances of over 120 miles.

Perhaps one of the most impressive examples was **Operation Bagration**, the Soviet offensive against Germany's Army Group Centre in mid-1944. By 1944, Soviet approaches to operational art had been informed by the lessons of experience. The weaknesses exhibited by Russian forces earlier in war, such as a lack of flexibility, lack of depth, problems in combined arms integration and a proclivity towards frontal assaults, had been addressed to varying degrees. Bagration commenced on 22 June 1944 and involved four Soviet fronts; through effective **deception operations** and the general quantitative advantages possessed by the Soviets at this stage of the war, they were able to build up an **enormous numerical advantage** at key points. Having broken through the German lines at several points, mechanised forces conducted **deep operations** to exploit local successes; reserve echelons were then employed to deal with surrounded German forces. By 4 July Minsk had been encircled; by mid-July the operation finally ground to a halt having outdistanced its logistic support. At its conclusion, Bagration had effectively destroyed Germany's Army Group Centre: 17 German divisions were eliminated and another 50 lost more than half their strength; 670,000 German troops were killed, wounded or captured.[40]

The importance of attrition

However, operational art could not guarantee success. Following close on the heels of the success of Operation Uranus, for example, was **Operation Mars**, intended by the Soviets to eliminate the German salient west of Moscow: but here the initial Soviet breakthroughs were contained and the Red Army suffered badly. In **Italy**, multiple Allied attacks over six months failed to achieve a decisive breakthrough and the theatre had cost them 320,000 casualties by the end of the war.

Nor could operational art eliminate attrition. Even successes in operational scale mobile warfare came at a cost. Bagration inflicted heavy losses on the victorious

Russians: 750,000 casualties, of whom 180,000 were killed. Indeed, the success of mobile operations tended to be built on the foundations of costly, positional battles of attrition. British and Commonwealth success at **El Alamein** in October and November 1942 resulted in a 900-mile pursuit of Erwin Rommel's Axis forces, but the battle itself comprised a carefully managed infantry and artillery assault in a series of what the British General Bernard Montgomery termed 'crumbling attacks' – essentially attritional assaults to thin the German and Italian lines.

The successes of **Operation Cobra** were preceded by fighting in June and July in Normandy every bit as bloody as the First World War. Thus, Allied operational level successes in the latter part of the war were hard won and more than casually related to the accumulated attrition of German forces that had occurred over previous years. Moreover, if attrition were ultimately a prerequisite for restoring manoeuvre, it could also be in itself an impediment to the exercise of effective operational art. In Normandy, for example, the losses inflicted on individual formations were often enormous: the US army's 90th division lost the equivalent of 100% of its soldiers and 150% of its officers in six weeks.[41] Losses of this scale had evident consequences for the morale and level of training of troops, making it more difficult to maintain the standards of enterprise, leadership, joint and combined operations required to support modern land warfare.

Thus, the 'evolution' of land warfare after 1941 was in essence built upon a strong element of continuity. Operational level land warfare was founded upon the application of principles developed in the First World War but applied on a larger scale: principles such as decentralisation, combined arms, and depth. Allied with major incidences of mobile warfare, positional warfare was still important: for this reason, the lessons of 1918 remained valid even for operational art. For example, artillery remained a crucial component of success. If technological developments widened the variety of artillery available to include self-propelled artillery and infantry assault guns, nevertheless the basic principles used in 1918, such as the 'hurricane bombardment', remained extant in the Second World War. Lavish use continued to be made of artillery preparation: Operation Uranus, for example, employed more than 13,500 guns in an 80-minute preparatory bombardment. Innovations such as the use of airborne landings provided additional means of realising existing approaches rather than revolutionising the practice of land warfare.

The operational level of war

Whilst the Allied armies of the Second World War in practice evolved an ad hoc form of operational art, driven by the **material** conditions of war, the **conceptual** basis was late in arriving. English-speaking armies were not formally introduced to operational art as a doctrine until the 1980s, but when it did arrive, the concept was tied heavily to a very specific definition of the operational *level* of war. Prior to this, the 'operational level' was simply an informal term describing the level at which operational art was applied at a given point in time. This might vary: it could be at an army or army group level; it could apply to a theatre or a portion of a theatre. However, in the 1980s the concept, developed in the main by the United States, created a more specific meaning for the operational level of war: it was tied to a geographic theatre; it was tied to a level of command (the theatre commander); and it was associated with a specific function (the planning, conduct and sustainment of campaigns).

The Cold War

Nuclear weapons provided a post-1945 challenge to the existing foundations of land warfare. Whilst the principles required to fight a conventional campaign in a nuclear environment echoed the tactical themes developed earlier in the century such as dispersion, fluidity and a mobile defence, the technological nature of the nuclear battlefield pushed a higher-level emphasis on bureaucratic and managerial approaches to war, rather than operational thinking. Because of their firepower, nuclear weapons, it seemed, had made operational art irrelevant because conventional warfare was irrelevant. However, tactical nuclear war-fighting seemed for many to be a ludicrously impractical concept, summed up by the US army's Davy Crockett mortar – a recoilless rifle firing an M388 nuclear round, which could, in theory, substitute for divisional-scale artillery, but which, in tests, had a range less than the blast radius of its own munitions.[42] By the early 1960s, it was evident that nuclear war was not inevitable and that nuclear parity between the two superpowers had recreated a spectrum of conventional warfare beneath the nuclear options. Indeed, **new technology** seemed to offer even more possibilities for conventional warfare and operational level activity: helicopters, especially, seemed to offer new opportunities for manoeuvre, deep exploitation and combined arms.

The Soviet Union continued to exhibit a commitment to the operational art through doctrine that remained formally codified. Soviet doctrine continued to enshrine within it principles developed in the inter-war period and refined in the Second World War, including deep and successive operations; combined arms integration; joint co-operation; the use of deception, surprise and mass to achieve breakthrough; and the use of rapid manoeuvre by reserves, supported by air attack, into the depth of the enemy to convert tactical successes into decisive victory. However, whilst the US army seemed receptive to new technology, creating an air cavalry concept and deploying the First Cavalry (Air Mobile) Division into Vietnam in 1965, there remained no formal US doctrine for the conduct of operational art in land warfare.

US operational level doctrine

The sources of the regeneration in US operational thinking arose from a number of factors. The lessons of the **1973 Yom Kippur/Ramadan War**, fought at a high tempo, with heavy casualties and extensive use of precision-guided munitions, pointed to a future land warfare environment in which features such as mass were less important and in which survivability would depend upon flexibility, agility, combined arms and competence with new technology.[43] **The Soviet threat** was another incentive: Soviet foreign policy activism was accompanied by new conceptual developments, including the creation of Operational Manoeuvre Groups (OMG), manoeuvrable, combined arms brigades that would be able to exploit and raid into the depth of North Atlantic Treaty Organization (NATO) forces.[44] **New technology** in the guise of laser-guided bombs, surveillance technology, and data processing and communications equipment seemed to be on the cusp of delivering extraordinary new capabilities.[45] Moreover the traumatic **defeat in Vietnam** created a powerful mood of introspection in the army and an atmosphere conducive to political and intellectual efforts to reform the institution. The institutional consequence of these factors was the creation of new organisations to oversee a process of reform. One was the Office of Net Assessment (ONA), set up in

1973 under Andrew Marshall, which concerned itself with analysing Soviet military strengths and vulnerabilities. Also established in 1973 was the army's own Training and Doctrine Command (TRADOC): focusing on knitting together the often disparate army training and doctrine programmes, TRADOC's purpose essentially was to change how the army trained and fought.[46]

Wider developments

Moreover, there were **parallel conceptual developments** in the 1980s and 1990s which supported thinking on the operational level of war, especially the advocacy of **manoeuvre warfare** and the **OODA loop** (Observe-Orient-Decide-Act; see Box 4.5). Proponents of a greater focus by the US army on **manoeuvre warfare** argued that the army's approaches to war were too attritional and too focused on taking and holding ground. Systematising historical 'best practice', enthusiasts argued that the most effective doctrine for success was a mode of warfare (manoeuvre warfare) founded upon creating a tempo of operations greater than that of the enemy: by accelerating the **relative rate of operations** through surprise, deception, speed, mobility, manipulating uncertainty and seizing the initiative, one could invalidate the enemy's decisions and undermine their cohesion.[47] The 'manoeuvre' component of manoeuvre warfare was thus about much more than mobility: it was in many respects a state of mind and a way of thinking (see Box 4.6). One of the consequences of the manoeuvre warfare concept was to reinforce the importance of operational art: only theatre level co-ordination of forces would impose coherence on the decentralised tactical activity inherent in manoeuvre warfare approaches.

Box 4.5 OODA loops

The OODA loop was developed as a concept by US Col. John R. Boyd. Built on Boyd's examination of the dynamics of tactical air combat in the Korean War, Boyd argued that major military advantages accrued to the side that could cycle fastest through the Observe-Orient-Decide-Act (OODA) process that underpinned effective decision making. Having a more rapid 'OODA loop' than one's opponent would ensure that one could act faster than the enemy accurately could respond, creating for the enemy a crisis of lost initiative and irrelevant orders.

Box 4.6 Manoeuvre warfare

Manoeuvre warfare was built as a concept on the theories of writers such as Liddell Hart, who argued that the most effective attacks were those formed like an 'expanding torrent': they flowed around obstacles and through gaps, thus maintaining the tempo of the advance. Manoeuvre warfare thinking drew heavily on German concepts such as *Aufstragstaktik* (mission command); *Schwerpunkt* (focus of effort or centre of gravity), and *Flachen und Luchen* (surfaces and gaps). Key enablers included decentralised decision making, risk-based exploitation based on 'reconnaissance pull', and the embrace of uncertainty and formlessness. Manoeuvre warfare was translated by the British in the mid-1990s into the concept of the 'manoeuvrist approach'. This concept was applicable to all forms of warfare and comprised a mode of thought focused on adaptability, unorthodoxy and a focus on exploiting enemy weaknesses: 'manoeuvre' was conceptualised in mental rather than physical terms.

AirLand Battle

Cumulatively, these factors led to the creation of the US army's **first formal doctrine** for operational level warfare. The first incarnation of this new doctrine was termed informally '**Active Defence**', and was contained in the 1976 edition of Field Manual (FM) 100-105, *Operations.* Active defence was controversial. The doctrine committed the US army to identifying and confronting the Soviet main effort, meeting it in a forward battle that would force the Soviets to mass their forces: this mass would then be vulnerable to NATO high-technology firepower and could be defeated through attrition. Critics argued that Active Defence was too conservative and too defensive; it allowed the Soviet follow-on echelons freedom to manoeuvre. However, the doctrine did focus on many familiar operational level themes: depth in attack; the importance of joint air-ground co-operation; shock; and rapidity.[48]

A revised doctrine entitled **AirLand Battle** was promulgated in the 1982 edition of FM 100-105. AirLand Battle focused explicitly on the operational *level* of war, a level that it defined as:

> planning and conducting campaigns. Campaigns are sustained operations designed to defeat an enemy force in a specific place and time with simultaneous and sequential battles. The disposition of forces, selection of objectives, and actions to weaken or outmaneuver the enemy all set the terms for the next battle and exploit tactical gains. They are all part of the operational level of war.[49]

It was this doctrine that formally associated operational thinking with a specific intermediate level of war between tactics and strategy; indeed, the doctrine did not mention operational art at all. Nevertheless, the new doctrine codified some familiar concepts:

- **Campaigns**: one concept was the role of the campaign in co-ordinating simultaneous and sequential tactical action.
- **Depth**: using emerging sensor and precision-strike technology, US forces would 'see deep, fight deep', pinning the first Soviet echelon and using deep attack to wreak havoc on the Soviet follow-up echelons.
- **Attributes**: AirLand Battle focused on key operational level attributes such as corps level co-ordination within the theatre, jointery, fluidity, combined arms, and integrated fire and manoeuvre.
- **Manoeuvrism**: moreover, FM 100-105 focused on a manoeuvrist approach to warfare. Manoeuvre was defined as 'the dynamic element of combat':[50] rapid, vigorous, unpredictable manoeuvre would disorientate the enemy – this, rather than attrition, would be decisive.
- **The German link**: there was in the new doctrine a notable focus on some key German concepts for operational level mobile warfare – the importance of *Aufstragstaktik*, for example, and defining the *Schwerpunkt*.

The 1986 version placed even more emphasis on jointery, simultaneity and integration; it also highlighted the importance formally of **operational *art*** as the means for using the operational level of war. Operational art was defined as 'the employment of military forces to attain strategic goals in a theatre through the design, organization, and conduct of campaigns and major operations'.[51]

The 'essence of operational art' was defined as the 'identification of the enemy's centre of gravity and the concentration of superior combat power against that point to achieve decisive success'.[52] The key elements of the US conceptions of operational art and the operational level of war were adopted by other Anglophone armies, including those of Canada, Australia and the United Kingdom, and remain extant today.[53]

Whether or not AirLand Battle was a revolutionary doctrine is a matter of debate. Nevertheless, AirLand was 'an intellectual symbol of an era'.[54] It reflected, amongst other things, the **formal codification** of a set of operational level ideas; the recognition that modern land warfare had a crucial **intermediate level** between tactics and strategy, and that effectiveness at this intermediate level was a *sine qua non* for military success in campaigns. AirLand Battle also reflected the fact that, notwithstanding its particular focus on contextual principles such as precision attack, there seemed to be a common set of foundation ideas inherent in planning and executing operational art, such as:

- Conceptualising the theatre as one integrated campaign.
- Planning operations that exploited the full depth of the theatre.
- The importance of combined arms.
- Sustaining the momentum of the attack.
- Decentralised command and control.

These ideas had their origins in the First World War, they were developed in the Second, and they continue to be reflected in contemporary land warfare doctrine.

Critiques and debates

The operational level of war and the concept of operational art have come to occupy a position of central importance in the planning and execution of modern military operations. However, whereas the tactics of modern land warfare are uncontroversial, the same cannot be said for land warfare in its operational form. One set of critiques focuses on the concept of operational art; a second set concerns itself with the weaknesses in the idea of a distinct operational level of war.

Debates on operational art

For some commentators, enthusiasts of the value of operational art are guilty of ignoring some of the difficulties associated with implementing the concept in real-world conflicts.

An inflated value?

One argument is that **operational art can be over-valued** as contributor to success: it is in danger of being seen by Western armies as a 'silver bullet' solution to military problems, especially the problem of sensitivity to casualties. In the 1970s, Colonel-General Andrian

Danilevich of the Soviet General Staff argued that 'the Soviets did not win the Great Patriotic War because Soviet generalship and fighting skills were superior to those of the Germans. The Soviet armed forces simply overwhelmed the Germans with superior numbers'.[55] Even if this view overstates the case, it is certainly true that some of the most spectacular operational level Soviet successes were facilitated by the significant quantitative advantage that they possessed. Equally, whilst for many the Gulf War of 1990–91 was a vindication of US operational art and of AirLand Battle, for others the Gulf War was a land-air battle, but one that did not feature the effective realisation of AirLand doctrine. As one US officer asserted, 'The U.S. Army that led the coalition forces to success was not a good army. It was simply a better army than its opponent'. Coalition victory, it was argued, was essentially one founded upon attrition.[56]

It is easy to see why doctrines of operational art that focus on shock and dislocation are attractive to Western armies, many of which are increasingly small in size and where casualty avoidance is a political benefit. For this reason it is not surprising to find commentators asserting that 'The very essence of operational art is to win decisively in the shortest time possible and with the least loss of human lives and materiel'.[57] However, such campaigns as that in Normandy from June–August 1944 demonstrate that attrition may be the essential prerequisite for operational level success. Tukhachevsky was clear on the importance of mass in operational art, and rejected the views of those like Fuller who argued that mechanisation and mobility would allow small forces to achieve decisive effects. Tukhachevsky commented: 'It is clear that discussions about small, but mobile and mechanized armies in major wars are a cock-and-bull story, and only frivolous people can take them seriously.'[58]

The complexity of operational art

Second, some argue that modern Western militaries have **over-intellectualised** operational art. Here, the argument is that Western armies fail to understand the degree to which the *Wehrmacht's* (German armed forces) success in 1940–41 was more about improvisation than it was about a formal doctrine. Western operational doctrines in consequence introduce a formality to operational art that can often be unhelpful. The experience of the Israeli Defence Forces (IDF) is interesting in this respect: having engaged in operational art informally during earlier wars, the IDF only finally formalised their thinking with the creation in the mid-1990s of the Operational Theory Research Institute (OTRI). However, having taken on board concepts such as deep operations, OTRI then constructed a formal doctrine built upon an arcane vocabulary and contested post-modern concepts. This doctrinal 'Tower of Babel' obscured more than it illuminated.[59] The danger of this is not just that doctrine can become unusable in practice, but that it can induce an anti-intellectual counter-culture – 'muddy-boots fundamentalism' – which rejects theorising and organises itself around such simplistic, 'back-to-basics' sloganeering as, 'The Army's job is to kill people and break things'.[60]

The challenge of implementation

Third, critics note the **problems of implementation**. Effective operational art may be difficult to realise for many reasons:

Military culture

There are the **constraints of culture**. The execution of operational art must overcome entrenched strategic cultures. For example, many argue that a focus on the overwhelming use of firepower lies at the heart of US approaches to warfare.[61] These 'habits of thought' can be reinforced by the demands of so-called 'post-heroic warfare' – wars in which receiving or inflicting casualties is politically problematic.[62] Under conditions such as these, an aversion to risk can make the principles of operational art difficult to implement.

Demanding requirements

Operational art also has **demanding requirements**. Operational art requires, according to current doctrine, 'commanders who understand their operational environment, the strategic objectives, and the capabilities of all elements of their force ... and are not bound by preconceived notions of solutions'.[63] Indeed, in contemporary contexts, operational art may require as much political creativity in a commander as it does military aptitude.

Changing contexts

Moreover, the form that operational art takes will **vary according to context**. For some observers, future operational art will be very different from that of the past because the conditions will be different. For example, information rather than manoeuvre may be the crucial determinant of success. Others question operational art's **applicability across the spectrum of war**. In essence, the argument here is that operational art was brought forth in the context of high-intensity conventional warfare: it is associated historically with certain physical and political conditions, such as the large scale of military forces involved, the essentially military nature of the participants, the clarity of objectives and the physical conditions for manoeuvre, which cannot be replicated in low-intensity conflicts. In complex counterinsurgency or stabilisation scenarios there is, as Chapter 7 explores, an absolutely vital need to co-ordinate the tactical actions of multiple actors across the theatre of operations in time and space in order to achieve the intended strategic goals. Operational art, so current doctrine argues, is therefore still relevant. However, as Chapters 6 and 7 discuss, counterinsurgency, peace support and stabilisation scenarios involve many complexities, including multiple actors; complex political factors; ambiguous and long-term objectives; and decisive non-military spheres such as diplomacy, economics and law. In these scenarios, the 'theatre of operations' may be defined as much by political relationships as it is by geography. In trying to reconcile military conceptions of operational art to these kinds of scenarios it may be that we are simply 'confusing operational art and purposeful action'.[64]

Overall, there is a suspicion in some quarters that our concept of operational art may be too ambitious – that operational art is simply 'the thoughtful sequencing of tactical action to achieve a subordinate objective within a campaign'.[65] It cannot guarantee decisive victory; it cannot easily be moulded to fit unconventional contexts. As Kelly and Brennan argue:

> Operational art is not the entirety of warfare. Operational art is not the design and conduct of campaigns. Operational art is not an interagency problem. Operational

art *is* the thoughtful sequencing of tactical actions to defeat a component of the armed forces of the enemy.[66]

The operational level

Debates on operational art tend not to question the basic utility of having the concept, but only the various difficulties in defining and implementing it. In recent years, however, there has been a growing debate on the validity of having a concept at all of the operational *level* of war.

Technology

For some, this is a question raised by the impact of **technology**. Advances in sensors and communications have, it is argued, resulted in the ability to generate a real-time picture of the theatre of operations. In conditions in which the strategic level leadership can access and influence tactical units, a 'compression in the levels of war' may result: the operational level may be in danger of becoming relegated to an administrative level of command where the crucial decisions regarding tactical activity are taken directly at the strategic level. Moreover, there may be a danger that the increasingly technological nature of modern armies is encouraging managerialism and centralisation at the expense of the qualities required for effective operational art.

Misinterpretation

More fundamentally, however, there is a body of thought that argues that the operational level of war is a perverse and self-defeating **misinterpretation** of the nature of operational art. Critics have noted that:

- The idea of having a distinct and formal operational *level* of war had no precedent before its introduction in the US army's FM 100-105 in 1982. As this chapter has already discussed, Soviet thinking, for example, formalised only a concept of operational art.
- The relationship in US army doctrine between the operational level and operational art remains ambiguous: 1982's version of FM 100-105 established an operational level of war, but did not mention operational art; the 1986 version introduced operational art as a concept, but did not associate operational art with a particular level of war, noting that 'No particular echelon of command is solely or uniquely concerned with operational art'. In contrast, by 2008 US doctrine argued that operational art was activity conducted only at the operational level of war. Yet by 2011, US army doctrine had returned to the earlier view that operational art 'applies to any formation that must effectively arrange multiple tactical actions in time, space, and purpose to achieve a strategic objective, in whole or in part'.[67]
- The army's original concept of the operational level of war was defined in terms of 'planning and conducting campaigns',[68] but traditionally, this had been the responsibility of strategy.
- In the 1980s, the operational level was focused more on co-ordinating tactical level action, because the Cold War strategic context seemed a given. However, the

> strategic uncertainty associated with the end of the Cold War has forced the operational level to consider increasingly ambiguous strategic level inputs.[69]

As a result of these factors, critics have argued that the concept of the operational level of war that had been adopted by the US army and which had then been adopted by other armies, is a dangerous doctrine. Having a specific level of military command that sits between the tactical and strategic levels and which is responsible for planning and conducting campaigns might have two related negative consequences.

For one group of critics, the danger is that the operational level **locks out inputs from the strategic level**. Planning and conducting campaigns should be activities shaped by a strategy defined by political inputs. Yet the idea of a distinct intermediate military level of command responsible for planning and conducting campaigns threatens to create conditions in which operational level commanders may seek autonomy from strategic level interference. In relation to the planning of the war in Iraq in 2003, for example, the operational level commander General Tommy Franks told Donald Rumsfeld's deputy at the Pentagon to 'keep Washington focused on policy and strategy. *Leave me the hell alone to run the war*'.[70] The operational level of war, reflected in a theatre level military command tasked with planning and executing campaigns, risks separating the conduct of war from its proper political context because of the risk that operational level military commanders might seek actively to resist political interference.[71]

For another group of critics, the operational level of war is problematic because it threatens to **substitute for strategy**. The assumption that the operational level acts as an interface between tactics and strategy is all well and good, but it presupposes that a clear and effective strategy exists and that there is effective strategic direction from above. In conditions in which the strategy for a campaign might be vague and/or contested, the operational level of war may be forced to perform the role of strategy. In doing so, it provides a 'strategic alibi' for politicians and higher-level military commanders: the operational level of war can mask the lack of strategy in its proper sense.[72]

Some have asserted that these problems, taken together, help to explain the problems experienced in the campaigns in Iraq and Afghanistan, noting that:

> In practice, the tactical acts in conflict, however successful in themselves, are conducted in a strategic vacuum, in which multiple sources of conflicting interests act more or less incoherently in an ill-defined theatre.[73]

Thus, for some at least, the operational level of war is an obstacle to effective military activity because it discourages strategic political intervention. What is defined by the military as a problematic 'compression of the levels of war' is actually strategy attempting to reassert its role. Instead of hiding behind an operational level of war, critics of the idea of a distinct operational level of war argue that the commander needs to engage with political leaders in order to ensure that the conduct of military activity genuinely meets the demands of higher-level political goals.

Conclusions

Modern land warfare at the tactical level created conditions in which offensive breakthroughs were possible even in the face of modern firepower, but the experiences of

1918 demonstrated that the scope of these successes was likely to be limited because of the difficulties in taking advantage of them. **Technological developments** in the realms of wireless communications and the internal combustion engine created new possibilities for exploiting tactical successes. Inter-war theorists used these possibilities to apply the tactical principles of land warfare on a wider scale, reflected in the development of a **concept of operational art**. The level of systematisation of operational art varied, but its importance as a means of restoring the power of decision to the modern battlefield was established during the early part of the Second World War. By 1945 the fundamentals of modern system land warfare had been established, but the period after 1945 was still significant for three reasons:

- First, a debate that had previously been carried out mainly in German and Russian was now conducted **increasingly in English**.
- Second, this debate was reflected in the adoption of **formal operational doctrines** by US and NATO armies.
- Third, these doctrines accorded a centrality not just to operational art but also to a **distinct operational level** of war.

Having established the conditions that have led to the emergence of modern system land warfare, it might be logical to assume that this system has been adopted by all armies. After all, the lethality of modern firepower affects all land forces, whoever and wherever they are. Logically, shouldn't all armies fight in a modern way if they wish to survive on the battlefield? In reality, though, the techniques of land warfare are subject to a great deal of variation. The next chapter explores the reasons why this is so, reinforcing the idea that many forms of land warfare can exist at any point in time.

Notes

1 Joint Publication 1-02, *Department of Defense Dictionary of Military and Associated Terms*, April 2001, 393.
2 Ibid., 548.
3 Ibid., 74.
4 Stuart Griffin, *Joint Operations: A Short History* (Training Specialist Services HQ, 2005), 12.
5 Ibid., 14.
6 AP 3000, *British Air and Space Power Doctrine* (4th edition), 1.1.2.
7 See G. Isserson, 'The Evolution of Operational Art', in Harold S. Orenstein, *The Evolution of Soviet Operational Art, 1927–1991: The Documentary Basis*, Vol. 1 (London: Frank Cass, 1991), 48–77.
8 Martin van Creveld, 'Napoleon and the Dawn of Operational Warfare', in John Andreas Olsen and Martin van Creveld (eds) *The Evolution of Operational Art: From Napoleon to the Present* (Oxford: Oxford University Press, 2011), 9–34.
9 Justin Kelly and Mike Brennan, *Alien: How Operational Art Devoured Strategy* (Carlisle, PA: Strategic Studies Institute, September 2009), 33.
10 Colin S. Gray, *Modern Strategy* (Oxford: Oxford University Press, 1999), 47.
11 Colin McInnes, *Men, Machines and the Emergence of Modern Warfare, 1914–1945* (Camberley: Strategic and Combat Studies Institute, 1992), 31.
12 For a discussion of inter-war theorists see Allan R. Millett and Williamson Murray (eds), *Military Effectiveness, Vol. 2: The Interwar Period* (Cambridge: Cambridge University Press, 2010).
13 Kelly and Brennan, *Alien*, 56.
14 For a discussion of Soviet perspectives see Jacob W. Kipp, 'The Tsarist and Soviet Operational Art, 1853–1991', in Olsen and van Creveld (eds) *The Evolution of Operational Art*, 64–95.
15 Milan Vego, *Joint Operational Warfare: Theory and Practice* (Newport, RI: US Naval War College, 2009) V-24.

16 Kipp, 'The Tsarist and Soviet Operational Art', 66.
17 Quoted in John Kiszely, 'Thinking About the Operational Level', *RUSI Journal* (December 2005), 38–43, 38.
18 Kelly and Brennan, *Alien*, 52.
19 See James S. Corum, *The Roots of Blitzkrieg: Hans von Seekt and German Military Reform* (Kansas, KS: University Press of Kansas, 1992), especially chapters 1–4.
20 Thomas G. Mahnken, 'Beyond Blitzkreig: Allied Responses to Combined-Arms Armored Warfare During World War II', in Emily O. Goldman and Leslie C. Eliason, *The Diffusion of Military Technology and Ideas* (Stanford, CA: Stanford University Press, 2003), 254.
21 Williamson Murray, *Military Adaptation in War: With Fear of Change* (Cambridge: Cambridge University Press, 2011), 121–22.
22 Vego, *Joint Operational Warfare*, I-22.
23 Dennis Showalter, 'Prussian-German Operational Art, 1740–1943', in Olsen and van Creveld (eds) *The Evolution of Operational Art*, 51.
24 Karl-Heinz Frieser, *The Blitzkrieg Legend: The 1940 Campaign in the West* (Annapolis, MD: Naval Institute Press, 2005), 4.
25 Kelly and Brennan, *Alien*, 34–35.
26 Gary Sheffield, '*Blitzkrieg* and Attrition: Land Operations in Europe 1914–45', in Colin McInnes and G.D. Sheffield, *Warfare in the Twentieth Century: Theory and Practice* (London: Unwin Hyman, 1988), 68.
27 Frieser, *The Blitzkrieg Legend*, 2.
28 Sheffield, '*Blitzkrieg* and Attrition', 70.
29 Mahnken, 'Beyond Blitzkreig', 253–54.
30 Kelly and Brennan, *Alien*, 39.
31 Jurgen E. Forster, 'The Dynamics of *Volksgemeinschaft*: The Effectiveness of the German Military Establishment in the Second World War', in Allan R. Millet and Williamson Murray (eds) *Military Effectiveness, Volume 3: The Second World War* (Cambridge: Cambridge University Press, 2010), 200.
32 Frieser, *The Blitzkrieg Legend*, 27.
33 Martin van Creveld, *Technology and War: From 2000BC to the Present* (London: Brassey's, 1991), 181.
34 David Glantz, *Barbarossa Derailed: The Battle for Smolensk, 10 July–10 September 1941, Volume 1* (Solihull: Helion and Company, 2010), 141.
35 Van Creveld, *Technology and War*, 161.
36 Albert Seaton, *The German Army*, 1933–45 (London: Weidenfeld and Nicolson, 1982), 186.
37 Van Creveld, *Technology and War*, 206.
38 Lloyd Clark, *Kursk: The Greatest Battle* (London: Headline, 2012), 207, 210–12.
39 Ibid., 204.
40 Kipp, 'The Tsarist and Soviet Operational Art', 84.
41 Richard Lock-Pullan, '"An Inward Looking Time": The United States Army, 1973–76', *The Journal of Military History* Vol. 69, No. 2 (April 2003), 496.
42 Brian McAllister Linn, *The Echo of Battle: The Army's Way of War* (Cambridge, MA: Harvard University Press, 2007), 176.
43 Lock-Pullan, '"An Inward Looking Time"', 498–99.
44 Kipp, 'The Tsarist and Soviet Operational Art', 88–89.
45 Keith L. Shimko, *The Iraq Wars and America's Military Revolution* (Cambridge: Cambridge University Press, 2010), 47.
46 Linn, *The Echo of Battle*, 197.
47 William S. Lind, *Maneuver Warfare Handbook* (Boulder, CO: Westview, 1985), 6.
48 Lock-Pullan, '"An Inward Looking Time"', 504.
49 Kelly and Brennan, *Alien*, 61.
50 Michael Evans, *The Primacy of Doctrine: The United States Army and Military Innovation and Reform, 1945–1995*, Army Occasional Paper No. 1 (August 1996), 20.
51 Field Manual (FM) 100-105, *Operations*, 1986, 6–2.
52 Antulio J. Echevarria II, 'American Operational Art, 1917–2008', in Olsen and van Creveld (eds) *The Evolution of Operational Art*, 155.
53 Kelly and Brennan, *Alien*, 62.

54 Lock-Pullan, '"An Inward Looking Time"', 509.
55 Kipp, 'The Tsarist and Soviet Operational Art', 88.
56 Linn, *The Echo of Battle*, 221.
57 Vego, *Joint Operational Warfare*, I-6.
58 Kipp, 'The Tsarist and Soviet Operational Art', 72.
59 Avi Kober, 'The Rise and Fall of Israeli Operational Art', in Olsen and van Creveld (eds) *The Evolution of Operational Art*, 166.
60 Linn, *The Echo of Battle*, 7.
61 Echevarria, 'American Operational Art,' 152.
62 Kober, 'The Rise and Fall of Israeli Operational Art', 183.
63 Army Doctrine Publication (ADP) 3-0, *Unified Land Operations*, 10 October 2011, 10.
64 Justin Kelly and Mike Brennan, 'The Leavenworth Heresy and Perversion of Operational Art', *Joint Forces Quarterly*, Vol. 56 (1st Quarter, 2010), 116.
65 Ibid.
66 Kelly and Brennan, *Alien*, 98.
67 FM 100-105, *Operations*, 1986, 10; ADP 3-0, *Unified Land Operations*, October 2011, 9.
68 FM 100-105, *Operations*, 1982, 2–3.
69 Hew Strachan, 'Strategy or Alibi? Obama, McChrystal, and the Operational Level of War', *Survival* Vol. 52, No. 5 (October–November 2010), 160–62.
70 Ibid., 166.
71 Kelly and Brennan, *Alien*, 67.
72 Strachan, 'Strategy or Alibi?', 168.
73 General Sir Rupert Smith, 'Epilogue', in Olsen and van Creveld (eds) *The Evolution of Operational Art*, 241.

Suggested reading

Howard G. Coombs and Rick Hillier, 'Planning for Success: The Challenge of Applying Operational Art in Post-Conflict Afghanistan', *Canadian Military Journal* (Autumn 2005), 5–14. This article provides an illuminating insight into the challenges of applying the concept of operational art in complex, unconventional environments.

Karl-Heinz Frieser, *The Blitzkrieg Legend: The 1940 Campaign in the West* (Annapolis, MD: Naval Institute Press, 2005). Frieser explodes the myth of *blitzkrieg* and highlights the informality of German approaches to operational art.

David M. Glantz, *Soviet Military Operational Art: In Pursuit of Deep Battle* (London: Frank Cass, 1991). Provides an incisive study of the development of Soviet approaches to operational art.

Justin Kelly and Mike Brennan, *Alien: How Operational Art Devoured Strategy* (Carlisle, PA: Strategic Studies Institute, September 2009). As well as surveying the development of operational art, Kelly and Brennan cast doubt on the utility of a formal concept of the operational level of war.

John Andreas Olsen and Martin van Creveld (eds), *The Evolution of Operational Art: From Napoleon to the Present* (Oxford: Oxford University Press, 2011). An excellent overview of the development of operational art through the lens of national case studies.

Lt Col. Trent Scott, *The Lost Operational Art: Invigorating Campaigning into the Australian Defence Force* (Canberra: Land Warfare Studies Centre, February 2011). An Australian perspective on their military's failure to employ effective operational art in contemporary operations and suggestions as to how this can be remedied.

Hew Strachan, 'Strategy or Alibi: Obama? McChrystal, and the Operational Level of War', *Survival*, Vol. 52, No. 5 (October–November 2010), 157–82. Strachan explores the dangers of an operational level of war in the context of poorly defined strategy.

David T. Zabecki, *The German 1918 Offensives: A Case Study in the Operational Level of War* (London: Routledge, 2009). A detailed examination of the 1918 offensive using new sources.

5 Land warfare

Context and variation

Key points

- In theory, processes of military competition might be expected to result in the emergence of a single type of land warfare. Armies should logically copy the methods of the most successful land forces.
- In practice, the challenges associated with military learning, combined with such influences as differing domestic contexts, organisational influences and cultural factors, mean that armies sometimes imitate other armies but that they often also respond in other ways: adapting to, compensating for, or even ignoring developments elsewhere.
- In consequence, despite the effectiveness of modern system land warfare, many different types of land warfare are likely to exist at any point in time. No single paradigm is likely to be able to explain the past, present, or future of warfare on land.

Why do armies fight differently from one another? If the system of modern tactics and operational thinking outlined in the two previous chapters is so effective, then surely all armies would adopt it: modern land warfare would comprise a single way of doing things and we could conclude that the way the British army fights would be the same as the US army or the Syrian army or the Chinese People's Liberation Army (PLA). It is clear, though, even in terms of more recent conflicts such as Afghanistan in 2001 and Iraq in 2003, that land forces often look different and fight differently; the Iraqi, US and Taliban forces displayed marked variations in their organisation, equipment and doctrine.

This chapter examines the processes that shape how armies adapt and change over time and what this means for our analysis of land warfare. First, we examine the different theories on military adaptation. In Chapter 2 we looked at the different views on the development generally of warfare on land. In this chapter, we refine those arguments by exploring those issues that explain how and why individual armies respond to the broader technological, conceptual, political, economic and social sources of military change. The important point to understand is that every army has a different context and that context can have a decisive impact on what that army looks like and how it fights. Second, we then consider what this means for modern land warfare. In this part of the chapter we look at why it is that some armies seem more able and others less able to implement the tenets of the modern system of land warfare. We also look at the impact that enemy activity and the political, economic and social context can have on the utility of a modern system approach to warfare. For some armies the modern system of land warfare is difficult to implement, inappropriate or simply irrelevant.

Examining these issues matters. The historian Jeremy Black notes the Western fascination with analytical 'metanarratives' – of imposing a single 'big idea' or model of doing things that can be used to explain complex phenomena.[1] Whilst the modern system of land warfare outlined in Chapters 3 and 4 presents a plausible model of how an army can conduct land warfare effectively, this does not mean that it constitutes the only way that land warfare is fought. At any point in time, many different kinds of land warfare will exist, and this will continue to be the case in the future. This has important ramifications for our discussion in Part III of this book on future land warfare.

What shapes adaptation in armies?

We begin by looking at what factors might shape how an army responds to military change. Chapter 2 covered the debates on the broad sources and structure of military change, but what those debates miss out is the question of how armies respond individually. In a general sense, for example, the German and French armies of the inter-war period were subject to the same agents of change, including new technology, and yet they approached war in 1940 in very different ways.

Military competition and the sources of military change

Eventually, shouldn't all armies fight in the same way? There is a ruthless competitive logic that should dictate that the 'modern system' of land warfare outlined in the previous chapters would become the dominant form of land warfare:[2]

- Success in land warfare matters. War, after all, is of **vital interest** to states and the importance of military success should ensure armies adopt more modern practices when they have an opportunity to do so, especially where perceived threats are high.
- Logically, there should be a strong **imitative** element in the development of armies: armies have an incentive to adopt and replicate the structures and values of successful armies because they have a proven efficacy.
- Even where armies are not at war, they have an incentive to copy armies that have proven themselves successful. Armies that are especially successful often become what has been termed '**paradigm armies**' – blueprints that other armies copy because they seem to embody in their values, structure, equipment and doctrine all that is effective.

Proponents of this view would point to the way in which, historically, successful military techniques initiated by one army have been adopted by others until they have become the norm: the Swedish system spread relatively quickly after its evident success in the Thirty Years War, for example, as did the techniques of the French army in the Revolutionary and Napoleonic Wars.[3] From this perspective, the development of land warfare should be driven by the competition of war, or the prospect of war, and should feature imitation of those armies that are deemed the most successful. As a competitive activity, war should weed out ineffective practices. The result should be that the modern tactical and operational techniques outlined in the previous chapters would become the norm in land warfare.

However, critics of this view note the difficulty in practice of this 'competitive learning' argument:

- Unlike commercial enterprises, armies are not in constant competition: they often operate in **prolonged periods of peace**, when the needs of future warfare can be uncertain.
- Even where it is clear that an army has been defeated by another, it is **not always clear why it has lost**: defeat may stem from many factors unrelated to the character of an army – it may be the result of poor strategy, for example. Even where defeat can be characterised as directly military in origin, the reasons can often be elusive, especially at the time (see Box 5.1).
- Moreover, critics also highlight that the likelihood of an army implementing substantive military change will be shaped by issues of **compatibility**. Compatibility constitutes the extent to which a given military change 'is perceived as being consistent with the existing values, past experiences, and needs of potential adopters'.[4]

Box 5.1 Understanding the sources of defeat

It is easy to misconstrue the nature of an adversary's advantages: in 1940, for example, Britain and France tended to characterise German *blitzkrieg* as a wholly new form of war, which acted to obscure some of the more mundane failings in the Allies' own conduct of the campaign. For these reasons, the empirical evidence suggests that there is no automatic relationship between defeat in war, and root and branch reform: the Russian army in the wake of the Russo–Japanese War provides an example of an army that suffered comprehensive defeat but failed to reform comprehensively as a result because it viewed the causes of its defeat as rooted in poor generalship rather than systemic failings in its own army.

(Stephen Peter Rosen, *Winning the Next War: Innovation and the Modern Military* (Ithaca, NY: Cornell University Press, 1991), 9)

Compatibility and adaptation

Focusing on compatibility highlights the relevance of a range of alternative explanations for the incidence and nature of major change in armies, including domestic politics, organisational interests and cultural factors.

Domestic politics

An alternative explanation for the dynamics of change in armies focuses on the **domestic circumstances** of a given state, rather than military competition, and the implications that this has for the scope of permissible change in its army. Theories on domestic politics focus on the ways in which factors such as the **social and political circumstances** of a state might influence whether and how an army responds to changing circumstances. For example, one writer, perhaps a little unfairly, characterises the

British army prior to the First World War as 'less an instrument for war … than as a state established institution to be maintained and perpetuated for its own sake'.[5] Advocates of this perspective note that the diffusion and adoption of new military ideas, techniques and technologies often take place in an **uneven and variable fashion**. Even when facing the same general problems, armies are likely to develop different responses because their states have different history, geography, values and so on. The acceptability of radical change cannot therefore be divorced from wider factors such as the structure of the society in question. Thus the Mongols, for example, had an extraordinarily effective military system, winning victories as far west as Hungary, but this system failed to have much direct imitative influence on other military systems at the time because the Mongol military system was the product of a very different kind of society from that of Western Europe.[6]

Organisational interests

The character of an army's development is also shaped by its status as a bureaucratic organisation: its behaviour is conditioned by a variety of **non-rational factors** such as structures within the army of power, authority, prestige, friendship, deference and so forth. Major military change is thus more likely if the proposed change is compatible with an army's **bureaucratic needs and outlook**. In general, theorists tend to highlight the resistance of bureaucracies to change. The development within bureaucracies of established routines and standard operating procedures, the creation over time of group loyalties, and vested interests, all mean that whilst major change within the military is clearly not impossible, it is often difficult because of the 'basic survival nature of large organizations'.[7] Armies compound the normal problems of bureaucracies because their leadership is chosen only from within the organisation, and the promotion path is dictated by criteria established by more senior officers. Thus, orthodoxy in armies can be **self-reinforcing**: military 'mavericks', fizzing with new and radical ideas, tend often not to be promoted to the highest levels of authority. Military change is thus a political issue for armies and breaking down the obstacles may require external help, in the shape of intervention by senior political figures outside of the organisation. Or it may rely on the activities internally of 'Young Turks', restless and not fully integrated, or of 'old timers' who are secure in their careers and thus able to take risks.

Cultural factors

Cultural perspectives focus on the important role that **beliefs and values** seem to play in shaping conceptions of what constitutes the 'proper' conduct of warfare. Major military change is more likely within an army if there is compatibility between the norms that underpin the new military techniques or ideas, and those of society and the army itself. Culture can be defined as 'acquired or learned characteristics'[8] that constitute a society's 'pattern of basic assumptions'.[9] These characteristics provide a 'toolkit of world views' and 'pre-fabricated action channels',[10] shaping the way that societies and their land forces view problems and the means to solve them. Advocates of this view argue that militarily effective reforms may be rejected by armies because they do not tally with the army's values. In this vein, the political scientist Peter J. Katzenstein observes that 'Collective identities can compel actors to make choices … that serve no apparent (rational) purpose'.[11] Many of the things that we regard as essential for military efficiency, such as a focus on professional officers, discipline and training are actually culturally conditioned and an expression of Western values and assumptions. Those values and assumptions may not accord with those of other societies. Societal and institutional cultures may also have important **trans-national dimensions**. Trans-national norms can shape conceptions of what a 'legitimate', modern army should look like: a large proportion of armies, for example, are standing forces that are standardised, technologically structured and state-based even where the state concerned might be better served by a less conventional force.[12] An army's own view of what a 'proper' army should look like and how it should act may therefore be important in dictating what aspects of change it embraces and what it rejects.

Conceptualising change in armies

The impact of military competition, domestic factors, organisational influences and culture varies depending on circumstances. For example, theoretically, one might expect organisational factors would be more applicable in times of low threat, like peacetime, and military competition theories more significant as an explanation where international threats are high. The complex interaction between these different influences also means, though, that armies can **respond to change in many different ways**. Rather than simply imitating one another, armies have adopted a range of responses to the political, economic, social and military conditions that have confronted them. These responses include:

- **Imitation**, the importation and recreation of values and ideas. In essence, armies may simply copy other armies.
- **Innovation**, the introduction of new techniques or ideas. An army may respond to change by developing wholly new techniques, force structures or technological solutions.
- **Resuscitation**, the repair of existing institutions that have fallen into decay. So, an army that has been defeated, for example, may take the view that that defeat was the result of not doing as effectively what it used to do.
- **Adaptation**, the contextualisation of imported values/ideas, doctrine or technology. An army may take on board new technology, techniques of structures, but shape them to meet its own particular circumstances.
- **Compensation**, the developing of counter-measures to invalidate enemy advantages.
- **Stasis**, maintaining the existing military system as it is.

The last of these categories can be a perfectly valid choice given that major military change can be disruptive, costly and perhaps irrelevant to a threat that might be better met through political or other means. It is also important to note the potential scope of military change, which can be applied, or not, across many areas, including:

- Doctrine.
- Values.
- Structures.
- Processes.
- Technology.

There tends to be a synergistic relationship between these dimensions: in other words, an army will often find that an adequate response to military change will often require parallel changes across a number of these categories. For example, realising a new doctrine will often require new force structures and a focus on different military values. Conversely, new technology or new force structures may drive the development of new doctrine. Even so, not all military change need result in comprehensive change across the board, and different kinds of military change need not be mutually exclusive and may exist in parallel across different categories.

Context and variation

What this all adds up to is that the modern system of land warfare identified in Chapters 3 and 4 may indeed be the most effective way theoretically of fighting large-scale, high-intensity, conventional warfare; however, this does not mean that all armies have adopted its tenets or applied them in the same manner, or look and fight in the same way. Variations in themes related to the interplay amongst military competition, domestic dynamics, organisational factors and culture have resulted in armies that have fought land warfare in different ways.

We can explore this important point further by examining three dominant manifestations that have often marked land warfare since the beginning of the twentieth century.

The variable application of the modern system

First, even where land forces have been locked into intense military competition, not all successful military change will result in straightforward imitation by other armies.

The German army in the First World War

For example, Chapter 3 explored the modern tactical system of land warfare that emerged during the First World War and spread amongst the belligerents through a process of competitive diffusion. However, this diffusion was uneven. It was the German army that was at the forefront of these new approaches and other armies appeared less effective at adopting them (see Box 5.2).

Box 5.2 The German army and military change

> When the war began we were all prepared for the Germans to be successful at first owing to their study of war and scientific preparation, but we argued that very soon we should become better than they, not being hidebound by a system. The exact contrary has been the case. The Germans with their foundation of solid study and experience have been far quicker to adapt themselves to the changed conditions of war and the emergencies of the situation.
>
> (British officer in 1915, in Samuels, *Doctrine and Dogma*, 110)

Some of the impetus for the development of modern system tactics by the Germans lay in ad hoc responses to deal with immediate military problems. Thus, flexible defence in depth was developed by the German army in response to the pressure of Allied attack; infiltration tactics were developed as an offensive means to make use on the Western Front of the time-limited opportunities provided by the defeat of Russia in 1917. However, **organisational factors** were also important in explaining the more robust embrace of new techniques by the Germans as compared with their adversaries:[13]

- The German army was an organisation **culturally more open to debate**, consultation, education and innovation – one in which the military hierarchy encouraged initiative from below on the basis that it was more junior officers who really understood the practical conditions of war.
- The German army also put mechanisms in place to **experiment with new ideas** and to disseminate new techniques: for example, in January 1917, a whole infantry division was established, the purpose of which was to test and train in new ideas.
- There was also a **social dimension** to the success of the German army in its more rigorous adoption of modern system tactics. The easier social relations between officers, NCOs and men, for example, provided a more solid foundation for the adoption of decentralised command and control than those found in the British army.

The Allies were not indifferent to the need for innovation. For example, in May 1915 French Captain André Laffargue published privately a pamphlet, *The Attack in Trench Warfare*, that suggested techniques that reflected many of the principles embodied in German infiltration tactics.[14] Moreover, the British army in 1917 embarked on a systematic study of German techniques, informed by experience, intelligence information and captured German documents. However, the armies of the Allies were simply **not as effective learning organisations** as the German army. The British army, for example, remained wedded to the importance of maintaining centralised control of operations; this rigidity in command and control was compounded by deficiencies in the training of British troops, and shortages of officers and NCOs that, cumulatively, led the British army to adopt the form of German techniques without much of the philosophical substance that underpinned it.[15] Overall, whilst all armies on the Western Front adopted aspects of the modern system, the German army was more comprehensive in its embrace of both the form and substance of the new techniques because, organisationally, it was more effective in debating, embracing, implementing and diffusing new ideas.

The modern system and the Second World War

The variable impact of military competition, and cultural, domestic and organisational factors was equally manifest in the Second World War. As the previous chapter has noted, one of the reasons for the gradual reduction in the effectiveness of German *blitzkrieg* was that techniques for effective operational scale attack and defence spread gradually amongst Germany's enemies. However, even in the relatively tightly knit system of Allies and adversaries that fought in the Second World War, modern system land warfare was not applied in the same way. There were many reasons for this.

Interpreting the lessons of war

One difficulty was that armies found it **difficult to draw unambiguous lessons from combat experience**. There were many reasons for this, including:[16]

The challenge of observation

For some participants the early German victories were **observed rather than experienced**. For the Russian army, their initial view of German techniques came through observation of the Polish and French campaigns. This lack of first-hand experience allowed the Soviets to reject the significance of German operational methods. A senior Red Army officer argued in September 1940 that there was 'no such thing as *blitzkrieg*', and in December 1940 that German victories did not provide a challenge to Soviet thinking.[17] The United States had the longest period to reflect on the lessons of 1939 and 1940. Moreover, as a neutral power, it had military attachés across Europe able to procure information from both sides. The United States' first armoured division was a systematic attempt to replicate German panzer divisions, with a similar structure and a doctrine that made an explicit attempt to embody key German features such as speed of manoeuvre, the seizing of the initiative, envelopment and encirclement. By 1942, the evidence of European fighting had led to changes: armoured divisions became more balanced, and German techniques such as the forming of sub-divisional all-arms battlegroups were taken on board by creating within US divisions two tactical headquarters to allow the forming of combined arms 'combat commands'. However, these peacetime initiatives still left the US army unprepared for the real challenges of the battlefield. In the first major operation against the Germans at Kasserine Pass in North Africa in 1943, US combined arms broke down under German pressure, and US forces were crushed piecemeal by the Africa Korps.

Multiple theatres of war

Another difficulty was that the **lessons from one theatre were not always applicable to others**. After the disaster of Dunkirk, Britain had, in effect, three armies, which would operate in very different theatres. One, rebuilt in Britain, was tasked first with home defence and then with operations on the European continent; another would operate in the deserts of the Middle East; another in the jungles of Asia. 'One-size-fits-all' approaches were clearly inappropriate because the methods and force structures applicable for one environment were less so for others. In the Normandy campaign, for example, experienced armoured units brought from the Middle East suffered heavily because the demands of cramped *bocage* fighting in the Normandy campaign were so different from those of the open desert. Processes of learning were also complicated by the confused lessons provided by different adversaries. British forces in the Middle East suffered heavily at the hands of the German Africa Korps in 1941 and much of 1942, not least because of German skills in combined arms warfare. However, the British approaches that proved so problematic against the Germans, such as a focus on tank-heavy armoured divisions, had been exactly the approaches that had earlier won extraordinary successes against the Italians in 1940–41. Like Britain, Europe was not the whole of the Second World War for the United States. In the Pacific, for example, the challenges facing US ground forces were very different from those in Europe, and mastery of amphibious operations, with the demands that they made on jointery, was a prerequisite for success.[18]

The impact of domestic circumstances

Domestic **political and economic factors** also shaped how armies responded to the perceived lessons of combat.

Political circumstances

The US army, for example, was never in sole **control of production or personnel decisions**: instead, these areas were shaped, amongst other things, by competition between the Congress and president, the influence of public opinion and the activities of lobbying groups. One consequence of this was a personnel policy in which the army struggled to obtain the best quality of recruits and in which the planned size of the army proved too small. As a result, by 1944, the US army was chronically short of trained officers and men in the teeth arms, a circumstance which encouraged a compensatory focus on firepower and technology. In the Soviet Union, the impact of political considerations had a direct influence on military decision making down to the tactical level through the **influence of the Communist Party and its structure of commissars**. Political purges in 1937–38 removed 90% of all general officers and 80% of all colonels, and the centralised authoritarian structure of the Soviet state encouraged political meddling in military considerations. Combined, these political factors tended to create a climate of fear in the Soviet army which undermined flexibility, and discouraged independent thought and decentralised approaches to planning and command.[19]

Economic conditions

In some countries economic conditions were simply **not conducive to supplying the technology or logistics support for modern mobile warfare**. In both Italy and Japan, for example, **poor economic administration**, **shortages in raw materials** and an **inadequate industrial base** for the mass production of heavy equipment made replicating German techniques impossible. For this reason, the Imperial Japanese Army's aspiration to form ten armoured divisions was wholly unrealistic. In the United States and Britain, military change had to be built on a foundation of relatively low peacetime military spending, creating early **bottlenecks in the production** of modern equipment. In 1939 US defence spending was just 2% of gross domestic product (GDP) on armed forces, compared with 42% by 1945. Even where the industrial base did provide an effective platform for change and modernisation, there were often problems caused by **dysfunctional procurement practices** such as the British practice of developing tank chassis and tank guns separately, with negative consequences for the effectiveness of British armour.

Cultural influences

Cultural factors were also manifest in explaining different approaches to modern land warfare techniques. For example, in the Italian army decentralised command and control was problematic because of a military culture that featured an unwillingness to shoulder responsibility. A focus on heroic individualism that rejected German techniques as 'the cult of organisation pressed to the extreme' contributed to a pronounced lack of professionalism in the officer corps.[20] This resulted in officers who exhibited marked deficiencies in training, administrative capability, staff work and technical competence. In the British army, the importance of combined arms was well understood, but its implementation was hampered amongst other things by a culture shaped by divisive influences such as **class differences** and the **regimental system**, one consequence of which was to entrench divisions between the combat arms and to encourage service narrow-mindedness. The Japanese army embodied values that focused on aggression and obedience. Whilst these values had many positive consequences in terms of encouraging risk taking and a determined pursuit of objectives, they also produced an **unbalanced focus on certain military activities at the expense of others**: defensive techniques were marginalised because they were regarded as passive; military administration and logistics were often neglected because they were regarded as secondary activities.

Organisational factors

Organisational factors also shaped the variable implementation of modern system land warfare. One recurrent problem was the difficulty experienced by Germany's adversaries in embracing military change in the midst of crisis. The organisational challenges embodied in the need simultaneously to fight the immediate battle whilst reflecting on the requirements for future change were enormous. Some organisational challenges related to the **inexperience of army personnel because of losses and/or massive expansion**. Once Operation Barbarossa was underway, the huge losses suffered by the Soviets led to an army made up increasingly of inexperienced and poorly trained officers and men, and weak in equipment – circumstances that made executing even basic combined arms techniques difficult, let alone implementing complex reforms. In both the United States and Britain heavy casualties in Europe in 1944, especially amongst officers and NCOs in the teeth arms, tended to encourage caution and a reliance on firepower to achieve goals.

In both the United States and Britain, mobilisation also created armies that were very inexperienced: it was problematic enough to get troops to master their specialism without then requiring troops to master complex combined and joint operations. Combined arms and joint warfare theory was therefore difficult to realise in practice: for example, air and ground forces radios often operated on different frequencies; and there was no means for infantry to communicate with tanks once the latter were closed up for action. Indeed, six months before the D-Day landings, 33 US and UK divisions had failed to conduct any joint air-ground training: 21 of them had not even seen friendly aircraft for air recognition purposes.[21]

Other organisational challenges stemmed from the **lack of formal mechanisms to learn and disseminate new techniques**. The British army, for example, lacked a dedicated structure to analyse, interpret, formulate and disseminate the lessons of the battlefield. British army training was often unimaginative and too decentralised, making it difficult to create and sustain new doctrinal approaches. In the Italian army, attempts to improve the quality of combined arms were undermined by an inadequate training infrastructure. **Service interests** also impeded implementation of aspects of modern land warfare, especially in the realms of air-land co-operation. Neither the US air force nor the Royal Air Force prioritised close air support of land forces, preferring to focus on independent activities such as air control, interdiction and strategic bombing. In Japan, land-maritime co-operation was heavily undermined by political struggles between the army and navy. Even within armies themselves, different arms clashed over how combined arms warfare should be conceptualised. In the US army, like many other armies, the role of armour was shaped by a 'cavalry outlook' that tended to focus armoured forces on the independent use of fast tanks for breakthrough and exploitation rather than on close co-operation with artillery and infantry.

Thus, whilst German *blitzkrieg* in theory provided a common learning experience for other armies, this did not result either in ease or uniformity in response. US and British forces, for example, tended to pursue techniques in land warfare that were often more cautious than their German adversary. One German commander serving in Italy noted that 'The conduct of the battle by the Americans and the English was, taken all round, once again very methodical. Local successes were seldom exploited'.[22] British operations often tended to be centralised and cautious, with a focus on firepower and the use of materiel. Aside from the impact of inexperience and casualties, one reason for this was a deliberate choice not to rely on the kind of mobile warfare in which the Germans had such a comparative advantage.

Variation and the relational nature of war

A second reason why variation in forms of land warfare is likely is because even if many armies make similar assumptions about the way that land warfare should be fought, an **adaptive enemy** can force a land campaign to be fought in ways that do not meet those assumptions. Perhaps the most recurrent manifestation of this problem is the **tension between conventional and unconventional operations**. It can be precisely because an army is seen as effective at modern system land warfare that an adversary may well choose methods that embody elements of adaptation and compensation that focus on unconventional tactics and techniques. In essence, rather than imitate, an adversary may choose **adaptation** and **compensation** to develop 'anti-strategies' and 'anti-tactics' designed to thwart modern system techniques (see Box 5.3).[23]

Box 5.3 The Sino–Vietnamese War of 1979

In the 1979 war between Vietnam and China, the Chinese were able initially to take several Vietnamese provincial capitals: Vietnam, however, then utilised a 'dynamic synergy of guerrilla and conventional warfare' to mitigate Chinese strengths and to focus pressure on Chinese weaknesses, especially their logistics and command and control. Road-bound Chinese forces were met by flexible Vietnamese local and regional forces, with regular Vietnamese forces generally held in reserve.

Where armies persist in applying conventional warfare mindsets against foes who have adopted effective counter-methods, success is likely to be elusive. This point is illustrated in such conflicts as the French struggle against Vietnamese nationalists in Indo-China, the Korean War, and the wars between Israel and its Arab adversaries.

Indo-China, 1946–54

The interaction between two adversaries can produce complex outcomes, as shown by the war in Indo-China between the communist Viet Minh and French from 1946–54.[24] There, French dominance in conventional mobile operations succeeded in defeating the Vietnamese when the Viet Minh attempted to fight the French in a conventional manner in the lowland areas of the north. However, this French conventional superiority seemed to matter little when it came to breaking Viet Minh control of the highland jungles, where the latter's guerrilla techniques nullified France's advantages. In response, the French engaged in **compensation**, developing a new approach that promised to allow them to bring their conventional superiority in technology and firepower to bear. This approach was based on the idea of the *base aero-terrestre*: creating heavily defended fixed positions across Viet Minh lines of communications that would be supplied by air. To restore their lines of communication, the Viet Minh would have to dispense with their normal advantages of mobility and agility and instead take up the burden of fighting an attritional battle of position against these French positions. The French first tried the technique at Na San in late 1953 and it worked, inflicting 3,000 casualties on the Viet Minh. A second base established at Dien Bien Phu in 1954 failed dismally.

In the interim, the Viet Minh had reflected on their failures at Na San and adapted their approaches to the next encounter: they invested heavily in anti-aircraft guns to neutralise French airpower; in artillery, to augment their firepower; in their logistics, to allow for a sustained attack on Dien Bien Phu; and in a sophisticated use of the jungle terrain to camouflage and fortify their own firing positions. In the end, the positional battle of attrition was fought on terms favouring the Viet Minh, leading to the eventual surrender of the Dien Bien Phu position. At Dien Bien Phu, both the French and the Viet Minh armies found themselves using approaches very different from their assumed norms of warfare, due to a process of competitive evolution.

The Korean War, 1950–53

This dynamic interplay between belligerents is also well illustrated by the techniques used by both sides during the Korean War (1950–53).[25] China intervened in the Korean War in November 1950 with an army that had demonstrated success, winning the Chinese Civil War against their Nationalist opposition in 1949. From guerrilla warfare origins, the Chinese PLA was built on early successes against the Nationalists and adopted a progressively more conventional approach to war, mixing guerrilla operations with offensive mobile operations. If the PLA lacked mechanisation and logistic capabilities by Western standards, its capabilities were effective enough compared with its Nationalist adversary, enabling it to fight a war of encirclement and annihilation.

In Korea, once the initial benefits of strategic surprise had worn off, the PLA found itself with a much more difficult proposition. United Nations (UN) forces (which were mainly drawn from the United States and South Koreans) had a significant advantage in firepower thanks to their artillery and air support; the latter advantage also posed a severe threat to PLA logistics and mobility. Moreover, the capacity of the PLA to exploit successes with rapid breakthroughs was severely limited by its lack of armoured and mechanised forces, and the lack of wheeled logistic transport. Over time, part of the Chinese response to these problems was founded on **imitation**: increasing the PLA's firepower, air support and logistics; improving the professionalism of the army; and developing such modern system techniques for mitigating enemy firepower as mobile defence in depth. Indeed, contrary to the 'human wave' stereotype of PLA attacks, the Chinese tended to use machine guns and mortars to pin down and suppress UN forces, with attacks conducted by assault groups that would creep forwards, searching for weak spots in UN positions.

However, the PLA response also featured important elements of **compensation**: on recognising, for example, that whilst the UN benefited from lavish firepower and modern technology, this also meant that UN forces were often overly dependent on roads, fire support and continuous, large-scale logistic support. The PLA also noted that UN forces were often hesitant at night, reluctant to engage at close quarters, and did not always exploit local terrain to its fullest effect. For this reason, the PLA often conducted attacks at night, utilising the terrain to try to infiltrate the depth of UN defences, bypassing strong points in order to threaten UN artillery positions and to allow attacks to be launched from unexpected directions. Alternatively, they would deploy their forces in as close proximity to UN forces as possible in order to complicate the use of UN air and artillery strikes.

Likewise, conditions in Korea often forced UN troops to adopt techniques rather different from those developed in the Second World War. Shortage of troops, the difficult terrain and PLA tactics that focused on infiltration by light forces posed serious difficulties for UN forces: how to deal with the real possibility of being isolated, of being attacked in the flanks or rear, or of sustaining logistic support far from roads. In response, UN forces placed less emphasis on defence in depth and instead relied on discontinuous lines of company positions entrenched on high ground for all-round defence: there could be hundreds, sometimes thousands, of yards between positions. The key to sustaining these positions was active patrolling and also, crucially, very high levels of artillery and air support. On occasion, a single company position might be supported by the artillery of an entire army corps.

Arab–Israeli conflicts

Armies may also be encouraged to develop their forces in particular ways because certain modes of warfare seem especially effective against primary adversaries. A useful example of this phenomenon is the Israeli Defence Forces (IDF).[26] In clashes with their Arab adversaries in 1948, 1956 and 1967, the IDF determined that victory could be achieved through aggressive, high-tempo operations in which tanks and aircraft played a primary role. For this reason, whilst the IDF reflected many aspects of a modern system army in their focus on principles such as professionalism, integration, co-operation and decentralisation, there was more emphasis on **joint warfare**, especially on the value of tanks as the primary instrument of ground manoeuvre, and less emphasis on **combined arms co-operation** between tanks and infantry. In certain cases, therefore, armies can develop rather specialised and unbalanced forms of modern system warfare if such approaches seem to have a proven efficacy in specific conditions.

Variation in the forms that land warfare can take should be expected, then, because armies develop techniques shaped, in part, by their adversaries. Elements of unconventional warfare have been adopted routinely by armies as a means to compensate for enemy advantages in conventional operations. However, this process is dynamic: regular armies may become more irregular; irregular armies may become more regular; armies may develop particular approaches designed to meet specific adversaries or particular conditions.

Incompatibility and pre-modern methods

Finally, variation in the ways that land warfare is fought should also be expected because, for many armies, political, economic, cultural or organisational conditions may render modern system land warfare **irrelevant** or **problematic to implement**.

The impact of local context

This chapter has already noted that even within Western states, armies may fight differently. Globally, conditions within states may result in very different models of land warfare:[27]

- Poor states may lack the **finance** or **economic infrastructure** to develop and maintain modern equipment.
- Authoritarian states may reject devolved systems of command and control for reasons of **political expediency**.
- States **lacking major external threats** may focus their armies on nation building or on defence diplomacy.

Crucially, the purpose of armies is a matter of **political choice**. Many armies do not exist primarily to fight conventional wars against external threats:

- In some states, **regime security** may be key. States such as Saudi Arabia, with its National Guard, may create military structures parallel to the regular army as a means to guard against coup attempts. Indeed, regime security may result in the removal of competent officers because they are a political threat.
- Important choices such as whether to create conscript or professional armies may be related as much to dynamics in **civil-military relations** as they are to military necessity. A state may want a conscript army, for example, because it distrusts large, regular forces.
- **Foreign policy choices**, such as whether to focus on regional or global interests, will have an important bearing on the nature of a state's army.
- In many cases, the **weakness of the state and central government** may render the notion of regular armed forces a practical fiction: this has been the case in many civil conflicts in Africa.
- More widely, armies reflect the **attributes of the societies that generate them**. They are likely to reflect attributes such as divisions in class, educational problems, weaknesses in professional military culture, ethnic cleavages, economic circumstances and so forth (see Box 5.4).

So, for many armies 'effectiveness' may not be measured necessarily in terms of military efficiency, but might instead be a function of their capacity to support a particular regime, to support their own interests, to maintain internal stability, to bind a disparate nation together, or to promote political ties with allies. For example, the armies of Latin America often have a broad range of modern conventional capabilities, yet their primary role has often been in counterinsurgency or public order roles.

Box 5.4 Culture and modern warfare

For some commentators, 'modern warfare' is synonymous with 'Western warfare'. Victor Davis Hanson[1] sees a 'Western Way of War' founded upon attributes such as consensual government, individualism and civic militarism, which together provide the essential foundations for Western military advantages in organisation, discipline, morale, initiative, flexibility and command. Geoffrey Parker[2] also identifies a 'Western

Way of War', built, in his opinion, on 'finance, technology, eclecticism, and discipline'. However, if 'modern' land warfare is culturally specific, then much of the land warfare that takes place globally must, by definition, be decidedly 'unmodern'.

1 Victor Davis Hanson, *Why the West has Won* (London: Faber and Faber, 2001), 13.
2 Geoffrey Parker, 'The Future of Western Warfare', in Parker (ed.) *The Cambridge History of Warfare* (Cambridge: Cambridge University Press, 2005), 417)

The intangible constituents of modern system land warfare are crucial. The fact that an army looks modern does not mean that it fights in modern ways; indeed, modern capital-intensive forces may not even be the most effective military solution for many states.[28] Looking globally, it is entirely possible to find armies that look modern in terms of the equipment that they deploy, but which are poor facsimiles of modern system armies when it comes to operations. The Iran–Iraq War of 1980–88 provides an excellent example of this.

The Iran–Iraq War, 1980–88

The Iran–Iraq War was an enormous conflict, involving by 1988, 2.5 million troops and causing 1 million casualties.[29] Each side was equipped with the panoply of modern equipment including tanks, aircraft, heavy artillery and ballistic missiles, although, owing to sanctions and external aid to the Iraqis, the Iranians suffered a marked disadvantage in numbers and serviceability. However, neither was in a position to execute modern system land warfare.

In autocratic states, the armed forces are often only as effective as the political leadership. In both states, the structures and dynamics of political leadership translated directly into military challenges. In Iraq, a focus on loyalty and compliance with Saddam Hussein led to over-centralisation in command and control, a lack of initiative and, often, a basic lack of competence in the officers chosen for command. In Iran, internal power struggles between the Mullahs and Iranian President Abolhasan Bani-Sadr, were reflected in the existence of semi-autonomous Pasdaran and Baseej militias which existed alongside the regular army. Indeed, the armies of the respective belligerents displayed an array of fundamental problems:

- Both sides lacked the **training** and **ethos** to manoeuvre effectively: whilst troops could fight well enough from prepared positions or in the early stages of attacks, once conditions became more fluid, command, control and cohesion tended to break down.
- Both sides exhibited a lack of training and competence in **combined arms** and **joint operations**.
- Both armies lacked **mobility**, **flexibility** and **initiative**.
- These problems were compounded by broader failings in **supporting services**: signals and electronic intelligence gathering was poor, for example, and there was also a lack of **logistic infrastructure** able to support rapid, mobile operations: even limited manoeuvre resulted in disorganisation and a breakdown in command and control.

Whilst on both sides there were serious efforts to learn lessons and adapt, which bore fruit particularly for the Iraqis, the reliance by both on conscript armies and the high attrition amongst combat-effective forces made it more difficult to identify and respond to the lessons of the battlefield.

In consequence of factors such as these, the Iran–Iraq War was a conflict that looked back to the early years of the First World War in terms of its conduct. Both sides built sophisticated defensive networks that featured extensive entrenchments, minefields, mortars, and major artillery complexes to the rear. In attacking these positions, the two armies tended to focus on deliberate, rigidly planned set-piece attacks. Assaults featured extensive preparatory firepower, followed by infantry-centric attacks: tanks, for example, were often dug in and used as artillery rather than for manoeuvre. Where breakthroughs were achieved, exploitation was difficult because forces found rapid and efficient manoeuvre difficult. Casualties in the war were enormous, but by 1988 the Iraqis were able to inflict a significant military defeat on the Iranians which acted as a catalyst to end the war.

Indeed, this last point is an important one. Just because an army does not apply the principles of the modern system of land warfare doesn't mean that it can't be successful.

Non-modern system land warfare and military effectiveness

The period since the Second World War provides many examples of armies that have obtained successes despite departing significantly from Western norms of conventional warfare. Somalia's conflict with Ethiopia over the Ogaden from 1977–78 featured familiar technology such as tanks, armoured personnel carriers and heavy artillery; it featured also familiar techniques such as air-mobile operations, combined arms warfare and the use of armour as breakthrough weapons.[30] Yet it was also a war in which regular forces were leavened with local militias; in which conventional warfare was accompanied by parallel guerrilla operations; and it was a war conditioned by the interplay of ideas between foreign advisers and indigenous forces. Likewise, in the so-called 'Toyota War' between Chad and Libya in 1987, the more conventional Libyan forces were roundly thrashed by adversaries who realised principles of aggression, mobility, tempo, and seizing the initiative through the unconventional use of a combination of light forces and light vehicles.[31]

The notion that non-modern system conventional armies can still be successful under certain conditions is well illustrated by the examples of the Indo–Pakistan War of 1971 and the Egyptian army in 1973.

The Indo–Pakistan War of 1971

In 1971, India won a stunning success against Pakistan in a 12-day assault into East Pakistan (now Bangladesh), which netted 100,000 prisoners.[32] It was a campaign that one author describes as 'a blitzkrieg, integrated into South Asian conditions', characterised by 'high tempo manoeuvre warfare'.[33] Yet the Indian army achieved this despite lacking significant mechanisation. A high **relative rate of manoeuvre** was achieved through leveraging local conditions: Indian **air superiority** enabled use of helicopter transport and air drops, and also interdiction operations against Pakistani units and logistics. Pakistani forces were deployed in often strong positions, but they were frequently widely dispersed. In East Pakistan, the local population often helped the invading forces with **intelligence and logistics support**. Therefore, despite their own logistic difficulties, and despite the many instances of frontal assaults against Pakistani positions, the Indian forces were, in relative terms, better positioned to seize and maintain the initiative despite the paucity of modern mechanised forces.

1973: the Yom Kippur/Ramadan War

In contrast to the lack of mechanisation in the Indian and Pakistani forces, it might be tempting to see the 1973 war as a clash between essentially US and Soviet armies, given Israel's extensive reliance on the equipment of the former, and the Arab reliance on equipment and advisers of the latter. However, each side fought the war using methods that reflected **local adaptation.**[34] The primacy given by the IDF to tanks and aircraft at the expense of armour-infantry combined arms has already been mentioned, and the IDF was confident that their experience had shown the decisive nature of aggression, flexibility and the seizing of the initiative. Likewise, the Arab armies were in many respects far from implementing Soviet practices. Cultural factors relating to conformity, deference and the avoidance of shame made it difficult to realise the principles of decentralisation, initiative and flexibility required to implement modern system warfare. The Syrian army, for example, was a poor facsimile of its Soviet mentor: rigid, inflexible, and inexpert at combined arms warfare. The Egyptian president, Anwar Sadat, recognised the limits of the capabilities of his own army. Recognising that the IDF focus on surprise, aggression, initiative, improvisation and mobility was something that the Egyptian army simply could not replicate, Sadat argued: 'We will simply have to use our talents and our planning to compensate.'[35]

The result was an Egyptian plan that in many respects reflected **'anti-modern' principles**: a focus on seizing ground and consolidation, not exploitation; a focus on infantry assault, with mechanised forces held in reserve; rigid planning of limited activities, exhaustively practised. Sadat hoped to mitigate problems in leadership, decentralisation and initiative by giving his forces a set of very limited and narrow tasks, and emphasising defence over attack. In doing this, he also sought to exploit IDF attributes, hoping that the IDF would respond to the initial Egyptian assault across the Suez Canal with a set of aggressive, improvised and essentially one-dimensional armour or air attacks that could each be dealt with by precision-guided munitions.

Indeed, in the initial stages of the campaign, Sadat's assumptions proved correct. Commenting on the wave of early IDF tank assaults against his forces, one Egyptian general noted: 'They have assaulted in "penny packet" groupings and their sole tactic remains the cavalry charge.'[36] The initial IDF response was marked by heavy losses, a lack of success and a great deal of confusion. The limits of the Egyptian approach became evident when Sadat made the decision to advance the Egyptian forces in order to help his Syrian allies. IDF forces had already begun to adapt to the threat posed by precision-guided munitions, but once the battle became more fluid, the IDF's advantages in mobile warfare quickly overwhelmed the Egyptian forces. Despite this, Egypt's early successes had shocked the Israelis and the peace settlement that Sadat obtained gave him broadly the peace that he wanted.

In the end, then, whilst features such as mechanisation, decentralisation, combined arms, jointery, depth in attack and defence, operational thinking, fire and manoeuvre, and military professionalism seem demonstrably to improve the military effectiveness of armies, these principles are relational: they can be realised in different ways; they may be problematic to implement; and in the end, military effectiveness may not be the primary *raison d'être* of an army anyway.

Conclusion

The purpose of this chapter has been to finesse the arguments provided in Chapters 3 and 4. The 'modern system' is a crucial development in land warfare: even so, land warfare is still a heterogeneous affair, characterised by many different approaches. We opened this chapter with the question 'Why do armies fight differently from one another?' Whilst, in theory, the processes of military competition should lead to the gradual spread to other armies of the single most effective model of land warfare, in practice many kinds of land warfare exist in parallel. The reason for this is that military competition is often a difficult and inefficient mechanism for the transfer of land warfare techniques, and the relevance of modern tactical and operational land warfare as developed during the First and Second World Wars to states with varying domestic, cultural and organisational contexts can often be marginal. Western-centric analyses of land warfare can often exclude large and important conflicts that exhibit important departures from Western norms.

This chapter has identified the sorts of themes that shape alternative approaches to land warfare: a state's political and economic conditions, for example; the organisational imperatives of armies themselves; and cultural influences that bound the definition of acceptable military practices. Because of this, armies do not always simply imitate the practices of other successful armies: they may adapt them to local circumstances; they may innovate and produce new approaches to land warfare; they may create compensatory initiatives that undermine the relevance of previously successful ways of war. An army may conclude that its basic techniques and structures are sound even in the face of military defeat, if it also concludes that the reasons for that defeat are due to factors outside of its own control, such as political mistakes.

Because of this, states may use variations on the modern system of war, as was the case amongst the Allies in the First and Second World Wars; like the French and UN in Indo-China and Korea, respectively, they may be forced to embrace less modern approaches; or, as with many other armies, they may be unable or unwilling to implement the modern system of warfare because of its political, economic or social costs. In any case, the important point is that despite the emergence of the modern system, many different forms of land warfare have existed in parallel in the past and present, and, linking forwards to Part III of this book, this will continue to be the case in the future. The development of the modern system of land warfare is a necessary, but not complete, element in the story of land warfare generally.

Notes

1 Jeremy Black, *War Since 1945* (London: Reaktion, 2004), 176.
2 Emily O. Goldman and Richard B. Andres, 'Systemic Effects of Military Innovation and Diffusion', *Security Studies* Vol. 8, No. 4 (Summer 1999), 82.
3 Geoffrey Parker, *The Military Revolution: Military Innovation and the Rise of the West* (Cambridge: Cambridge University Press, 1996), 23–24, 37.
4 Everett M. Rogers, *Diffusion of Innovations*, 3rd edn (New York: Free Press 1983), 16–17, quoted in Goldman and Andres, 'Systemic Effects', 90.
5 Williamson Murray, *Military Adaptation in War: With Fear of Change* (Cambridge: Cambridge University Press, 2011), 60.
6 Goldman and Andres, 'Systemic Effects', 80.
7 Barry R. Posen, *The Sources of Military Doctrine: France, Britain, and Germany Between the World Wars* (Ithaca, NY: Cornell University Press, 1984), 58.

8 Colin Gray, *Another Bloody Century: Future War* (London: Weidenfeld and Nicolson 2005), 58.
9 Alastair Iain Johnston, *Cultural Realism: Strategic Culture and Grand Strategy in Chinese History* (Princeton, NJ: Princeton University Press, 1995), 33.
10 Peter J. Katzenstein, *Cultural Norms and National Security: Police and Military in Postwar Japan* (Ithaca, NY: Cornell University Press, 1996), 19.
11 Ibid., 25.
12 Theo Farrell, *The Norms of War: Cultural Beliefs and Modern Conflict* (Boulder, CO: Lynne Rienner, 2005), 33–38.
13 Colin McInnes, *Men, Machines and the Emergence of Modern Warfare, 1914–1945* (Camberley: Strategic and Combat Studies Institute, 1992), 28; Bruce I. Gudmundsson, *Stormtroop Tactics: Innovation in the German Army 1914–1918* (Westport, CT: Praeger, 1995), 172–76.
14 See Jonathan Krause, *Early Trench Tactics in the French Army: The Second Battle of Artois, May–June 1915* (Farnham: Ashgate, 2013) for an examination of early French innovation.
15 Martin Samuels, *Doctrine and Dogma: German and British Infantry Tactics in the First World War* (London: Greenwood Press, 1992), 130, 137–48, 149–68.
16 See Thomas G. Mahnken, 'Beyond Blitzkreig: Allied Responses to Combined-Arms Armored Warfare During World War II', in Emily O. Goldman and Leslie C. Eliason, *The Diffusion of Military Technology and Ideas* (Stanford, CA: Stanford University Press, 2003), 258–63; and the relevant chapters in Allan R. Millet and Williamson Murray (eds), *Military Effectiveness, Volume 3: The Second World War* (Cambridge: Cambridge University Press, 2010).
17 Mahnken, 'Beyond Blitzkrieg', 256.
18 Jonathan M. House, *Combined Arms Warfare in the Twentieth Century* (Lawrence, KS: University Press of Kansas, 2001), 108; Mahnken, 'Beyond Blitzkrieg', 254–55.
19 John E. Jessup, 'The Soviet Armed Forces in the Great Patriotic War, 1941–45', in Millet and Murray (eds) *Military Effectiveness, Volume 3*, 260, 269.
20 MacGregor Knox, 'The Italian Armed Forces, 1940–43', in Millet and Murray (eds) *Military Effectiveness, Volume 3*, 165.
21 House, *Combined Arms Warfare*, 170.
22 Williamson Murray, 'British Military Effectiveness in the Second World War', in Millet and Murray (eds) *Military Effectiveness, Volume 3*, 127.
23 Black, *War Since 1945*, 11–12.
24 See Martin Windrow, *The Last Valley: Dien Bien Phu and the French Defeat in Vietnam* (London: Cassell, 2004).
25 Yu Bin, 'What China Learned from its "Forgotten War" in Korea', in Mark A. Ryan, David M. Finkelstein and Michael A. McDevitt (eds) *Chinese Warfighting: The PLA Experience Since 1949* (London: M.E. Sharpe, 2003), 127–30.
26 Robert M. Citino, *Blitzkrieg to Desert Storm: The Evolution of Operational Warfare* (Lawrence, KS: University Press of Kansas, 2004), 173–74.
27 See Murray, *Military Adaptation in War*, 27; Jeremy Black, *War: Past, Present and Future* (New York, NY: St Martin's Press, 2000), 76–80.
28 Farrell, *The Norms of War*, 53.
29 Anthony H. Cordesman and Abraham R. Wagner, *Lessons of Modern War: The Iran–Iraq War* (Boulder, CO: Westview Press, 1991), 412.
30 Black, *War Since 1945*, 74.
31 Kenneth M. Pollack, *Arabs at War, 1948–1991* (Lincoln, NE: University of Nebraska, 2004), 391–97.
32 Citino, *Blitzkrieg to Desert Storm*, 188.
33 Ibid., 209
34 Michale J. Eisenstadt and Kenneth M. Pollack, 'Armies of Snow and Armies of Sand: The Impact of Soviet Military Doctrine on Arab Militaries', in Emily O. Goldman and Leslie C. Eliason, *The Diffusion of Military Technology and Ideas* (Stanford, CA: Stanford University Press, 2003), 87–90.
35 Ahron Bregman, *Israel's Wars, 1947–93* (Abingdon: Routledge, 2009), 72.
36 Citino, *Blitzkreig to Desert Storm*, 178.

Suggested reading

Jeremy Black, *War Since 1945* (London: Reaktion, 2004). Black argues for the need to view war in less ethno-centric terms and to consider global, rather than just Western, developments.

Theo Farrell and Terry Terriff (eds), *The Sources of Military Change: Culture, Politics, Technology* (Boulder, CO: Lynne Rienner, 2002). Topped and tailed by chapters examining the general issues associated with military change, this volume examines a broad range of historical case studies.

Kenneth M. Pollack, *Arabs at War, 1948–1991* (Lincoln, NE: University of Nebraska, 2004). A fascinating case study of the problems experienced by Arab armies in absorbing Western and Soviet techniques.

Stephen Peter Rosen, 'Military Effectiveness: Why Society Matters', *International Security*, Vol. 19, No. 4 (Spring 1995), 5–31. A compelling case for the importance of social structures in determining military power.

Martin Samuels, *Doctrine and Dogma: German and British Infantry Tactics in the First World War* (London: Greenwood Press, 1992). This comparative analysis of the German and British armies in the First World War sheds an interesting light on the organisational obstacles to military reform.

Lawrence Sondhaus, *Strategic Culture and Ways of War* (London: Routledge, 2006). An excellent examination of culture and the debates on its military impact.

Part II

What is victory?

'War's very object is victory, not prolonged indecision. In war there is no substitute for victory', asserted General Douglas MacArthur in 1951.[1] At that time commander of United Nations forces in the Korean War, MacArthur reflected in his comment the traditional view that success in warfare stems from the military defeat of the enemy's armed forces; that political victory and military victory are, in essence, the same thing. By inflicting a crushing military defeat upon the enemy, one can then dictate the sort of peace that one wants.

Yet, increasingly, there have appeared to be significant substitutes for victory defined in this way. As the next three chapters illustrate, achieving victory through the use of land forces may require accepting outcomes very different from the decisive battlefield victories of the past. Chapter 1 noted that armies are flexible but not infinitely so, and one recurring problem seems to be that even where armies are excellent at high-intensity conventional warfare, they may struggle to succeed in other kinds of conflict. Indeed, in complex counterinsurgency, peace and stability operations, the military defeat of the enemy may be only one, and not even the most important, role for land forces. In these environments, the modern system of land warfare may be of only marginal relevance in determining success or failure. In operations that have variously been termed 'small wars', 'low-intensity conflicts' and 'operations other than war', the control over ground exerted by land forces must be translated not only into neutralising armed adversaries, but into political influence over the local population. In these circumstances, the conventional use of force can become an obstacle to success and not an enabler.

The next chapter, Chapter 6, explores this issue in the context of counterinsurgency operations. Chapter 7 then considers the issue of peace and stability operations. Chapter 8 draws the discussion together by highlighting some of the generic difficulties that have faced armies in the twentieth and twenty-first centuries in translating land power into victory in complex, non-traditional scenarios.

Notes

1 Quoted in Richard H. Rovere and Arthur Schlesinger Jr, *General MacArthur and President Truman: The Struggle for Control of American Foreign Policy* (New Brunswick, NJ: Transaction, 1992), 276.

6 Counterinsurgency operations

Key points

- Insurgency is an asymmetric strategy that utilises varied tactics including subversion, guerrilla warfare and terrorism.
- Contemporary counterinsurgency (COIN) has its roots in the imperial policing operations of the nineteenth century and the so-called 'golden age' of COIN theory that emerged in the 1950s and 1960s.
- The principles of effective counterinsurgency have a long history and include at their heart such ideas as the primacy of political over military action, intelligence, unity of action, and the importance of winning the support of the local population.
- These principles have often proven difficult in practice to apply. Having principles for the conduct of COIN is no guarantee that they can be implemented effectively.

In previous chapters we have focused our attention on conventional land operations. Land warfare, however, has important unconventional dimensions, not least in relation to insurgency and counterinsurgency. Insurgency is one of the most widespread types of armed conflict. Yet despite its ubiquity, the history of efforts to defeat insurgency is a history marked by profound problems and controversies. Indeed, given experiences over the last ten years or so in Iraq and Afghanistan, an observer might be forgiven for thinking that insurgencies are almost impervious to military action. They are not: insurgencies can be, indeed often have been, defeated, and over time the methods by which this can be done have been codified by many armies into formal doctrines.

This chapter concerns itself primarily with one key question: what are the principles of modern counterinsurgency operations? Our investigation begins with a brief overview of the phenomenon of insurgency – its origins, objectives and methods. The chapter then considers the development of classic counterinsurgency techniques according to three broad chronological headings: imperial policing operations, charting responses by colonial powers to indigenous resistance; the so-called 'golden age' of counterinsurgency that emerged after the Second World War and which was associated with attempts to quell ideologically motivated insurgent groups; and contemporary COIN efforts, focused on the period after the end of the Cold War. The final part of the chapter explains the difficulties associated with counterinsurgency, especially the difficulties in defining success, the debates surrounding the nature of contemporary insurgency, and the problems involved with the execution of COIN operations themselves.

What we will see through this analysis is that counterinsurgency can succeed but that success is far from guaranteed: when it comes to counterinsurgency, it would seem that practice does not make perfect. In particular, two problems are important to note. First, counterinsurgency is **often marginalised** by conventional armies; regular land forces have a tendency to define their standards of military excellence in relation to high-intensity warfare against other regular military organisations. Second, whilst it isn't difficult to define a range of plausible general principles that should guide efforts at counterinsurgency, it is much more **difficult to apply these principles in practice** because each COIN context varies. In counterinsurgency, the devil is always in the detail.

Insurgency

In order to examine counterinsurgency we should first define its subject: insurgency. The latest iteration of US COIN doctrine, Field Manual (FM) 3-24, *Counter-insurgency*, defines insurgency as 'an organized movement aimed at the overthrow of a constituted government through the use of subversion and armed conflict' (see Box 6.1 for some others).[1] Insurgency can perhaps be understood through the metaphor of a virus: it attempts to take over a body (the state) and then alter it to meet its own requirements (a set of political objectives).[2]

Box 6.1 Definitions of insurgency

[A] clandestine political movement to gain control of a country from within.

(Thomas R. Mockaitis, *Iraq and the Challenge of Counterinsurgency* (Westport, CT: Praeger, 2008), 16)

[A] struggle between a non-ruling group and ruling authorities in which the non-ruling group consciously uses political resources (e.g. organisational expertise, propaganda, and demonstrations) and violence to destroy, reformulate, or sustain the basis of legitimacy of one or more aspects of politics.

(Bard O'Neill, *Insurgency and Terrorism* (Washington, DC, 2005), 15)

[An] armed movement aimed at the overthrow of a constituted government, or separation from it, through the use of subversion and armed conflict. It is a protracted politico-military struggle designed to weaken government control and legitimacy while increasing insurgent control. Political power is the central issue in an insurgency.

(Joint Publication 1-02, *Department of Defense Dictionary of Military and Associated Terms*, April 2001)

There is often a tendency to use such terms as insurgency, guerrilla warfare and terror interchangeably, but the latter two terms better describe some of the **methods** that are used by insurgents to achieve their goal (see Box 6.2). The tactics of guerrilla warfare and terror have a long history; indeed, even regular armies have made use of both. Terror, in the sense of exemplary violence directed towards the psychological intimidation of a population, has been a routine part of warfare throughout history reflected from the Roman use of crucifixion through to strategic bombing. Similarly, though the term *guerrilla* ('little war') was first coined during the French occupation of Spain in the Napoleonic Wars, armies have long made use of the tactics of irregular warfare either as an adjunct to regular warfare (through the use of special forces, partisans, or local auxiliaries), or as a primary technique (as used by many tribal warriors). Insurgencies often begin with political, rather than military action. Insurgents may later utilise, often in parallel, a variety of tactics: guerrilla warfare, subversion, terrorism, constitutional politics (through proxy political organisations), propaganda, political mobilisation and even conventional warfare if they have become strong enough.

Box 6.2 The definitional morass

Terms such as 'irregular warfare', 'insurgency', 'guerrilla warfare' or 'terrorism' are often contested, value-laden terms with complex political, legal and moral implications. Definitions often vary depending upon whether a focus is placed on who is prosecuting it, or the methods that they are using, or the purposes to which the activities are being directed.

For example, in the early stages after the invasion of Iraq in 2003, there was a marked Coalition reluctance to describe the rising tide of violence there as an insurgency because of the implication that the Coalition might be confronting popular Iraqi resistance rather than banditry and the activities of desperate pro-Saddam Hussein remnants; instead, terms such as 'no-hopers' or 'terrorists' were preferred. In the battle between insurgent and counterinsurgent, vocabulary is another weapon and language is used to isolate, de-legitimise, motivate and rally. Insurgents may be labelled 'guerrillas', 'bandits' or 'terrorists' in order to de-legitimise them; they often refer to themselves in terms reminiscent of regular forces. So, for example, whilst anti-government forces in Syria struggling against the regime of Bashir al-Assad might meet the doctrinal definition of insurgents, they are referred to by the Syrian government as terrorists, and they term themselves the Free Syrian Army. Political and moral factors may be as important as objective reality in the labels employed in counterinsurgency.

The characteristics of insurgency

Insurgencies arise from three phenomena. First, insurgencies rest upon a condition of **desperation**: that is, they are formed by the existence of grievances – political or economic discrimination, poverty, corruption, persecution and so forth. Second, insurgencies require **motivation**. This motivation is typically conditioned by a political ideology that provides an idiom of struggle: a grammar, ideas and symbols that define the nature of the problem against which the insurgents are struggling, and which defines also the solution that needs to be implemented to remedy this. Ideology provides the inspiration, determination and coherence that sustain insurgency. For this reason, insurgency is sometimes also called revolutionary warfare. Third, insurgency also depends upon a condition of **opportunity**. This opportunity may exist because the state will not address grievances (perhaps because the government reflects the interests of a particular ethnic or political group), or because the state cannot do so because it is too weak.[3] Every insurgency is unique, but insurgency as a strategy has a number of common elements:

- **Weakness**: fundamentally, insurgency is a strategy of the weak: it is used in circumstances where more direct methods would expose the insurgents to the full force of government conventional countermeasures.
- **Asymmetry**: insurgency is asymmetric. The insurgent seeks to shift the terms of the conflict from the material political and economic sphere in which the government is stronger, to those dimensions where the insurgent can compete on equal or stronger terms: the political and psychological dimensions.
- **Common functions**: in this competition insurgent groups share three core functions: to survive, to strengthen themselves, and to weaken the government.[4] Classically, this process involves protracting the conflict. In his *The War of the Flea* (1965), the journalist Robert Taber argued that the insurgent 'bites, hops, and bites again, nimbly avoiding the foot that would crush him'[5] (see Box 6.3).
- **Key target**: Last, the insurgent's key target is the local population, and the tactics of insurgency are designed to influence the population to support the insurgents and to cease supporting the government (see Box 6.4). For this reason, insurgency has sometimes been referred to as 'armed theatre': material effects are often less important than political and psychological ones.

Box 6.3 Time and the insurgent

Taber argues that the insurgent 'fights the war of the flea, and his military enemy suffers the dog's disadvantages: too much to defend, too small, ubiquitous, and agile an enemy to come to grips with. If the war continues long enough – this is the theory – the dog succumbs to exhaustion and anaemia without ever having found anything on which to close his jaws or rake with his claws'.

(Robert Taber, *The War of the Flea: A Study of Guerrilla Warfare Theory and Practice* (New York: L. Stuart, 1965), 20)

Box 6.4 The essence of insurgency

> A population will give its allegiance to the side that will best protect it … Accordingly, the highest imperative of the insurgent is to deprive the population of that sense of security. Through violence and bloodshed, the insurgent seeks to create a climate of fear by demonstrating the authorities' inability to maintain order and thus highlight its weakness … here, the fundamental asymmetry of the insurgency/counterinsurgency dynamic comes into play: the guerrillas do not have to defeat their opponents militarily; they just have to avoid losing. And, in that respect, the more conspicuous the security forces become and the more pervasive its operations, the stronger the insurgency appears to be.

(Bruce Hoffman, Insurgency and Counterinsurgency in Iraq (Santa Monica, CA: RAND, 2004), 15–16)

Models of insurgency[6]

Broadly, the strategies pursued by insurgents can be divided into three categories. One strategy is a **proto-state** approach that focuses on **emulation**, in which the insurgent group seeks to weaken the government through irregular attack, build its own capabilities and then win through developing a greater military and political strength. The insurgents seek to create and expand liberated areas. The second strategy is that of **malignancy**, which focuses on **destruction**: here, the insurgents lack the strength to create liberated areas and instead seek first to promote the collapse of the existing state. The third strategy focuses on **limited control**: more focused on criminality, the insurgents do not seek control of the whole country but rather only selected areas that they exploit for financial and political gain.

The development of insurgency

Insurgencies therefore have a core of common features and they routinely adopt one of three strategies to achieve their objectives, but how is it that insurgency has developed? Insurgency rests on a combination of **ideological motivation** and **irregular warfare tactics**, directed towards the overthrow of the existing political order.

Ideological motivation

Modern insurgency arose from the emergence and dissemination of vigorous new political philosophies.[7] The origins of these ideologies rested on such factors as:

- The Enlightenment of the mid-seventeenth and eighteenth centuries.
- The Industrial Revolution with its attendant social and economic consequences.
- European imperialism.
- Technological developments such as railways, the electric telegraph, mass printing, steamships.

These new philosophies included **Liberalism**, which emerged in the late eighteenth and nineteenth centuries as a potent force: such nationalist leaders as Simón Bolívar in Latin America, Wolfe Tone in Ireland and Giuseppe Mazzini in Italy agitated for the liberation of their peoples from the domination of foreign powers. Liberalism helped to provoke wars of liberation in America and Latin America, Europe-wide upheavals in 1848, the 'Year of Revolutions', and the emergence of such new states as the United States, Greece and Italy.

However, in the second half of the nineteenth century other ideologies were also promoting revolutionary violence: **Anarchism**, influenced by writers such as Pierre-Joseph Proudhon (1809–65) and Mikhail Bakunin (1814–76); **Nihilism**, shaped in part by the ideas of Ivan Turgenev (1818–83); and **Socialist movements**, influenced by Karl Marx (1818–83) and Friedrich Engels (1820–95), co-authors of *The Communist Manifesto* (1848). As the twentieth century dawned, these new political forces had spawned a wave of revolutionary violence: anarchist bombings and assassinations across Europe; Social Revolutionary activity in Russia. Nationalist violence was also increasing, reflected in the formation of such groups as the Serbian Black Hand (1901) struggling for the formation of a Greater Serbia. Governments struck back with repressive measures in what *The New York Times* described in 1881 as 'The War on Terror'.[8]

Irregular warfare tactics

Whilst earlier revolutionary theorists argued for the immediate implementation of armed struggle, others argued that existing governments were simply too strong: such revolutions as those of 1848, the Paris Commune in 1871, and the revolutionary unrest in Russia in 1905 were crushed by government forces; clearly, **alternative methods** were required to side-step the conventional military inferiority of revolutionary movements. Revolutionaries therefore began to consider in detail not just protest and resistance, but the problems and solutions associated with promoting change in the context of a strong asymmetry in power. In his *Rules for the Conduct of Guerrilla Bands* (1832), Giuseppe Mazzini outlined a forward-looking strategy for revolutionaries: a **hit-and-run guerrilla warfare** conducted until the revolutionaries were strong enough to form a national army. Later, theorists such as Marx and Engels argued for the need to mobilise support through **propaganda and education**, and to strike against the government using unconventional techniques including **assassinations**, **strikes** and **acts of terror**.

Theories of insurgency

Insurgency blossomed in the second half of the twentieth century. This development was the result of several factors, including growing nationalism and anti-colonialism; the presence, thanks to the Soviet Union, of an ideology of revolution; growing global interconnectedness; and a powerful copy-cat effect.[9] In a world marked by increasing connectivity and access to education, insurgents were in a stronger position than ever before both to codify their techniques and for these to diffuse to other insurgent groups. This prompted the development and dissemination of theories of insurgent warfare. Perhaps the two most significant theories were those of **rural-based communist insurgency** developed by Mao Zedong in his struggle against the Chinese Nationalists, and those of rural and urban-based **Focoist strategies** that were developed in Latin America in the 1960s and 1970s.

Maoist thinking

Mao Zedong led the Chinese Communist Party to victory in its military struggle against the Chinese Nationalists between 1927 and 1950. Mao's victory gave his ideas wide power and publicity; indeed, for some, Mao's thinking is at the heart of 'nearly all contemporary insurgency theory'.[10] Perhaps Mao's most significant statement of his views was in *On Protracted War* (1938). Mao's theories were important for laying out a number of core ideas:

- For Mao guerrilla warfare was not the crucial element in a revolutionary struggle: **political action was the foundation of success**. Mao noted that, 'The richest source of power to wage war lies in the masses of the people'.[11] This power could be unlocked only through political action. Revolutionary warfare was therefore as much about political mobilisation, propaganda, winning the hearts and minds of the peasant, and political organisation, as it was about military action: it was political action that gave military activity its meaning. The centrality of political mobilisation was reflected in Mao's famous metaphor characterising the relationship between the revolutionary army and the people as that of the relationship between the fish and the sea.

- In relation to the military struggle, Mao articulated the essence of **effective guerrilla warfare strategy**. Guerrilla warfare should focus on manipulating space, time and geography: guerrillas should use difficult terrain and avoid holding ground and instead create a fluid fight in which they held the initiative. The essence of this approach was contained in Mao's dictum that: 'the enemy advances, we withdraw; the enemy camps, we harass; the enemy tires, we attack; the enemy retreats; we pursue.'[12] However, Mao was also clear that guerrilla warfare alone would be insufficient. He conceived of the revolutionary struggle as comprising **three phases**: the enemy strategic offensive and revolutionary strategic defensive; enemy consolidation and revolutionary strategic defensive; revolutionary strategic counter-offensive and enemy's strategic retreat. Whilst guerrilla warfare might suffice for the first two stages, by stage three the revolution would need forces capable of conventional positional warfare: it was regular warfare that would provide the last and decisive element in the revolutionary struggle.
- Last, despite the mythologising of Mao's theories that later took place as part of the hardening of China's communist orthodoxy, he was actually very pragmatic. Mao noted that **theory had to be contextualised** to take into account actual circumstances: success therefore depended upon adaptation; on study, evaluation and reflection.

Focoism

Whilst South-East Asia was the focus of rural-based communist insurgency, an alternative theory developed in Latin America, where revolutionaries were struggling against right-wing authoritarian governments. **Ernesto 'Che' Guevara**, a key lieutenant of Fidel Castro, developed an alternative perspective on revolutionary war based on an interpretation of the success of Castro's seizure of power in Cuba in 1959. In his 'Guerrilla Warfare' (*La Guerra de Guerrillas*) of 1960, Guevara argued that political preparation and mobilisation of the rural peasantry was not a prerequisite for revolutionary success. Instead, a **revolutionary vanguard of fighters** acting as the *foco* (focus) of resistance could, through violence, create the conditions for revolution. Armed struggle could therefore precede the mobilisation of the masses; indeed, it would be a vital enabler for it. Violence would stir and energise the population; it would precipitate over-reaction from the government which would begin to alienate society.[13]

Guevara's ideas were adapted by **Carlos Marighella**, a Brazilian Marxist who penned the *Minimanual of the Urban Guerrilla* (1970). Marighella believed that the concept of the rural *foco* could be adapted to the **urban environment**. From 1967, using such tactics as kidnappings, bombings and robberies, Marighella's National Liberation Action organisation attempted to create the conditions that would precipitate a revolutionary struggle in Brazil.

Experience demonstrated that neither Maoism nor Focoism constituted a foolproof strategic tool. For example, Guevara's ideas misinterpreted his own experience in Cuba, and his attempts to apply his theories in the Congo in 1965, and Bolivia in 1966 both failed; in the latter enterprise Guevara was killed. However, effective or not, these theories had **immense influence**. They encouraged **other revolutionary movements**. Maoism, for example, provided the inspiration for insurgencies in Malaya, Burma, the

East Indies and the Philippines. Most prominently, perhaps, Mao's ideas provided the strategic framework for Ho Chi Minh's struggle in Vietnam, first against the French (1941–54) and then against the South Vietnamese government supported by the United States (1963–75). Moreover, Mao's success in China, and the apparent influence of his ideas elsewhere, combined with Guevara's profile as the poster boy of left-wing revolution, conspired to generate a heightened awareness of insurgency as a global ideological phenomenon. Combined with the growing incidence of insurgency, the period from 1945–70 saw a significant series of insurgent successes. British defeats in Palestine (1945–47) and Aden (1962–67), Dutch withdrawal from Indonesia (1945–49), French defeats in Indo-China (1941–54) and Algeria (1954–62), and Castro's success in Cuba (1959), amongst others, gave an impression that insurgency was growing in potency. How, then, did counterinsurgents respond?

Counterinsurgency

Counterinsurgency embodies all of the measures required to thwart insurgency. FM 3-24 defines COIN as: 'military, paramilitary, political, economic, psychological, and civic actions taken by a government to defeat insurgency' (see Box 6.5 for some alternatives).[14]

Box 6.5 Defining COIN

[T]he attempt to confound organised armed challenges to established authority.

(Jones and Smith, 'Whose Hearts and Whose Minds?', 81)

[M]easures taken by a threatened state and its supporters to defeat an insurgency.

(Mockaitis, *Resolving Insurgencies*, 11)

Those military, law enforcement, political, economic, psychological and civic actions taken to defeat insurgency, while addressing the root causes.

(British Army Field Manual Vol. 1 pt 10, *Counter-Insurgency Operations* (Ministry of Defence, 2010), 1-6)

Implicit in this definition is that counterinsurgency is not solely a military activity or responsibility. Counterinsurgency is **multi-dimensional**: it will require political, military, economic and social activity; these activities may be executed by a wide variety of military and non-military agencies and organisations. However, counterinsurgency is also a **relatively new concept**. That irregular warfare might require techniques that differed from regular warfare was something that had been recognised since ancient times. The Byzantine Emperor Maurice wrote in his *Strategikon* of *c.* AD 600 of the sorts of stratagems that might be required to defeat forces of irregular horse nomads. Modern counterinsurgency techniques have their roots in the imperial policing operations of the nineteenth century. During this period, colonial armies developed techniques to meet the challenges of unconventional warfare that were at variance to those required for conventional warfare against symmetrical opponents. These imperial policing techniques were often bottom-up solutions developed over time to meet local problems. However, by the end of the century these techniques were beginning to be codified in the works of a number of soldier-theorists.

Imperial policing

The gradual emergence of insurgency as a threat was not something wholly ignored by regular armies. In terms of practice, for example, armies had often responded to the tactical and political needs of countering irregular adversaries. As mentioned, it was during the Peninsular War (1808–14) that the term *guerrilla* emerged, and the French responded to Spanish guerrillas by using *contra-guerrillas* in an attempt to counter the insurgents with their own irregular forces. There also emerged some written reflections on the difficulties of dealing with revolutionary warfare. In his *Treatise on Partisan Warfare* (1784), **Johann von Ewald**, a Hessian officer of jägers and a veteran of the American War of Independence, mused on the unique conditions that seemed to have emerged during the war, including the power of nationalism and self-determination in shaping a high degree of motivation and commitment on the part of the colonists.[15]

However, whilst colonial powers had extensive experience of putting down rebellions, modern counterinsurgency really only began in the late nineteenth and early twentieth centuries, when the lessons of the past began to be codified and disseminated, often in written form.

Theories of imperial policing

Theorists, often soldiers, began to articulate ideas on how properly to defeat guerrilla opposition and did so evoking ideas that would form many of the foundations of future thinking: the importance of gaining the support of the local population; the pitfalls in focusing too much on force; the significance of political and economic activity; and the peculiar demands of irregular warfare which meant that the techniques of conventional warfare would not suffice. Some of these works are today largely ignored: the prolific British barrister Thomas Miller Maguire's *Guerrilla or Partisan Warfare* (1904) and General Sir William Heneker's *Bush Warfare* (1907) fall into this category, but there were others whose works continue to be referenced even today.

Hubert Lyautey and Joseph Gallieni

Two of the key exponents of specialist counter-guerrilla techniques were the French officers Marshal Hubert Lyautey (1854–1934) and Marshal Joseph Gallieni (1849–1916). Gallieni's ideas, added to and developed by Lyautey, drew on their experiences in North Africa, Madagascar and Indo-China. French approaches to colonial warfare had long understood the need both for tactical military adaptation and parallel non-military activity (see Box 6.6). In particular, Gallieni and Lyautey saw the key battle-ground as the struggle for the **loyalty of the population**: they had to be protected and encouraged to view French control as preferable to the alternatives. Gallieni used the metaphor of the *tache d'huile* ('oil spot') in which, through promoting economic development and security, French control would spread from secure areas outward. Lyautey used the metaphor of the plant and soil: the guerrillas, like plants, could only thrive in a particular kind of environment; thus defeating the insurgent required principally measures to secure the sympathy of the population, which constituted the 'soil' from which insurgents grew. The military organisation required to conduct anti-guerrilla operations required, in consequence, an ability to **administrate as well as to fight**.[16]

Box 6.6 French colonial policing

Shaped from the 1830s onwards by their experiences in Algeria, French techniques of imperial policing were in many respects remarkably advanced. Facing an irregular enemy with no traditional centre of gravity, lines of communications or interest in holding ground, the French made use of an adaptation of the Bedouin tactic of the *razzia*, or raid. This comprised the use of mobile columns, targeted by intelligence, to intimidate recalcitrant villagers and to threaten the one thing that the locals could not make mobile: their villages. Booty taken from the villages also allowed the columns to be self-sustaining.

However, the French colonial army recognised that it was not enough to drive off the guerrillas: something had to be put in place that would consolidate French influence. What the French termed 'material conquest' was insufficient: it had to be accompanied by 'moral conquest'. In 1833, the French first experimented with what would later become the *Bureaux Arabe*. Staffed by military officers, and focused initially on intelligence gathering, this organisation was later tasked specifically with political and economic action to win the loyalty of the population. With a role in local development, justice and the managing of political relationships with the tribes, the Bureaux tried to work through established political structures and respect indigenous customs, values and language in order to avoid imposing at a local level an alien form of government.

Charles Callwell

Another notable theorist from this period was the British officer Major-General Sir Charles Callwell (1859–1928), author of *Small Wars: Their Principles and Practice*, first published in 1896 but revised in 1899 and 1906. 'Small wars' Callwell defined as 'operations of regular armies against irregular, or comparatively speaking irregular, forces'.[17] Callwell's book is interesting for many reasons. One feature was its recognition that whilst there were some parallels between small wars and conventional wars, **small wars required a unique approach**. Callwell noted that:

> the conditions of small wars are so diversified, the enemy's mode of fighting is often so peculiar, and the theatres of operations present such singular features, that irregular warfare must generally be carried out on a method totally different from the stereotyped system … The conduct of small wars … is an art in itself.[18]

Callwell identified **the particular difficulties in defeating irregular adversaries**: their mobility; the difficulty in obtaining information; the problem of sustaining communications and logistics when chasing down highly mobile adversaries. Irregular adversaries had the advantages both of time and of the initiative. In consequence, Callwell also recognised the need for training and preparation based on an understanding of the specific conditions to be faced, noting that an army must know 'what nature of opposition they must expect, and should understand how best to overcome it'.[19]

Lawrence of Arabia

Another seminal soldier-theorist from this period was T.E. Lawrence (1888–1935), the famed 'Lawrence of Arabia', termed by one author 'the progenitor of modern irregular warfare in the West'.[20] Lawrence wrote the famous *The Seven Pillars of Wisdom* (1922), an account of his involvement in the Arab revolt of 1917–18. Lawrence's perspective was unique, because as a key actor in mobilising the Arab revolt against the Ottoman Empire during the First World War, he was fighting on the side of the irregulars. Lawrence's analysis of the revolt provided an important insight into the advantages of irregular warfare. Anchoring his analysis in the concept of what he called 'desert power', Lawrence identified the importance of **maintaining the initiative** in irregular warfare, and the consequent ability to **harness hit-and-run tactics, intelligence, psychological warfare and propaganda**. Lawrence concluded: 'Our cards were speed and time, not hitting power, and these gave us strategical rather than tactical strength.'[21] Lawrence's analysis of the strengths and weakness of irregular warfare led him to conclude that it was foolish for his irregulars to focus on taking enemy outposts or cities: instead, the focus should be on exploiting Ottoman logistic weaknesses through attacks on their lines of communications. Irregular warfare could be a powerful tool: 'granted mobility, security … time, and doctrine', noted Lawrence, the insurgents would win.[22]

Charles Gwynn

In another significant publication, *Imperial Policing* (1934), Major-General Charles Gwynn (1870–1962) focused on a theme that would become central to later counter-insurgency thinking: the problems surrounding **the role of force**. Gwynn argued that examples from 1919–31 demonstrated that a difficult paradox permeated the conduct of imperial policing operations. On the one hand, the excessive use of force by the government could have a profoundly negative impact on the views of neutral or hitherto loyal elements of the local population. Gwynn commented that 'Excessive severity may antagonise this element, add to the number of the rebels, and leave a lasting feeling of resentment and bitterness'. On the other hand, however, a lack of a willingness to use force might be interpreted as a sign of weakness. It was therefore important, noted Gwynn, that 'the power and resolution of the Government forces must be displayed'.[23] To manage this paradox Gwynn advocated that the need for firm and timely action needed to be balanced by adherence to the principles of **minimum force**; **close civil-military co-operation**; and **civilian control of policy making**.

The US Marine Corps

Shaped by long experience of involvement in expeditionary operations, the US Marine Corps' *Small Wars Manual* (1940) was in many respects extraordinarily forward looking. It included chapters not just on the tactical handling of military operations, but on such **broader themes** as military co-operation with the State Department and with civilian agencies, disarming the local population, the handling of local auxiliary units, the principles of military government, and the supervision of elections. The manual noted that 'The motive in small wars is not material destruction. It is usually a project dealing with the social, economic, and political development of the people'.[24]

The limitations of imperial policing

However, if this corpus of material on imperial policing provided much that would re-surface later in modern doctrine, it still had important limitations:

An apolitical focus

First, techniques of imperial policing tended to **see colonial opposition in apolitical terms**: they were viewed as local malcontents, rebels, mutineers. They were not viewed as the vanguard of movements fighting for revolutionary political change. This meant that the basis of successful imperial policing tended to be seen in terms of military organisation and techniques rather than more holistic approaches that would require significant civilian inputs or any great effort at understanding the political goals of local adversaries. In these contexts, the enemy were often characterised in quite simplistic terms. For example, Callwell saw 'small wars' as comprising 'expeditions against savages and semi-civilized races'.[25]

Limited diffusion

Second, the methods developed to deal with colonial policing were still seen by regular armies as auxiliary techniques: the lessons of imperial policing **tended not to percolate through armies as a whole**. In France, for example, imperial policing was something conducted by the colonial army – the French army more generally continued to focus its efforts on meeting the threat of conventional warfare against such adversaries as Prussia/Germany. Whilst such writers as Lyautey were widely read, they nevertheless had little concrete impact on mainstream military doctrine.

Brutality

Third, the colonial context of the nineteenth century **permitted techniques that involved great brutality**. Indeed, Callwell argued explicitly that 'small wars' would require methods that might be unacceptable in regular warfare. These techniques began to cause controversy towards the end of the period: British methods in South Africa, US techniques in the Philippines, French approaches in Madagascar – all began to cause domestic and sometimes international unease.[26]

Continued problems

Indeed, fourth, the age of imperial policing also highlighted **recurrent difficulties** in the practice of defeating irregular opponents. Despite the cerebral approach of French techniques, for example, Lyautey was still very critical of many of the features of the French army. He criticised the too rapid rotation of troops into and out of the theatre of operations; the failure of the army to provide enough officers trained in such key skills as relevant languages; the lack of unity of command; the division that emerged over time between the colonial army and the metropolitan army.

Thus the period up to the Second World War saw a gradual leavening of more traditional forms of irregular warfare with the newer phenomenon of insurgency. The latter was shaped by conditions of **material weakness** that created a reliance on avoiding conventional military clashes, using instead other approaches such as guerrilla warfare; however, at its heart, insurgency was shaped by the **impact of ideology** that gave a new order of purpose, motivation and organisation to irregular warfare. In parallel, imperial armies involved in countering local rebellions began to **systematise their thinking** on the lessons of dealing with irregular warfare. Both of these strands laid the foundations for developments in the so-called 'golden age' of counterinsurgency thinking.

The golden age of counterinsurgency

The period from the end of the Second World War until the 1970s saw an **unprecedented incidence of insurgency**. Anti-colonial rebellions, separatist struggles in weak states, leftist guerrillas struggling against authoritarian regimes, proxy battles fuelled by Cold War superpower rivalries all conspired to make insurgency the dominant form of armed conflict. Insurgencies also benefited from the attention of an **increasingly pervasive international media**. Increased scrutiny of counterinsurgency campaigns was one consequence, a **greater transparency** that encouraged debates on the methods, goals and relative costs of what were often brutal campaigns. However, the media also gave a **greater voice** to the insurgents: even, in the case of charismatic individuals such as Che Guevara, making them into celebrities and, for some, heroes.

The result of this challenge was the emergence of what has been termed a 'golden age' in counterinsurgency thinking: an effusion of written theory on how the challenge of modern insurgency could be met.

Classical counterinsurgency

That insurgency was not, in fact, as powerful as might be supposed is evident from a number of **high-profile counterinsurgency successes** during the period: communist insurgency was defeated in Malaya between 1948 and 1960; Indonesian efforts to undermine Malaysia were defeated from 1962–66; US advisers helped the Philippines deal with the communist Hukbalahap insurgency; the Bolivians dealt with Che Guevara's Marxist revolutionaries. Nevertheless, the scale of the insurgent problem and the evident difficulties in dealing with it prompted a **flowering of thought** on the principles of effective counterinsurgency. This thinking included Roger Trinquier's *Modern Warfare: A French View of Counterinsurgency* (1961) and John McCuen's 1966 work *The Art of Counter-revolutionary War: A Strategy of Counter-insurgency*. Perhaps the most significant writers, though, were David Galula, Sir Robert Thompson and Sir Frank Kitson. The views of these writers crystallised what is termed 'classical' counterinsurgency theory: principles that would exert a powerful impact into the present.

David Galula

David Galula (1919–67) was a French officer who served as a company commander in Algeria from 1956–58. Galula codified his thoughts in his *Pacification in Algeria* (1963) and *Counterinsurgency Warfare: Theory and Practice* (1964). Three ideas, in particular, were especially important themes in Galula's works. One was that successful COIN was not fundamentally about killing insurgents; it was about **protecting the population**. In COIN, it was the population that was 'the prize'.[27] Only if one could succeed in separating the population from the insurgents could one achieve victory. A second feature of Galula's writing was the importance he attached to the need for the military to **master a wide variety of skills and activities**. Galula argued that the soldier 'must be prepared to become a propagandist, a social worker, a civil engineer, a schoolteacher, a nurse, a boy scout'.[28] Nevertheless, a third recurrent theme was that counter-insurgency primarily is a **political rather than a military activity**. Galula noted '[that] the political power is the undisputed boss is a matter of both principle and practicality. What is at stake is the country's political regime, and to defend it is a political affair'.[29] Moreover, even if the military had to master a variety of non-military tasks, this should be 'only for as long as he cannot be replaced, for it is better to entrust civilian tasks to civilians'.[30]

Sir Robert Thompson

Sir Robert Thompson (1916–92) was adviser to the British Director of Operations during the Malayan Emergency (1948–60) and later head of the British Advisory Mission to South Vietnam. Thompson's experiences led him to write *Defeating Communist Insurgency: Experiences from Malaya and Vietnam* (1966). Thompson articulated **five basic principles of counterinsurgency**:

- The government should have a clear political aim for its COIN campaign.
- There should be an overall plan for the campaign.
- In the guerrilla phase, the counterinsurgents should first secure their own base areas.
- The government should operate within the law.
- Finally, counterinsurgency should focus less on defeating the enemy guerrillas and should instead prioritise defeating political subversion.

Thompson was also significant for first using the terms '**clear and hold**', reflecting the importance not just of removing or suppressing insurgent activity but also of then being able to restore normal government so that an area could be held in the long term. Like Galula, Thompson was also population-focused, arguing that '"Winning the population" can tritely be summed up as good government in all its aspects'.[31]

Sir Frank Kitson

General Sir Frank Kitson served with the British army in Malaya, Kenya, Oman, Cyprus and Northern Ireland. His publications included *Gangs and Counter-Gangs* (1960), *Low Intensity Operations: Subversion, Insurgency and Peacekeeping* (1971), and *Bunch of Five* (1977). Kitson's writings reflected his judgement that insurgencies **could not be defeated through military operations**. Kitson's view was that 'there can be no such thing as a purely military solution because insurgency is not primarily a military activity'.[32] Kitson placed particular emphasis on the importance in counterinsurgency of **good intelligence**, noting: 'If it is accepted that the problem of defeating the enemy consists very largely of finding him, it is easy to recognise the paramount importance of good information.'[33] Obtaining general information on the insurgent was not enough: this 'background information' had to be turned into 'contact information' – information that could be used to improve the chances that guerrillas could be found and engaged successfully.

The principles of counterinsurgency

The writings of theorists were also now accompanied by official publications that attempted to codify established 'best practice'. In 1952, for example, Britain published *Conducting Anti-Terrorist Operations in Malaya*, and in 1969 *Land Operations: Counterrevolutionary Warfare*. In 1953, the United States produced FM 31-15, *Operations Against Irregular Forces*, and followed with a slew of other publications relevant to counterinsurgency, including: National Security Action Memorandum 182 (1962), FM 31-16, *Counterguerrilla Operations* (1963), DA-PAM 550-100 and the *US Army Handbook of Counterinsurgency Guidelines for Area Commanders* (both 1966), and FM 8-2 *Counterinsurgency Operations* (1967).

Core themes in counterinsurgency

Whilst official publications, and those of military theorists, might emphasise different elements of counterinsurgency, they all tended to emphasise **the same sorts of core themes**:

- One recurring idea was that of **separating the population from the insurgent**: insurgents relied on the support, or at least acquiescence, of the local population to survive. Where support amongst the population for the insurgent was low and support for the government was high, the insurgents would be deprived of supplies, recruits and hiding places; government forces would be able to obtain effective local intelligence and would be able to rely on locally recruited forces. Effecting this separation required military action: the government must be able to provide meaningful security. However, political and economic action was also crucial: dealing with local grievances and providing effective governance. Insurgents espoused a cause: action needed to be undertaken that would undermine the basis of the insurgent cause. As one commentator observed in the 1960s, insurgency was 'not out-fought but out-governed'.[34]
- Another theme was the importance of **political primacy**. COIN may have had important military dimensions, but overall control needed to be exerted by government not the military because only this would ensure that political rather military priorities prevailed.
- Political primacy was also associated with the importance of an **effective overall strategy**: having a clear and attainable goal for the campaign, as well as a long-term plan to ensure that any early successes could be transformed successfully into a stable peace.
- **Unity of effort** was also a recurrent idea. Effective counterinsurgency required that all of the elements of national power, diplomatic, military, political, economic, intelligence and so on were all co-ordinated. This requirement for unity of effort extended to co-operation between national contingents in multi-national operations and to the various component parts of a campaign: military units, government agencies, international organisations, non-governmental organisations.
- **Restraint** was often seen as crucial: COIN might well require the use of force, but it should be proportional and discriminate; the government should also act within the law.
- **Intelligence** was often an important contributor to restraint because it was successful intelligence that allowed force to be focused. Intelligence was also a central means through which a proper understanding of the context could be achieved: an understanding of the enemy, for example, and of the local population.
- Counterinsurgency forces needed **flexibility**: approaching COIN operations using the techniques associated with regular warfare was unlikely to bring success. Most

writing on counterinsurgency argued that the security forces needed to adapt to use small unit tactics, with such features as decentralised command and control. This often led to the formation of specialist COIN units: the British Ferret Force in Malaya, for example, or the French *Groupement de Commandos Mixtes Aeroportes* (GCMA) used in Indo-China.

All of these themes **tended to be related**: for example, separating the population from the insurgent was helped by avoiding indiscriminate violence, and by having intelligence that facilitated an understanding of the population's grievances and aspirations. The strong continuities that tended to exist in approaches to counterinsurgency are evident in contemporary COIN doctrine.

Contemporary doctrine

By the end of the twentieth century the counterinsurgency doctrines of such countries as the United States and the United Kingdom exhibited two related features. First, their **core principles were very similar** (see Box 6.7). Second, those principles were **evolutionary** and drew heavily on the ideas and themes developed by classical counterinsurgency theorists, and, indeed, on their colonial past.

Box 6.7 Pre-2001 US and UK COIN principles

Table 6.1

US	*UK*
Political primacy	Political primacy and political aim
Unity of effort	Coordinated government machinery
Intelligence	Intelligence and information
Isolating the insurgent	Separating the insurgent from his support
Long-term commitment	Longer-term post-insurgency planning
Legitimacy	Neutralising the insurgent
Rule of law	Understanding the environment

In fact, though, even the latest iterations of doctrine, built on experiences learnt from campaigns in Iraq and Afghanistan, **amplify but do not substantially change** these principles. The difficulties in stabilising Iraq and Afghanistan prompted a notable process of reflection and debate, especially in the United States. The result was **FM 3-24**, *Counterinsurgency*, a document that one commentator argued represented 'probably the most comprehensive operational document for COIN operations ever produced'.[35] Reflecting the lessons of Iraq and Afghanistan, FM 3-24 had at its heart a **'population-centric' approach** to counterinsurgency that drew explicitly on the theorists of the 1960s, especially David Galula, but it also referenced earlier writers such as T.E. Lawrence. FM 3-24 argued for an approach to counterinsurgency that stressed the primacy of political objectives – thus, it argued that 'The primary objective of any COIN operation is to foster development of effective governance … by the balanced application of both military and non-military means'.[36] FM 3-24 also stressed the importance of a variety of other themes that resonated with classical approaches:

- The importance of **civil-military unity of effort**.
- The importance of the **non-military aspects of COIN**, including economic development, political activity and information operations.
- The need for the **constrained use of force**.
- The importance of **providing security** for the local population.

Further publications reinforced these themes: the *U.S. Government Counter-insurgency Guide*, published in 2009, highlighted such themes as the importance of **interagency co-operation**, **economic development**, and a political strategy to promote **political reconciliation**, reform and **effective governance**.[37] Indeed, the principles at the core of FM 3-24 still had at their heart the same sorts of themes identified in earlier doctrines (see Box 6.8).

Box 6.8 FM 3-24 principles of COIN

- Legitimacy is the main objective: the key objective is to encourage the development of legitimate and effective government, legitimacy defined in terms of the consent of the local population.
- Unity of effort is essential: assure the synchronisation of actions and messages across the spectrum of actors involved in the COIN effort.
- Political factors are primary: overall success in COIN requires a political solution. Military activity should be governed by political considerations so that it does not hinder a political resolution.
- Understand the environment: success requires a strong understanding of local culture and society, including ideologies, political structures and motivations.
- Intelligence drives operations: only with timely, reliable and specific intelligence can COIN operations be successful.
- Isolate the insurgents from their cause and support: physical and political initiatives are required in order to marginalise support for the insurgents amongst the local population.
- Security under the rule of law: without security, political and economic development is impossible, but security needs to be provided in a lawful manner if it is to be viewed as legitimate.
- Long-term commitment: because insurgencies tend to be protracted, counter-insurgency campaigns must demonstrate the physical and political willingness to persevere.
- Manage information and expectations: unrealistic expectations amongst the local population regarding the pace of political and economic development damage the credibility of the counterinsurgents.
- Use the appropriate level of force: high levels of force may sometimes be necessary but careful thought needs to be given to who applies it, when and how.
- Learn and adapt: the counterinsurgent force must create the structures and values to adjust to meet changing circumstances.
- Empower the lowest level: 'mission command' approaches are the ideal approach in COIN campaigns.

- Support the host nation: the long-term goal in COIN is to leave a viable local government. Success depends upon getting the host nation to take responsibility as soon as is practicable.

Again, there remains much similarity between countries. British doctrine has ten principles: primacy of political purpose; unity of effort; understand the human terrain; secure the population; neutralise the insurgent; gain and maintain popular support; operate in accordance with the law; integrate intelligence; prepare for the long term; learn and adapt.

In many respects, FM 3-24 was actually **a rediscovery of the past**: Galula had been largely ignored for many years and yet, in the wake of this new process of reform, he became essential reading on some officer training courses. However, Galula himself was developing ideas that had been extant in the French army since the colonial period. If Galula was, according to some, the 'Clausewitz of counterinsurgency', then Lyautey was, for others, 'the godfather of population-centric counterinsurgency theory'.[38] Indeed, the sense that the basic themes of effective counterinsurgency had been long identified was evident in Johann von Ewald's comment, which he wrote in his 1785 *Treatise on Partisan Warfare*, that 'I know that I am not writing anything new'.[39]

However, if the basic themes of effective counterinsurgency were identified so very long ago, why is COIN often so difficult to do well? This question leads us to the final part of this chapter and a focus on those issues that explain the difficulties in translating theory into practice.

The challenges of counterinsurgency

The problems experienced by Coalition forces in Iraq and Afghanistan are in danger of giving the false impression that insurgencies always tend to win. This isn't so. Even in the difficult struggle against Islamist insurgent groups, counterinsurgency efforts have had some effect. In 2004, for example, the Al Qaeda strategist Abu Musab as-Suri admitted: 'The Americans have eliminated the majority of the armed jihadist movement's leadership, infrastructure, supporters and friends.'[40] Whilst insurgencies have, since 1945, become more successful than previously, they still fail more than they triumph. Out of 443 insurgencies since 1775, the counterinsurgents won 63.6% of those wars that ended, the insurgents won 25.5%, and the remainder were draws. Since 1945 the insurgents have become more successful, winning 40.3% of those wars that ended, but the counterinsurgents have still triumphed in 50.8% of cases.[41]

Yet, despite the statistics, **counterinsurgency remains an area of deep controversy** for land forces. The reasons for this relate partly to the difficulty in defining success in counterinsurgency, partly in the problems associated with implementing the principles of COIN, and partly also because of recent debates surrounding whether the character of the insurgent threat is changing.

What is success in counterinsurgency?

When Lt General David Petraeus took command of Coalition forces in Afghanistan in 2010, he asserted that 'we are in this to win',[42] but one of the difficulties in counterinsurgency is determining what victory means. Even where insurgents do not achieve all of their primary goals, a counterinsurgency campaign may still not be accompanied by a clear-cut sense of success. Three issues, in particular, tend to cause difficulties:

Rating outcomes

First, rating outcomes is complex. Even where insurgents are deemed to have lost, they have historically **often extracted concessions** from the government and created political settlements that give them some of what they want. For example, this was the case with the outcome of Britain's struggle against the Provisional Irish Republican Army (PIRA), where some suspected terrorists were able to enter democratic politics and occupy important political positions in Northern Ireland politics. Indeed, the grievance settling that often forms part of trying to win the support of the local population may require such concessions. Historically, governments that refuse to compromise at all have often lost counterinsurgency campaigns.[43]

Controversies

Moreover, COIN campaigns, even where successful, are often protracted and, over time, can **create intense controversies and deep divisions** within domestic politics. One set of controversies relates to fears regarding the use of **special legislation** and the impact on individual rights and freedoms. These difficulties have been illustrated by the controversies surrounding the United States' use in recent years of 'enhanced interrogation techniques' on suspects, wiretapping without warrants, detention without trial at Guantanamo Bay, the use of drone strikes for targeted assassinations in such places as Pakistan and Yemen, and the rendition of detainees to countries like Algeria and Egypt where they may be subject to torture.

Another area of controversy is the **behaviour of security forces** engaged in counterinsurgency. From the conditions in British concentration camps during the Boer War (1899–1902), to the behaviour of French paratroopers in the Casbah of Algiers in 1957; from the massacre by US troops of Vietnamese villagers at My Lai in 1968, through to the activities in Abu Ghraib prison in Iraq: the exposure by the media of the extreme acts that can occur during counterinsurgency campaigns can have a profoundly damaging effect on society, polarising opinion, poisoning civil-military relations and de-legitimising the existing political system.

Creating long-term success

Finally, even where an insurgent group is crushed, this **short-term COIN success may be squandered** if political, economic and social measures are not put into place to establish the conditions for long-term, sustainable peace. Without these measures, counterinsurgency success may only be short lived and may create nothing more than a pause before a later resumption of insurgency. The struggle in Peru against *Sendero Luminoso* (the Shining Path) illustrates this point. After a brutal campaign from 1980 to 2009 that cost the lives of more than 70,000 people, the Peruvian government succeeded in marginalising the Shining Path. Yet the Peruvian government then failed to address the crushing rural poverty and exclusion that had provided the catalyst for the emergence of the Shining Path in the first place. In consequence, the Shining Path was able to maintain a measure of support, and this, combined with involvement in coca production, has enabled it to continue to conduct attacks.[44]

Collectively, the points above mean that even success in counterinsurgency can seem curiously opaque: it can be militarily indecisive, politically ambiguous, costly, controversial and also short-lived. However, there is another set of difficulties, and this relates to the application of the principles of counterinsurgency.

Applying the principles of COIN

One of the (many) depressing things about the counterinsurgency phases of the wars in Iraq and Afghanistan has been the evident difficulty experienced by coalition forces in planning and executing an effective COIN campaign. These difficulties seem perverse given that the land forces of such powers as the United States, Britain and France had established counterinsurgency doctrines built on principles with a long historical provenance. However, experience in COIN campaigns suggests that the real difficulty in counterinsurgency operations is not in developing guiding principles. As this chapter has noted already, such themes as unity of effort, minimum force and political primacy have long been recognised as important even when they were not formally codified in doctrine. The real difficulty seems to be in **applying COIN principles to a given counterinsurgency context**. History suggests that there are many recurrent problems.

The influence of strategy

Counterinsurgency cannot escape the primacy of strategy outlined in Chapter 1. Counterinsurgency principles are **tactical and operational level techniques**: unless they are harnessed by an effective strategy to a set of realisable goals the principles are unlikely to produce success (see Box 6.9). Lack of resources, lack of domestic support, lack of strategic level unity of effort, improper or non-existent objectives, strategic level failures to understand the reality of the local context – these can all undermine attempts to realise the principles of counterinsurgency, yet they are outside of the gift of tactical and operational level military commanders to remedy easily. In Iraq, strategic level decisions like the one to limit the invasion force to only 140,000 men had a profoundly negative impact on the ability of US forces to conduct effective counterinsurgency operations once organised resistance began to emerge. Indeed, the US failure to realise that the campaign in Iraq was changing into an insurgency began at the top, with the US Secretary of Defense, Donald Rumsfeld, characterising the rising violence merely as the actions of 'pockets of dead enders'.[45]

Box 6.9 Vietnam

The war fought by the United States from 1965–75 in support of the South Vietnamese government is widely (though not wholly) regarded as a disastrous enterprise. Direct US involvement actually began in 1950, under a programme of support for the French, but US ground troops were formally committed in 1965. Many reasons have been advanced for US defeat: poor strategy; the weakness of the South Vietnamese government; the effectiveness of the North Vietnamese; media scrutiny; the ebbing away of US domestic support; the US military's 'regular war' focus; the lack of a unified COIN campaign.

In fact, though, even in circumstances of strategic defeat, there were many innovative US ideas on the conduct of COIN. For example, in the early 1960s, US Special Forces ran Civilian Irregular Defence Groups (CIDGs), in which local teams helped provide weapons, medical and economic support to local villages. From 1965, the US marines began creating Combined Action Platoons, a village-level blending of US and Vietnamese forces. In 1967, the US created unified civil-military teams at province and district level under the Civilian Operations and Revolutionary Development Support (CORDS) programme.

In the end, though, the US military's capacity to overcome the North Vietnamese in tactical military engagements did not translate into equivalent success in the other critical dimensions of counterinsurgency; the focus on firepower and attrition, expressed through a 'search and destroy' focus, proved counterproductive. One US Colonel, meeting a Vietnamese officer in 1975, commented to him: 'You know, you never defeated us on the battlefield'; the North Vietnamese officer replied: 'That is correct; it is also irrelevant.'

Organisational culture

Land forces themselves have often been complicit in creating problems in implementing COIN for reasons that stem from their values, not least that armies tend not to prioritise counterinsurgency operations. Given limited resources, many modern armies often 'train high to fight low': they prepare for conventional operations rather than unconventional operations because it is assumed to be easier to adapt conventional forces to an unconventional fight than it is to do the reverse. Most armies tend to define their professional standards by their ability to conduct conventional operations. The British General Sir John Kiszely summed up the problem in 2007, arguing that counterinsurgency embodies:

> features with which the pure warrior ethos is uneasy: complexity, ambiguity, and uncertainty; an inherent resistance to short-term solutions; problems that the military alone cannot solve, requiring cooperation with other highly diverse agencies and individuals to achieve a comprehensive approach; the need for interaction with indigenous people whose culture it does not understand; and a requirement to talk to at least some of its opponents, which it can view as treating with the enemy.[46]

For this reason, even where armies have a doctrine for COIN, there is often a reluctance to provide the necessary training, equipment and force structures to prepare in peacetime for unconventional operations.

Variations in context

However, even where a counterinsurgency campaign is guided by effective strategy and implemented by an adaptable military organisation, basic difficulties still remain.

One of the most important is posed by variations in context. In essence, and reflecting the comments on strategy in Chapter 1, **every COIN campaign will have a different internal and external context**.

- **Internally**, each campaign will involve populations with different histories, cultures and political attributes; each campaign will involve different geography and patterns of settlement; each campaign will involve different political actors with their own dynamics; the counterinsurgents will vary in composition and experiences.
- **Externally**, each campaign will be defined by different levels of international interest that will shape a different scale and nature of international involvement. These factors can have a profound impact. For example, the difficulty in defeating an insurgency is materially altered by the degree of external support on which it can draw: whilst some insurgents can win without such support (Fidel Castro in Cuba, for example), for others such as the United States in the American War of Independence, the Spanish in the Napoleonic Wars, or the Taliban in Afghanistan, external support has provided marked increases in resilience to counterinsurgency techniques.

This variation in context ensures that methods from one counterinsurgency campaign cannot simply be applied as a template in another. Variations in context can allow **radically different approaches** to counterinsurgency. Russian efforts in the Caucasus, Israeli efforts in Gaza and the West Bank, Chinese COIN in Tibet: none of these conform to the sorts of norms outlined in FM 3-24. However, in contexts like these in which the international community is unwilling to intervene, in which domestic audiences support a hard-line approach, in which the military have a considerable influence on COIN strategy, in which the governments are unwilling to compromise, and in which the local population may be irreconcilable, quite brutal counterinsurgency efforts may provide a workable enough solution for a state (see Box 6.10).

Box 6.10 Chinese COIN

A long-running Chinese counterinsurgency campaign has been conducted in Xinjiang province in the north-west of China, home to the Muslim Uyghurs. There has been a long history of tension between the Uyghurs and the Chinese government, but this escalated in the 1990s. Chinese COIN efforts in Xinjiang have developed over time and have come to include many features that reflect established counterinsurgency verities: a focus on establishing efficient intelligence networks; the co-option and integration of local elites by offering them jobs, privileges and so forth; heavy spending on economic development; the creation of local auxiliary security forces; re-shaping local education and religious structures. However, China's approach to COIN in Xinjiang has also been built consistently on a foundation of coercion, featuring such techniques as mass resettlement, summary trials, executions, torture and beatings, as well as special legislation that widened the definition of terrorism to criminalise virtually all expressions of dissent. The government has also encouraged ethnic Han Chinese immigration into Xinjiang so that they now comprise around half of the population.

By Western standards China's focus on coercion and the continued lack of a political compromise might be defined as a failure, but the Chinese context is different: international criticism is muted because there is little outside knowledge of the Xinjiang problem; the government's COIN campaign is uncontroversial domestically, because the state-controlled media control the flow of information and because the bulk of the Han Chinese population is unsympathetic to the Muslim Uyghurs anyway. The political will of the Chinese government to persevere is very high because the costs of failure are profound: with a history of social upheaval, crushing the insurgency is not just about securing Xinjiang, it is also about demonstrating the Chinese government's willingness to crush any other forms of internal dissent. Given the extent to which China has contained the problem and suppressed the violence, and given the broader political obstacles to a negotiated solution based on concessions, China's policies in Xinjiang might be construed as having provided a solution to the Xinjiang insurgency.

The misuse of history

The problems posed by variable context are compounded by the way in which the historical lessons of counterinsurgency are often generated. Rather than engaging in extensive and in-depth historical analysis, historical case studies have often been used by the military in a superficial way in order to support doctrinal predispositions (see Box 6.11). In other words, the **evidence has been shaped to fit the doctrine** and not the reverse.[47] For example, it was long believed that Britain was especially effective at counterinsurgency because of its reliance on two key principles: minimum force and political primacy. However, new scholarship has thrown into doubt the extent to which, (a) these principles were ever applied systematically, and (b) whether they were really the key to success (see Box 6.12). A re-examination of such case studies as the British success against the Mau Mau insurgency in Kenya (1952–60) has concluded that it actually depended to a considerable extent on a quite ruthless application of exemplary violence. The difficulty, then, is not only that armies can make the mistake of taking methods from the past and attempting to apply them subsequently to very different contexts; it is also that the evidence for the effectiveness of past techniques is itself contestable.

Box 6.11 *Guerre revolutionnaire*

In its war against the Algerian *Fronte de Liberation Nationale* (FLN) from 1954–62, the French army utilised an informal counterinsurgency doctrine known as *guerre revolutionnaire*. This doctrine was the fruit of a searching analysis by influential officers of the reasons for French defeat in Indo-China from 1946–54. In particular, the French army saw this defeat as the consequence of the ideological power of Marxism, and a lack of commitment on the part of the French state. In response, in Algeria the French army pursued a counterinsurgency strategy that focused on an equivalent ideological commitment, on the involvement of French society in the war through the use of conscripts, and on psychological, social and economic action to undermine the appeal of Marxist ideology.

But *guerre revolutionnaire* failed. First, it failed to understand that the challenge in Algeria was not Marxism, but nationalism, and no programme of local economic and social development could address the appeal to Algerians of national independence. Second, using conscripts merely exacerbated the domestic controversies in France over the war. Third, viewing the war as a decisive battle in the wider war against communism legitimised the use by the French army of extreme methods that alienated international opinion. The FLN were in many respects defeated militarily, but *guerre revolutionnaire* was a political disaster.

At the heart of these flaws were the values of the French army itself: *guerre révolutionnaire* was informed in part by the army's contempt for the decadence of liberalism; its simplistic understanding of the global politics of the time; its disdain for civilian society. The semi-mystical sense of mission that informed the army was exemplified by the comments of one officer who argued: 'We want to halt the decadence of the West and the march of communism. That is our duty, the real duty of the army. That is why we must win the war in Algeria.'

(Alistair Horne, *A Savage War of Peace, Algeria 1954–1962* (New York: Penguin Books, 1974), 49)

Box 6.12 Malaya

Unequivocal counterinsurgency successes are rare, but perhaps one of the most famous was the British defeat of the Malayan Communist Party (MCP) during the Malayan Emergency (1948–60). Traditional narratives highlight the clumsy nature of the early British COIN campaign, which featured a lack of unity in direction, poor intelligence, and the use of rather crude large-unit sweeps of the jungle. Over time, Britain developed a much more effective approach known as the 'Briggs plan' after the Director of Operations, Sir Harold Briggs. The British approach included the creation of a unified civilian-military effort through the appointment of a unified commander of the campaign, Sir Gerald Templar, and the creation of combined civil-military security committees at each level of administration. The campaign also featured such themes as minimum force, small unit tactics, a focus on winning the hearts and minds of the local population, and decentralised command and control.

However, the traditional narrative can be subjected to two criticisms. First, some have argued that a proper reading of the evidence suggests that the early, military-focused, stage of the campaign was actually successful and that it laid the essential foundations for the later success of British hearts and minds activities. Second, it has also been argued that British success owed much to the local context: that the insurgents were concentrated amongst the Chinese minority in Malaya; that Britain had already committed itself to withdrawing from the colony; that the MCP had very limited external support and no 'safe havens' across the border; that the domestic public was largely unconcerned over the use of what today would be considered very repressive measures. In either case, the Malayan Emergency may be more problematic as a template for counterinsurgency success than is often assumed.

The brutalisation of counterinsurgency

Another difficulty is that protracted counterinsurgency operations can often have a **brutalising impact on armies**. COIN campaigns can often generate in the security forces profound **frustration and de-sensitisation**: the enemy attacks using such unconventional and 'cowardly' techniques as ambushes, bombings and Improvised Explosive Devices; responses are heavily curtailed by often restrictive rules of engagement; the enemy are difficult to find, often hiding amongst the local population; indeed, many of the local population may provide passive or active support to the insurgents (or may even be part-time insurgents), leading security forces to develop an active fear and dislike of local civilians. In this context counterinsurgency can develop a **mimetic quality**: the security forces begin to use the same methods as the insurgents as a way of releasing their fear and anger and as a way of avoiding constrained rules of engagement. Under these pressures, the security forces may themselves begin to use terror as a tactic: shooting rather than capturing insurgents; threatening the local population; using torture and exemplary violence. Manifestations have included controversies surrounding an alleged British policy of 'shoot to kill' against IRA suspects in Northern Ireland; the use made by the Colombian security forces of paramilitary 'death squads' in their struggle against the Fuerzas Armadas Revolucionarias de Colombia (FARC, or the Revolutionary Armed Forces of Colombia); and acts of brutality by US and British forces in Iraq. Mimetic violence has been a common feature of counterinsurgency operations, illustrating that under the often extraordinary psychological pressures of COIN, land forces may begin actively to resist or subvert the implementation of their own doctrine.

Theory into practice

Finally, even where COIN principles are valid, they are often difficult to implement. For example:

- Contemporary COIN doctrine places a great deal of emphasis on understanding the **local 'human terrain'** (local politics, economics, societal dynamics and so forth), but this is easier to do in theory than it is in practice. Often, attempts to identify and understand 'the other' simply become exercises in crude cultural stereotyping.[48] Moreover, it can often be difficult to reconcile the requirement to develop a detailed understanding of local politics and culture with the limited tours of duty (often six months) of military units.
- Establishing **legitimacy is vital** in COIN, but external intervening forces find it difficult to avoid acting as a catalyst for the spread of local nationalism. It is easy for a COIN operation to look quickly like an army of occupation, a circumstance that is exploited politically by insurgents.
- Population-centric COIN is not as gentle as is often supposed. Winning 'hearts and minds' has usually required the use of sticks as well as carrots, but how much force is enough? The difficulty is that what constitutes 'minimum force' is often simply a matter of perspective: what military forces define as minimum force may not be the same as the local population or the international media.

The future insurgent threat

A third difficulty in conducting COIN is that the consensus on the value of the classical principles of counterinsurgency has begun to break down as a result of **debates on the nature of contemporary insurgency**.

Especially since the terrorist attacks of 11 September 2001 (9/11), much of the focus of Western counterinsurgency thinking has focused on the threat posed by **radical Islamist insurgent groups**. Insurgent threats in the first decade of the twenty-first century seem to be dominated by themes evident in Iraq and Afghanistan:

- Insurgencies waged by a multitude of fragmented groups, lacking an overall leader or unifying ideology.

- Groups that seemed less interested in controlling territory.
- Groups that seem to have little coherent conception of what might replace the existing state structure if it were overthrown.
- Groups that utilise urban environments.
- Insurgents who seem often to play as much to global as to local opinion.

However, there is **no consensus** on the extent to which these features reflect substantive or cosmetic change. Crudely, perspectives on contemporary and future counterinsurgency can be divided into **three viewpoints**: the neo-classicists, the global insurgency school, and the security policy perspective.

The neo-classicists

This first school of thought argues that in relation to the insurgent threat less has changed than meets the eye. In terms of the **causes** of insurgency, neo-classicists argue that whilst insurgents might cloak their cause in the rhetoric of Islamism, fundamentally the root causes of insurgencies still lie in the realms of unmet grievances. Islamism is merely an effective mechanism for mobilisation of popular support and legitimacy. For example, analysis of the motivations of Taliban fighters seemed to show that only a minority were motivated by religious zeal and that three quarters fought for more prosaic reasons such as monetary gain or tribal politics.[49] Thus, in terms of **responses**, the same framework used to meet insurgencies of the classical variety can be applied today and in the future because at its heart the demands of counterinsurgency have not changed: the key battleground remains the hearts and minds of the local population.

Neo-classicists recognise that traditional principles might require some adaptation or augmentation. In particular, experience seems to show that a greater emphasis on such themes as 'cultural knowledge' might be required: only through a nuanced understanding of local society and politics can effective counterinsurgency techniques be framed for separating the population from the insurgent. However, this was reflected in FM 3-24, which emphasised the importance of the 'human terrain' in COIN. FM 3-24's heritage was evident in its principles, which reflected and augmented traditional approaches.

Global COIN

The global COIN perspective points to the wider developments taking place in an increasingly globalised international system, and argues that **insurgency has undergone a paradigm shift**. Old-style, left-wing insurgencies are grinding to a halt: Marxism has been discredited by the collapse of the Soviet Union and China's shift towards market economics. Many long-running insurgent groups, such as the IRA and Maoists in Nepal, have ended their armed struggle and entered mainstream politics.

The global insurgency school argues that a new kind of insurgent, the 'global insurgent' has emerged as a consequence of several developments. **New grievances** have emerged founded on perceptions of global economic inequalities, increasing stratification within and between societies, and Western hegemony, creating intense alienation and frustrated aspirations.[50] **Traditional insurgency is less effective** because of the decline in levels of external support for insurgents relative to the Cold War and the increasing effectiveness of state security forces. Globalisation has also **empowered non-state actors**, such as insurgent groups: for example, communications technology has given transnational actors an ability to speak locally and vice-versa. Globalisation has resulted in a diffusion of resources and information that have weakened some states and allowed other sources of power to flourish. In particular, technological developments have strengthened the **importance of global networks**. The internet and new media have increased dramatically the ability of insurgents to mobilise financial and political support globally.[51] The 'infosphere' constitutes a new arena of COIN.

The effect of globalisation, in the form of dynamic international networks, has created **a new form of insurgency**: **global insurgency**. Insurgency is no longer a national phenomenon, waged by groups seeking to overturn a government, using staged methods focused on influencing the local population. The characteristics of the new global insurgent, of which Al Qaeda is an expression (see Boxes 6–13), include:

- Expansive goals driven by transnational religious purpose.
- Extreme violence.
- Decentralised control.
- International level organisation.
- An international, rather than just local, constituency.

Box 6.13 Al Qaeda as a global insurgency

Al Qaeda was the quintessential expression of this new model of insurgency: globalisation allowed Al Qaeda to craft a narrative of outrage and resistance that was able to tie Islamist ideology to a variety of local struggles. In its objective of establishing an Islamic caliphate, Al Qaeda pursued a goal well beyond simply changing the character of a single national government; indeed, its objectives were couched in terms of an almost civilisational struggle between Islam and predatory Western 'crusaders'. Al Qaeda's methods, not least the 9/11 attacks, seemed to embrace extreme violence as an end, not just a means, and one that seemed to invalidate the traditional assumption that insurgents placed politically motivated constraints on the scale and nature of their killing. Al Qaeda also established international links to local conflicts through the emergence of such 'franchised' affiliates as Al Qaeda in Iraq, Al Qaeda in the Islamic Maghreb and Al Qaeda in the Arabian peninsula. Al Qaeda's financial resources were also international in nature, and their audience similarly broad, reducing in importance the insurgents' traditional reliance on the local population for support.

The global COIN school argues that the **COIN response** must be to replace a focus on counterinsurgency at a local or national level, with a focus instead on breaking the linkages between the national and international sources of insurgent strength. Only by isolating local insurgents from their network of international finance, ideology and organisation can success be achieved. This strategy can be characterised as 'global disaggregation'. Measures are required to interdict and deny communications, finance, technology, personnel and other exchanges between global organisations and local insurgent groups within theatres.[52]

Security policy

Our third perspective on contemporary COIN is what can be termed a security policy perspective. This perspective criticises the neo-classical and global COIN schools because, for all their differences, they have one key similarity: both make an **assumption that political ideology is not central** to contemporary insurgencies. For both perspectives, such ideologies as Islamism are not a root cause of insurgency; the roots of contemporary insurgency still lie in negative social conditions that create the circumstances for radicalisation. For this reason both COIN perspectives have the same basic remedy for insurgency: grievance reduction. COIN is centrally about 'hearts and minds'.

The security policy perspective takes issue with this assumption. If insurgency really is simply a manifestation of political, social and economic grievances then all the political ideologies that have motivated political violence, including Anarchism, Liberalism, Nationalism, Marxist-Leninism and Islamism, have simply been insubstantial veneers: none of the groups that have espoused those ideologies has believed substantively in them. The security policy perspective argues that this is a flawed assumption: **political ideology matters to insurgents**; it creates COIN realities that cannot easily be wished away.[53] From a security policy perspective, counterinsurgency against groups like Al Qaeda is not simply a war against poverty or political disenfranchisement; it is a struggle against 'militant, ideologised Islam: or Islamism'.[54] The current insurgent challenge may be international in scope, but its roots lie in the domestic activities of Islamist groups. Because of this flawed assumption, a security policy perspective argues that neo-classical and global perspectives cannot answer key questions:

- First, how far is it possible to **compromise** with jihadists? Is there actually a programme of grievance reduction that would appease them, given the expansive nature of the goals that they pursue and their antipathy towards the West?
- Second, **where** should counterinsurgency efforts be focused? Traditional COIN has tended to see counterinsurgency as something that is practised abroad: it is done to others; however, work in counter-terrorism has come to a different conclusion. The key arena, increasingly, lies with the domestic politics of Western states.
- Third, **how far** should one be willing to compromise the interests of the majority domestic population to undercut Islamism? Domestic political support is the foundation for prosecuting any extended COIN campaign, but how can this support be assured in liberal, multi-cultural states if important political values have to be compromised in order to try to meet illiberal Islamism?

Thus, the security policy perspective argues that the neo-classical and global COIN approaches fail to engage sufficiently with the motives of insurgents; counterinsurgency is reduced instead to a set of techniques. The security policy perspective argues that in defeating Islamism, political ideology matters; in defeating this ideology, COIN needs to **embrace domestic counter-terrorism**, particularly approaches to domestic de-radicalisation. COIN cannot simply be done 'over there'; it must be part of a broader security approach that focuses on the 'here' as well.

Conclusions

Insurgency is the most common form of armed conflict in the contemporary period. Utilising a variety of techniques, including subversion, terrorism and guerrilla warfare, insurgency grew on the back of **developments in political ideology** to become a pervasive challenge to the organised military forces of states. Over time, land forces have developed counterinsurgency techniques to meet the insurgent challenge. These techniques had at their core a remarkably similar corpus of **principles and themes**, including: the primacy of politics; the importance of long-term planning; the need for effective intelligence; a focus on the use of the minimum necessary force; and especially, the vital importance of winning the hearts and minds of the local population.

Despite the flourishing of thinking on counterinsurgency operations, the practice of COIN has often been problematic. It would be wrong to think about the historical development of COIN as if it were a linear progression over time of methods of ever-increasing effectiveness. Land forces often seem to forget the lessons of previous campaigns, or learn the wrong lessons, or apply templates learnt from one campaign into other, rather different, contexts. **COIN remains a difficult task** because every context requires adaptation and innovation. Assessing Joseph Gallieni's counterinsurgency techniques, Hubert Lyautey wrote in 1897:

> There is no method, there is no cliché of Gallieni; there are ten, twenty – or, if there is a method, its name is suppleness, elasticity, adaptability to place, time, and circumstances.[55]

The difficulties inherent in COIN continue. Whilst new thinking on counter-insurgency has emerged in the opening years of the twenty-first century, this thinking has been built on contested foundations: Has globalisation created a new form of insurgent threat? How useful are traditional approaches? Where should COIN focus its efforts? COIN, however, has not suffered these uncertainties in isolation. As the next chapter demonstrates, peace operations have also been the subject of a great deal of innovative new thinking. However, whether that new thinking is relevant is the subject of intense debate.

Notes

1 US Army Field Manual (FM) 3-24, *Counterinsurgency* (Chicago, IL: University of Chicago Press, 2007), I-2.
2 Thomas R. Mockaitis, *Resolving Insurgencies* (Carlisle, PA: Strategic Studies Institute, 2011), 7.
3 Steven Metz, 'Insurgency', in Ole Jurgen Maao and Karl Eric Haug (eds) *Conceptualising Modern War: A Critical Inquiry* (London: Hurst, 2011), 119–22.
4 Steven Metz, 'The Internet, New Media, and the Evolution of Insurgency', *Parameters* Vol. 42, No. 3 (Autumn 2012), 80.
5 Max Boot, *Invisible Armies: An Epic History of Guerrilla Warfare from Ancient Times to the Present* (New York, NY: Liveright, 2013), 398.
6 Metz, 'The Internet, New Media, and the Evolution of Insurgency', 80.
7 Boot, *Invisible Armies*, 229–34.
8 Ibid., 232.
9 Metz, 'Insurgency', 126–27.
10 Boot, *Invisible Armies*, 342.
11 Mao Zedong, *Selected Military Writings of Mao Tse-Tung* (Beijing: Foreign Language Press, 1966), 260–61.
12 Boot, *Invisible Armies*, 333.
13 John Shy and Thomas W. Collier, 'Revolutionary War', in Peter Paret (ed.) *Makers of Modern Strategy: From Machiavelli to the Nuclear Age* (Princeton, NJ: Princeton University Press, 1986), 850.
14 FM 3-24, I-2.

15 James D. Kiras, 'Irregular Warfare', in David Jordan, James D. Kiras, David J. Lonsdale, Ian Speller, Christopher Tuck and C. Dale Walton, *Understanding Modern Warfare* (Cambridge: Cambridge University Press, 2008), 241–42.
16 Thomas Rid, 'The Nineteenth Century Origins of Counterinsurgency', *The Journal of Strategic Studies* Vol. 33, No. 5 (October 2010), 753.
17 C.E. Callwell, *Small Wars: Their Principles and Practice* (Lincoln, NE: University of Nebraska Press, 1996), 21.
18 Ibid., 23.
19 Alexander Alderson, 'Britain', in Thomas Rid and Thomas Keaney (eds) *Understanding Counterinsurgency: Doctrine, Operations, and Challenges* (Abingdon: Routledge, 2010), 32–33.
20 Kiras, 'Irregular Warfare', 252.
21 T.E. Lawrence, 'Evolution of a Revolt', *The Army Quarterly* Vol. 1, No. 1 (October 1920), 63.
22 Shy and Collier, 'Revolutionary War', 831.
23 Alderson, 'Britain', 33.
24 *Small Wars Manual* (New York, NY: Skyhorse Publishing, 2009), 1–10.
25 Shy and Collier, 'Revolutionary War', 830.
26 Even before then, too. See, for example, Douglas Porch, 'Bugeaud, Gallieni, Lyautey: The Development of French Colonial Warfare', in Paret (ed.) *Makers of Modern Strategy*, 381–82; Gil Merom, *How Democracies Lose Small Wars* (Cambridge: Cambridge University Press, 2003) charts the broad dimensions of the problem.
27 Daniel Marston and Carter Malkesian, 'Introduction', in Daniel Marston and Carter Malkesian (eds) *Counterinsurgency in Modern Warfare* (Oxford: Osprey, 2010), 14.
28 Boot, *Invisible Armies*, 399.
29 Marston and Malkesian, 'Introduction', 14.
30 Rid, 'The Nineteenth Century', 729.
31 Marston and Malkesian, 'Introduction', 14–15.
32 Frank Kitson, *Bunch of Five* (London: Faber and Faber, 1977), 283.
33 Marston and Malkesian, 'Introduction', 15.
34 John A. Nagl and Brian M. Burton, 'Thinking Globally and Acting Locally: Counterinsurgency Lessons from Modern Wars – A Reply to Smith and Jones', *Journal of Strategic Studies* Vol. 33, No. 1 (February 2010), 125.
35 D.M. Jones and M.L.R. Smith, 'Whose Hearts and Whose Minds? The Curious Case of Global Counterinsurgency', *The Journal of Strategic Studies* Vol. 33, No. 1 (February 2010), 84.
36 FM 3-24, I-113.
37 Mockaitis, *Resolving Insurgencies*, 2.
38 Boot, *Invisible Armies*, 399.
39 Ibid., 63.
40 Ibid., 527
41 Ibid., 559.
42 Mockaitis, *Resolving Insurgencies*, 1.
43 Marston and Malkesian, 'Introduction', 16.
44 Mockaitis, *Resolving Insurgencies*, 43.
45 Boot, *Invisible Armies*, 538.
46 Quoted in Jones and Smith, 'Whose Hearts and Whose Minds?' 82.
47 Karl Hack, 'The Malayan Emergency as Counter-insurgency Paradigm', *The Journal of Strategic Studies* Vol. 32, No. 3 (June 2009), 383.
48 See Patrick Porter, 'Good Anthropology, Bad History: The Cultural Turn in Studying War', *Parameters* Vol. 37, No. 2 (Summer 2007), 45–58.
49 Nagl and Burton, 'Thinking Globally and Acting Locally', 133.
50 Jones and Smith, 'Whose Hearts and Whose Minds?' 111.
51 Metz, 'The Internet, New Media, and the Evolution of Insurgency', 86.
52 David J. Kilcullen, 'Countering Global Insurgency', *The Journal of Strategic Studies* Vol. 28, No. 4 (August 2005), 609–10.
53 Jones and Smith, 'Whose Hearts and Whose Minds?' 104–5.
54 Ibid., 103.
55 Rid, 'The Nineteenth Century', 756.

Suggested reading

Ian F.W. Beckett (ed.), *The Roots of Counter-insurgency: Armies and Guerrilla Warfare, 1900–1945* (London: Blandford Press, 1988). Beckett's book provides a firm grounding in the imperial policing and counterinsurgency techniques of the US, French, British, German and Soviet armies, with additional material on the Chinese Civil War.

Max Boot, *Invisible Armies: An Epic History of Guerrilla Warfare from Ancient Times to the Present* (New York, NY: Liveright, 2013). A sweeping, engaging history of the development of insurgency and counterinsurgency.

Paul Dixon, *The British Way in Counterinsurgency: From Malaya and Northern Ireland to Afghanistan and Iraq* (Basingstoke: Palgrave Macmillan, 2012). This book provides a comprehensive reassessment of Britain's supposed excellence in COIN operations.

David Martin Jones, Celeste Ward Geventer and M.L.R. Smith (eds), *The New Counter-insurgency Era in Critical Perspective* (Basingstoke: Palgrave Macmillan, 2014). Provides an incisive and critical analysis of orthodox perspectives on counterinsurgency theory and practice.

David Kilcullen, *Out of the Mountains: The Coming Age of the Urban Guerrilla* (London: C. Hurst and Co., 2013). Another view of the future insurgent threat: here, Kilcullen sees insurgencies that are increasingly urban, interconnected and meshed with criminality.

John Mackinlay, *The Insurgent Archipelago: From Mao to Bin Laden* (London: C. Hurst and Co., 2009). Mackinlay provides an eloquent examination of the phenomenon of globalised insurgency.

Bard O'Neill, *Insurgency and Terrorism* (Washington, DC: Potomac Books, Incorporated, 2005). This book provides a comprehensive introduction to the definitions, concepts and theories associated with insurgency.

David H. Ucko, *The New Counterinsurgency Era: Transforming the U.S. Military for Modern Wars* (Washington, DC: Georgetown University Press, 2009). A revealing examination of the challenges experienced by the US army in adapting itself to the conditions of counterinsurgency in Iraq.

The U.S. Army/Marine Corps Counterinsurgency Field Manual (Chicago, IL: University of Chicago Press, 2007). Widely available online, FM 3-24 makes instructive reading and indicates just how sophisticated contemporary approaches to COIN have become.

7 Peace and stability operations

Key points

- Since the end of the Second World War especially, land forces have been involved increasingly in complex peace and stability operations.
- These operations have challenged many of the conventional war-focused and military-centric assumptions of traditional army campaigning. Victory is a more ambiguous concept, and such methods as a focus on firepower and the military defeat of adversaries can be counter-productive.
- The complexity of these operations has multiplied since the end of the Cold War, with operations taking place in difficult intra-state environments marked by multiple belligerents and ambitious political goals often focused on nation building.

Land forces don't just fight wars. Reflecting the versatility of armies that was explored in Chapter 1, the period since the end of the Second World War has seen their increasing employment in such tasks as the prevention of conflict, its management and its resolution. In performing these roles, land forces exploit their ability to control ground and to translate this control into opportunities to exert **political influence** over local populations. Controlling ground allows armies to provide security, provide humanitarian relief and to set the conditions for development in other spheres by permitting the safe development of local infrastructure, economic activity and political processes.

However, peace operations are challenging: traditional approaches to land warfare focusing on such themes as manoeuvre, firepower and decisive military victory may be at best irrelevant to peace operations and at worst entirely counter-productive. Whilst the military contribution to peace operations can provide the crucial foundation for success, that contribution must be linked harmoniously to activity in the political, economic and humanitarian spheres. At the same time, the use of land forces in such roles is likely to pose a significant challenge to conventional war doctrines, structures and ethos.

This chapter examines the debates surrounding such activities as peace support operations and stabilisation. In it, we will look at the basis of such operations, examine their requirements, and chart the development in methods and doctrines designed to perform those tasks more effectively. This chapter is divided into two parts: first, the chapter looks at the development of peace operations, examining the peacekeeping heritage of the Cold War and developments thereafter; second, the chapter examines the emergence of doctrines of 'stabilisation', especially in the context of the wars in Iraq and Afghanistan.

Peace operations

Peace operations have often had a high political profile, especially since the end of the Cold War, but why have they become so significant a role for land forces, and why have armies adapted themselves to meet the demands of tasks that are often very different from core war-fighting functions?

The origins of peace operations

Peace operations are often associated in the public consciousness with the United Nations (UN). In reality, the origins of peace operations lie in three key developments that took hold, especially, in the nineteenth and twentieth centuries:

- The first of these was **developments in international organisation**, reflected in the creation periodically of general frameworks to prevent or mediate conflict, like European congresses (such as the Congress of Berlin in 1878 to discuss conflict in the Balkans), and the creation of a variety of military missions designed to promote peace including multi-national operations to suppress the Atlantic slave trade.
- The second were **normative developments** in views regarding war: the First World War strengthened the perspective that war was an abhorrent activity and that humanity's best efforts should be directed towards trying to eliminate it. This was reflected in the creation of the flawed League of Nations in 1919.
- The third was **superpower rivalries** between the United States and Soviet Union after the Second World War, which created an environment of often intense strategic competition between nuclear-armed superpowers and their allies. In this context, conflict prevention and/or amelioration became a desirable activity.[1]

The United Nations

These three developments came together in the wake of the Second World War with the creation in 1945 of the UN, an organisation created expressly to promote international peace and security. Though peace operations were never specifically mentioned in the UN Charter, their development emerged from a flexible interpretation of the UN Charter, especially:

- **Chapter VI**, 'Pacific settlement of disputes': Chapter VI focuses on peaceful means of establishing or maintaining the peace through reconciliation, mediation, adjudication and diplomacy.
- **Chapter VII**, 'Action in respect to threats to the peace, breaches of the peace and acts of aggression': Chapter VII focuses on a wide range of enforcement actions from diplomatic and economic measures to the extensive application of armed force by member states.
- **Chapter VIII** 'Regional arrangements': Chapter VIII authorises regional organisations such as the North Atlantic Treaty Organization (NATO), Organization of American States (OAS), African Union (AU), Economic Community of West

African States (ECOWAS) and European Union (EU) to act to prevent, halt or contain conflict in their respective regions.[2]

Peace operations during the Cold War

If the Cold War acted as an imperative for the development of conflict prevention tasks by the UN, it also acted as a constraint on the development of peace operations. Considerations of **national interest** in the context of the **Cold War** tended to limit the number of operations undertaken by the UN to those on which both superpowers could agree. Between 1945 and 1988, the UN undertook only 13 peace operations (see Box 7.1). The Cold War also tended to **limit the nature** of such operations as well. Most peace operations undertaken during the Cold War were what became known as 'traditional peacekeeping' operations. Traditional peacekeeping can be defined as an operation 'undertaken with the consent of all major parties to a dispute, and one designed to monitor and facilitate implementation of an agreement to support diplomatic efforts to reach a long-term political settlement'[3] (see Box 7.2).

Box 7.1 Cold War UN peace operations

1948: UN Truce Supervision Organization (UNTSO), Middle East
1949–present: UN Military Observer Group in India and Pakistan (UNMOGIP), Kashmir
1956–57: UN Emergency Force I (UNEF I), Syria and Israel
1958: UN Observation Group in Lebanon (UNOGIL), Lebanon
1960–64: UN Operation in the Congo (ONUC), Congo
1962–63: UN Security Force in West New Guinea (UNSF), Dutch West New Guinea
1963–64: UN Yemen Observation Mission (UNYOM), Yemen
1964–present: UN Peacekeeping Force in Cyprus (UNFICYP), Cyprus
1965–66: Mission of the Representative of the Secretary-General in the Dominican Republic (DOMREP), Dominican Republic
1965–66: UN India-Pakistan Observation Mission (UNIPOM), India and Pakistan
1973–79: UN Emergency Force II (UNEF II), Egypt/Israel
1974–present: UN Disengagement Observer Force (UNDOF), Golan Heights
1978–present: UN Interim Force in Lebanon (UNIFIL), Lebanon

Box 7.2 Peace operations: alternative definitions

The UK's Joint Warfare Publication (JWP) 3-50 uses the term Peace Support Operations (PSO), a concept it defines as: 'An operation that impartially makes use of diplomatic, civil and military means, normally in pursuit of United Nations Charter purposes and principles, to restore or maintain peace. Such operations may include conflict prevention, peace making, peace building and/or humanitarian operations.'

British doctrine defines traditional peacekeeping as: 'A peace support operation following an agreement or ceasefire that has established a permissive environment when the level of consent and compliance is high, and the threat of disruption is low. The use of force by peacekeepers is normally limited to self-defence.'

(JWP 3-50, *The Military Contribution to Peace Support Operations*, June 2004, 1–2)

Traditional peacekeeping operations had a number of key features:

- First, they tended to be **inter-state** operations – that is, they were deployed in border areas between states.
- Second, they were deployed **after a ceasefire or settlement** had been agreed between the belligerents; in other words, they were deployed after fighting had stopped and there was a peace to keep.[4]
- Third, traditional peacekeeping operations tended to involve very **limited missions**. In general, these missions fell into two categories: **supervision tasks**, which might require the peacekeeping force to supervise the disengagement of belligerent forces, assist in the partition of territory and help in maintaining local order; and **monitoring tasks**, requiring activities such as verifying compliance with ceasefire agreements, reporting and investigating suspected breaches, liaising with local officials or non-governmental organisations (NGOs).

Overall, the purpose of traditional peacekeeping operations was not to solve conflicts or to play a direct role in a political process; rather it was to **de-escalate** and **manage** conflict in order to allow other actors to negotiate a political settlement.[5]

It was these sorts of limited operations that shaped the development of three key peacekeeping principles, the so-called 'Holy Trinity' of peacekeeping:

- **Consent**: the peacekeeping operation should be deployed only with the consent of main parties to the conflict. Without this consent, the peacekeeping force would be unable to fulfil its mandate and would risk being drawn into armed conflict.
- **Impartiality**: the peacekeeping operation should execute its mandate 'without favour or prejudice'. This did not mean neutrality – peacekeepers might find themselves acting more against one belligerent than another – but if they did, this would be conditioned by their mandate, not by political expediency.
- **Self-defence**: no use of force except in self-defence. This had implications for the kinds of forces deployed, which tended to be lightly armed. Later, the term 'self-defence' was extended to include defence of the peacekeepers' mission.[6]

The UN Military Observer Group in India and Pakistan (**UNMOGIP**) provides a useful example of this kind of traditional peacekeeping operation. UNMOGIP was deployed in 1949 into Kashmir along the ceasefire line agreed by India and Pakistan after their dispute over who should control Kashmir. It was deployed with the consent of both sides; its mission was limited to observing the ceasefire line and reporting and investigating any infractions; it was a small mission of some 40 observers. The purpose of UNMOGIP was thus never a coercive one – it was to reassure each of the belligerents that their opponents would abide by the ceasefire agreement and in consequence to facilitate the development of a longer-term political settlement.

It is important to remember, however, that not all UN operations during this period were of the traditional peacekeeping type. The US and allied forces deployed during the **Korean War** from 1950–53 did so under a UN Chapter VII enforcement mandate. Likewise, the United Nations Operation in the Congo (ONUC)[7] deployment into **Congo** from 1960–64 operated eventually under a Chapter VII mandate; it also operated within the host state and conducted activities that went well beyond separation and monitoring tasks. Thus, although traditional peacekeeping dominated the Cold

War period, there were already precedents for more expansive kinds of peace operations.[8]

From peacekeeping to peace operations

In terms of peace operations, the period up until the late 1980s was dominated by traditional peacekeeping. However, the period from 1988 to 1993 constituted an important **transitional period** in the evolution of peace operations. During this period, peace operations underwent significant development.

The sources of change

The sources of these changes lay in the impact of a number of interrelated factors:

Globalisation

Globalisation (see Box 7.3) was a central motor for change. The processes of globalisation had a number of important consequences including empowering an increasingly diverse range of such **international actors** as regional associations, trans-national corporations and NGOs; these would play an increasing role in the causes and/or amelioration of violent conflict. An increasingly **globalised media** also had a significant impact in raising consciousness of conflict and suffering in hitherto marginalised areas of the world.

Box 7.3 Globalisation

'Globalization refers to a process – still ongoing – through which the world has in many respects been becoming a single place.'

Though often associated with growing economic interdependence, globalization is multi-faceted with important political, cultural, social, technological, military and environmental dimensions.

(John Baylis and Steve Smith (eds), *The Globalization of World Politics: An Introduction to International Relations* (Oxford: Oxford University Press, 1999), 19)

Internal conflicts

Another factor prompting developments in peace operations was the relative decline in the incidence of inter-state conflicts and the rise to **prominence of intra-state wars**. In consequence, traditional peacekeeping, with its focus on ameliorating conflict between states, seemed less relevant to many of the conflicts that began to emerge in the early 1990s.

Normative changes

A further development lay in the realms of the normative debates prompted by questions regarding whether some key features of the traditional, or **'Westphalian' state system**, such as the centrality accorded to state sovereignty, were increasingly anachronistic (see Box 7.4). It is worth noting that in respect of the factors outlined above, the

Cold War was less of a watershed than might have been supposed. Globalisation, for example, is a phenomenon with a long historical pedigree. Similarly, normative debates surrounding the extent to which the international community had a right to involve itself in the domestic affairs of states were not peculiar to this period: for example, as early as 1946 the UN felt it necessary to debate the iniquities of South Africa's apartheid-based political system.

Box 7.4 The 'Westphalian' and 'post-Westphalian' system

Signed in 1648, the Treaty of Westphalia ended the Thirty Years War. Its broader significance was that the treaty itself embodied ideas which would form many of the foundation assumptions of the modern international system: the equality of states, the principle of non-intervention in the affairs of other states, and the consequent idea of national sovereignty. Latterly, some have argued that the processes of globalisation and changes in international norms have eroded the centrality of the state in the international system, creating a new 'post-Westphalian' order.

The end of the Cold War

Without the Cold War, the UN was **less constrained** by the competitive use of vetoes in the Security Council. Thus, whilst from 1945 to 1990 there were 238 vetoes cast in the Security Council, between 1990 and 2002 there were only 12. The demise of the Soviet Union also removed an important constraint on US foreign policy action; though far from making it omnipotent, the United States was placed in a strong position to pursue the notion of 'liberal peace' (see Box 7.5). Moreover, the end of the Cold War created a **positive mood** conducive to international activism. This was reflected in the galvanising of an international coalition in 1990 to meet Iraq's invasion of Kuwait. The Gulf War of 1990–91 prompted US President George Bush, Sr, to articulate the notion of a 'new world order': one in which international relations would be built on values of justice and freedom.[9]

Box 7.5 'Liberal peace'

Emerging in the eighteenth century with philosophers such as Immanuel Kant (1724–1804), the idea of 'liberal peace' argues that democratic states are much less likely to fight one another. This is because of attributes inherent in liberal regimes including the ability to hold governments to account. Thus, in theory, to encourage the spread of democracy in the international system is also to encourage the spread of peace. Needless to say, the concept is a contested one with critics arguing that the theory ignores problems over how one defines such concepts as 'democracy' and 'peace'.

This transitional period saw demonstrable developments in the **quantitative** and **qualitative** aspects of peace operations.

Quantitative developments

The period from 1988–93 was marked by increases in three areas:

- **The number of operations**: whereas the period from 1946 to 1988 had seen only 13 UN peace operations, in the period from 1988 to 1993, 20 new missions were initiated, including the UN Iraq–Kuwait Observation Mission (UNIKOM); the UN Good Offices Mission in Afghanistan and Pakistan (UNGOMIP); the UN Angola Verification Mission I (UNAVEM I); the UN Transition Assistance Group Namibia (UNTAG); the UN Observer Mission in El Salvador (ONUSAC); and the UN Observer Group in Central America (ONUCA).
- **The number of troops involved**: three of the new operations were especially large: UNTAC (the UN Transitional Authority in Cambodia); UNOSOM (the UN Operation in Somalia); and UNPROFOR (the UN Protection Force, deployed in the Former Yugoslavia). All three were initiated in 1992. These three operations, in particular, resulted in a massive increase in the numbers of troops serving on peace operations: some 75,000 during this period.
- **The number of participating states**: at the same time, states traditionally wary of involvement in UN peace operations, such as the United States and the United Kingdom, began to participate: indeed, this transitional period saw more than 40 countries participating in peace operations for the first time.

Qualitative developments

More important, perhaps, were the **qualitative developments** in peace operations that were associated with this period. Whilst many of the new operations were traditional peacekeeping operations, the UNTAC, UNPROFOR and UNOSOM operations, especially, expanded the remit of peace operations well beyond the interposition and monitoring tasks of traditional peacekeeping. In particular, these operations featured a number of characteristics that would be repeated in future operations:

- One key feature of these large operations was that they were **intra-state deployments**: rather than being deployed between states, the operations were deployed within states. This shift from an inter-state to an intra-state focus placed such operations squarely within the complexities of civil wars, state collapse and intense humanitarian crises.
- Another important feature was that some operations, such as UNOSOM, were deployed **before ceasefires** had been constructed and **without the consent** of all of the belligerent parties.
- Crucially, many of the new peacekeeping operations also had **expansive missions** that went well beyond traditional monitoring and supervision. These new missions involved ambitious long-term goals associated with building the capacity of governments, building peace and resolving conflict – not just interposition and observation, but helping to facilitate peace settlements, implement peace settlements and peace building afterwards. These new missions included election supervision, democratisation, humanitarian assistance, disarmament, demobilisation and reintegration (DDR) of militias and armed groups, and the protection of human rights.
- These new missions also often involved very important **non-military dimensions**, with significant numbers of civilian, police and non-government personnel (in Bosnia, for example, there were more than 200 NGOs involved).[10]

Sometimes termed '**second generation peacekeeping**' or 'new peacekeeping', the operations that emerged during this transitional period had clearly expanded well beyond the bounds of traditional peacekeeping. UNTAC, for example, was deployed in Cambodia from 1992–93 after the signing of a ceasefire in 1991. The Khmer Rouge, however, continued to pose a recurrent problem so that the operation had to confront periodic armed conflict. UNTAC had a broad remit, including protecting human rights, civil administration, maintenance of law and order, repatriation and resettlement of refugees, rehabilitation of militias, repair and development of infrastructure, and organisation of free and fair elections. It meant, for example, that UNTAC ran Cambodia's foreign affairs, security, finance and communications systems for a period of time. To do this, UNTAC had a very large non-military component including human rights officials and civil administration personnel; 1,400 police were deployed and more than 140 NGOs were involved in the operation.

Doctrinal evolution

In many respects, however, the peacekeeping environment changed much faster than peacekeeping techniques and approaches. States such as the United States and the UK, for example, had had only limited involvement in peacekeeping operations during the Cold War. Both approached peacekeeping according to principles shaped by generic low-intensity warfare doctrine. **Specific doctrine** did develop, in the form of the UK's 'Peace Operations' of 1987 and the United States' Field Manual (FM) 100-120, *Low Intensity Conflict* of 1990, which contained specific peacekeeping guidance; however, this doctrine essentially reflected the principles expressed in the 'Holy Trinity'. The **limitations** of this doctrine became evident in operations in places such as **Somalia**, 1992–95, in which there was no formal government from which to obtain consent for the deployment of UNOSOM; where the multitude of warring militias were never likely to view foreign intervention as impartial; and where the proliferation of violence made extremely vulnerable lightly armed peacekeepers operating according to the principles of self-defence.

New approaches

The need for approaches better tailored to the complex challenges of this transitional period were evident, but military approaches evolved only slowly. In 1994, a specific peace operations doctrine was published by the United States, **FM 100-123, *Peace Operations***. This doctrine had a number of important features:

- It recognised modes of peace operations beyond traditional peacekeeping, such as peace enforcement.
- It recognised that the character of an operation would be shaped by varying levels of consent and impartiality.
- It also recognised that force might have an important role to play in peace operations.

Thinking was constrained, however, by a reluctance on the part of both the military and politicians to embrace tasks associated with **nation building**. Thus, despite this new doctrine, and despite developments in other areas, such as the creation by the US army of a Peacekeeping Institute, peace operations doctrine remained heavily bounded. This left the US military woefully **unprepared** for operations in **Bosnia** in 1995, in which realities on the ground forced the United States to perform the nation-building-type tasks that were largely excluded from its doctrine. In what one US officer described as 'a "roll-your-own" situation', the military found itself having to improvise to meet the needs of tasks such as disarmament and demobilisation of militias, supporting elections, establishing the rule of law, assisting civil governance and rebuilding infrastructure.[11] The lessons of Bosnia, allied with those of Somalia, Haiti and Kosovo also combined with a degree of cross-fertilisation with such militaries as those of the UK, Canada, Norway and Sweden. Subsequent doctrine embodied in documents such as 2001's FM 3-0, *Operations*, and 2003's FM 3-07, *Stability Operations and Support Operations*, reinforced the status of peace operations as a **distinct and complex set of activities**, with a widening array of tasks and principles.

Contemporary peace operations

One consequence of this period of rumination was that the term 'peacekeeping' now seemed inadequate to describe the complex range of operations that were being performed: in consequence, the term '**peace operations**' evolved (often also referred to as '**peace support operations**') as an over-arching term to describe a number of sub-categories of activity of which traditional peacekeeping was now only one element. Peace operations comprise:

> … crisis response and limited contingency operations, and nominally include international efforts and military missions to contain conflicts, redress the peace, and shape the environment to support reconciliation and re-building and to facilitate the transformation to legitimate governance.[12]

Peace operations embody a number of sub-categories of activity:

- **Peacekeeping**: these are military operations 'undertaken with the **consent** of all major parties to a dispute, and are designed to monitor and facilitate implementation of an agreement to support diplomatic efforts to reach a long-term political settlement'.[13] Peacekeeping might be of the traditional kind, involving observation and interposition missions between states, but it might also include more complex missions comprising military, police and civilians, conducting tasks such as humanitarian relief, overseeing disengagement of forces, restoring critical infrastructure, and maintaining law and order in the interim.
- **Peace enforcement**: generally conducted under Chapter VII of the UN Charter, these are operations that are 'generally **coercive** in nature and rely on the threat of the use of force'.[14] This might include activities such as: forcing entry into a conflict zone; forcibly separating belligerents; enforcing sanctions or exclusion zones; protecting humanitarian operations; and re-instating an established civil authority. In peace enforcement, consent is clearly not a requirement, but impartiality and restraint are crucial factors: coercion should be used only on behalf of the operations mandate or an agreed peace settlement, and force should be used in a restrained and precise manner.
- **Peace building**: these are operations that cover '**post-conflict actions**, predominantly diplomatic, legal, and security related that support political, social, and military measures aimed at strengthening political settlements and legitimate governance and

rebuilding governmental infrastructure and institutions'.[15] Peace building is a long-term process designed to create a self-sustaining peace, to address the root causes of conflict, and to prevent a slide back into war. It may involve activity in up to six general sectors: security; humanitarian assistance and social well-being; justice and reconciliation; governance and participation; and economic stabilisation and infrastructure. Peace-building tasks often focus on 'capacity building': improving the ability of the host-nation government to function more effectively.[16] Activities might include: restoring law and order; training local security forces; providing for humanitarian needs; helping to establish a functioning police and judicial system; training civil servants; encouraging a free and open media; and initiating market reforms.

- **Peace making**: this comprises 'a **diplomatic process** aimed at establishing a cease fire or an otherwise peaceful settlement of a conflict'.[17] As such, peace making is used to address a conflict that is already ongoing. It may include diplomatic initiatives by a variety of actors including the Secretary-General of the UN, special envoys, third-party states and regional organisations.
- **Conflict prevention**: this category of activity is made up of 'diplomatic and other actions taken in advance of a predictable crisis to **prevent or limit** violence, deter parties, and reach an agreement short of conflict'.[18] Activities may include political initiatives by the UN, interested states or regional organisations or military activities such as preventative deployments, surveillance or embargoes.

'Multi-dimensional peacekeeping'

Another fruit of developing thinking on peace operations was the recognition that these categories of operation could **not be neatly bounded**, nor be seen as strictly linear. Operations such as those in Bosnia (1992–95) or Darfur (2004–present) were deployed initially in the wake of severe violence and in support of only weakly embedded ceasefires or political agreements. In consequence of developing events, these operations might find themselves undertaking several categories of peace operation at the same time – what the UN termed 'multi-dimensional peacekeeping', depending on the extent to which local political actors did, or did not, adhere to the political agreements that they had signed up to. Especially in intra-state environments, the nature of a peace operation was **highly dependent on local conditions**, including the degree of ongoing violence, the capacity of the host-nation government and the level of destruction wrought by previous conflict. A peace operation might initially be deployed in a Chapter VI peacekeeping mode, but a breakdown in the political settlement that it was deployed to support might require it to engage in peace enforcement in defence of that settlement and in active peace-making efforts to broker a new ceasefire. Peace operations might find themselves cycling back and forth between conflict, ceasefire and post-conflict phases of an operation, depending on local circumstances. This also complicates the military contribution to such operations because the military's input into peace operations can vary depending amongst other things on the phase of a conflict: in relation to peace building, for example, the military can have a central role to play in the period immediately following a cessation of hostilities; in the longer term, however, peace-building tasks are likely to be undertaken primarily by civilian organisations and the host nation.[19]

Box 7.6 Principles of peace operations

1 Consent: promotion of consent is central to long-term success. Forces and mandates need to be tailored to the level of consent.
2 Impartiality: this does not preclude coercion, but the operation must treat belligerents according to what they are doing in relation to the operation's mandate, not who they are.
3 Transparency: local and international actors must be made aware of the purpose, objectives and methods of the operation.
4 Credibility: the operation must be seen to have the capability to execute its mandate effectively.
5 Freedom of movement: the operation must have the freedom of movement to execute its mandate.
6 Flexibility and adaptability: the operation needs to be able to adapt quickly to changing circumstances.
7 Civil–military co-operation: the need to co-ordinate, integrate and synchronise civilian and military efforts to ensure harmony between them.
8 Restraint and minimum force: military force should be carefully disciplined to minimise negative consequences.
9 Clear objectives: objectives need to be clearly defined and attainable.
10 Perseverance: forces must be prepared for a long-term commitment.
11 Unity of effort: the wide variety of local, national and international actors need to be directed towards a common purpose.
12 Legitimacy: local, domestic, national and international actors need to believe in the legality and justice of the operation.
13 Security: the need for the ability to secure the resources, the mission and the local population.
14 Mutual respect and cultural awareness: the need to develop a mutually positive perspective amongst all actors.
15 Current and sufficient intelligence: crucial to determining the correct nature of operations conducted and in measuring their success.

The fundamentals of peace operations

Thinking on the requirements of peace operations has also led to reflection on the sorts of guiding principles that might enhance success in more complex post-Cold War peace operations (see Box 7.6). Of particular note is the growing importance attached to themes such as unity of effort, understanding and respect, legitimacy, the nature of objectives, and more nuanced understandings of consent:

Unity of effort

Modern peace operations are potentially very complex, involving **a wide array of local, national and international actors**, both military and civilian, governmental and non-governmental. The UN, for example, recognises some 50,000 different NGOs. One recurring lesson is the importance of trying to ensure unity of effort amongst these actors; to ensure that all are working with one another towards a common purpose. Different national and organisational priorities, organisational cultures and objectives can create serious difficulties in ensuring this, creating uncertain command arrangements and inter-agency tensions. Considerable effort needs to be put into providing solutions. These can include:

- Establishing military-civilian **co-ordinating structures** such as committees, action groups and joint liaison.
- Developing effective theatre-level **joint interagency co-ordination**.
- Developing inter-agency and multinational **doctrine and procedures**, and improving co-ordination with host nations.

Understanding

Understanding and respect amongst all participants, local and international is another factor identified as fundamental for success. The sources of this may be partly knowledge based, such as an understanding of local customs, history and language, and partly conduct based, requiring respect and professionalism in the conduct of activities.

Legitimacy

The principle of 'understanding' is also key to establishing and maintaining the **legitimacy** of an operation: the perception locally, nationally and internationally that the operation is right and just. Legitimacy is also a function of other attributes, such as perceptions of the impartiality of an operation; the success, or otherwise, of attempts at strategic communication to create transparency surrounding the goals and conduct of an operation; and the behaviour of intervening forces. In supporting the legitimacy of an operation, contemporary doctrine also places much emphasis on encouraging local ownership of the peace-building process by encouraging local participation as early and extensively as possible.[20]

Objectives

Much is also made in contemporary doctrine of the need for **clear and attainable objectives**. The recurrent criticism is that political objectives expressed in the mandates for peace operations have often been vague. This has had two effects: first, vague mandates are difficult to translate into operational political and military tasks and objectives. Second, vague mandates allow scope for 'mission creep' (a gradual escalation in the level of military force used within an operation), and 'mission cringe' (a situation in which the operation does less than it might). FM 3-07 notes, for example, that 'every military operation should move toward a clearly defined, decisive, and attainable objective', and that officers 'must understand strategic aims, set appropriate objectives, and ensure these objectives contribute to overall unity of effort'.[21] End states also require useful and quantifiable metrics that allow a peace operation to determine its progress in achieving its goals.

Consent

These sorts of fundamentals need to be put into the context of the operational environment, especially the degree of **consent**. In particular, the degree of consent dictates the level of opposition the peace operation might face, the degree of force it might have to use and, in consequence, the kinds of mandates, force structures, rules of engagement and tasks that it might have to perform. Because of this, a peace operation should seek broadly to increase the level of consent over time. The degree of consent may depend on many factors:

- The legitimacy of an operation.
- Its competence in providing security and delivering on local expectations.
- The degree of transparency in its operations.
- The degree of control that strategic level actors can exert over operational and tactical level actors.

Consent can often be challenged by specific actors, 'spoilers', who have little or no interest in existing or proposed peace settlements: these may be warlords, political entrepreneurs, splinter groups, militias and so on.[22]

Stability operations

It's clear that since 1945 there has been a growing sophistication in approaches to peace operations, and a growing consensus on the sorts of general concepts, ideas and principles surrounding such operations. In recent years, however, an additional concept has emerged: that of 'stability operations'.

The emergence of stability operations

The development of stability operations was shaped powerfully by the context provided by the terrorist attacks on the US World Trade Center on 11 September 2001 (9/11), and the wars in Afghanistan and Iraq that followed. The **9/11 attacks** helped to shape a perception that **failed or failing states** would constitute one of the key sources of future insecurity because of the opportunities that weak or ungoverned spaces provided for terrorists and criminals; because of the humanitarian and political crises that were often associated with state failure; and because of the ripple effects that such internal conflicts had regionally and globally. For some, failed or failing states were a locus of instability in a future marked by 'persistent conflict', driven by religious extremism, ethnic conflict, ideology, terrorism and global competition.[23] From this observation followed the logic that a capacity to **reverse state failure**, and to create conditions within a previously failed or failing state conducive to long-term stable peace, would be a crucial means of pre-empting future security threats.

Afghanistan and Iraq

However, experiences in Afghanistan from 2001 onwards and in Iraq from 2003 demonstrated **serious deficiencies** in the ability of governments and militaries to re-build states (these controversies are dealt with in more detail in the next chapter). It was evident that combinations of conventional military success allied with humanitarian relief could not alone create the conditions for political stability within a country. In both Iraq and Afghanistan the early conventional military success was overwhelming; however, the protracted security, political, economic and social crises experienced in both countries illustrated that winning battles decisively, the essence of traditional conceptions of victory, did not in any sense guarantee success overall. In the United States, especially, which bore the brunt of operations in both these countries, it seemed evident that there was a serious deficiency in abilities to conduct peace-building activities.

It is important to note that the term **stability operations** was not new: the US FM 3-0, *Operations*, for example, had long made reference to the importance of stability operations, but they were, nevertheless, an auxiliary element in military activity. Stability operations as understood in FM 3-0 embodied the sorts of tactical tasks – such as humanitarian relief, ensuring law and order, re-building infrastructure and training local security forces – that were familiar to peace operations doctrine, but which FM 3-0 recognised as being applicable to other operational environments as well: effective counterinsurgency (COIN) campaigns, for example, would often embody these kinds of tasks. However, Afghanistan and Iraq seemed to show that the US military had not accorded stability operations the priority necessary to do them well: in the triumvirate of Offensive, Defensive and Stability operations, the last was a poor third place in terms of the training, equipment and thinking accorded to it.

As a consequence, in November 2005 Department of Defense Document 3000.05 made the decision that stability operations would become a **core mission** and should be made of **equal importance** to combat operations.[24] Operations could now encompass Offense, Defence and Stability activities, and in some circumstances the last of these could predominate. In theory, stability operations could be applied across a **whole spectrum of activity**, from supporting partners in peacetime engagement, supporting a friendly government in fighting irregular warfare, through to post-conflict activities (see Figure 7.1).[25]

An essential reflection of this was the development of specific stability operations **doctrine**, building on the lessons identified from operations in Iraq and Afghanistan, reflected in publications such as FM 3-07 *Stability Operations*, published in 2008. This new doctrine defined stability operations as:

> [V]arious military missions, tasks and activities conducted outside the United States in co-ordination with other instruments of national power to maintain or re-establish a safe or secure environment, provide essential government services, emergency infrastructure reconstruction, and humanitarian relief.[26]

Though the new doctrine asserted that stability operations had applicability across the spectrum of conflict, much of its actual focus was on nation building – unsurprising, given the influence of the campaigns in Iraq and Afghanistan, and the broader identification of failed or failing states as a problem. In consequence, whilst stability operations might be tactical tasks, much of the focus was on the broader purpose of **stabilisation**:

> the process by which underlying tension that might lead to resurgence in violence and a breakdown in law and order are managed and reduced, while efforts are made to support conditions for successful long-term development.[27]

Stabilisation aimed to effect **conflict transformation**:

> the process of reducing the means and motivation for violent conflict while developing more viable, peaceful alternatives for the competitive pursuit of political and socio-economic aspirations.[28]

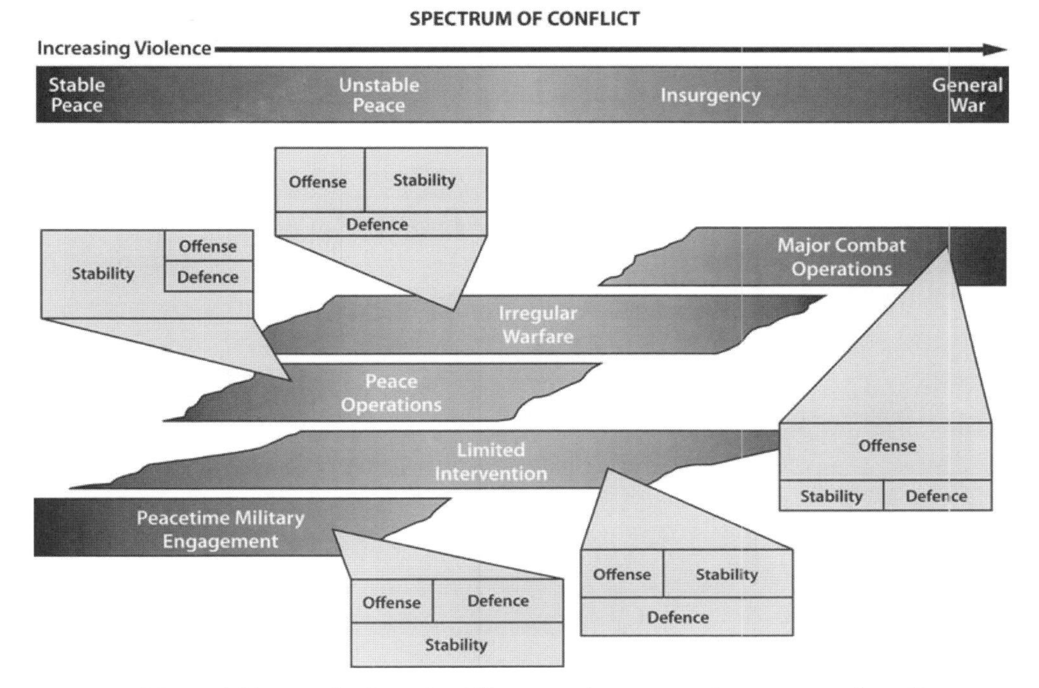

Figure 7.1 The variable contribution of stability operations across the spectrum of conflict

Stabilisation might have **different imperatives**, depending on the point at which it was executed:

- As part of an **initial response** to a crisis, it might need to focus on meeting immediate needs.
- In the **transformation phase**, the focus would need to be on establishing conditions for long-term development.
- If **sustainability** was the priority, operations would need to focus on embedding permanent solutions (see Figure 7.2).

Just as with peace operations, stabilisation had **objectives** very far indeed from traditional conceptions of military victory. FM 3-07, for example, identified five conditions for success:

- A **safe and secure environment**, defined in terms of establishing public security, enforcing an end to large-scale violence, and returning central government control over the means of legitimate violence..
- An **established rule of law**, in regard to establishing a just legal framework, enforcing law and order, ensuring access to justice, promoting a culture of lawfulness, and establishing public security.
- **Social well-being**, through encouraging access to delivery of basic needs, the right of return to previous conflict areas, and the promotion of peaceful co-existence.

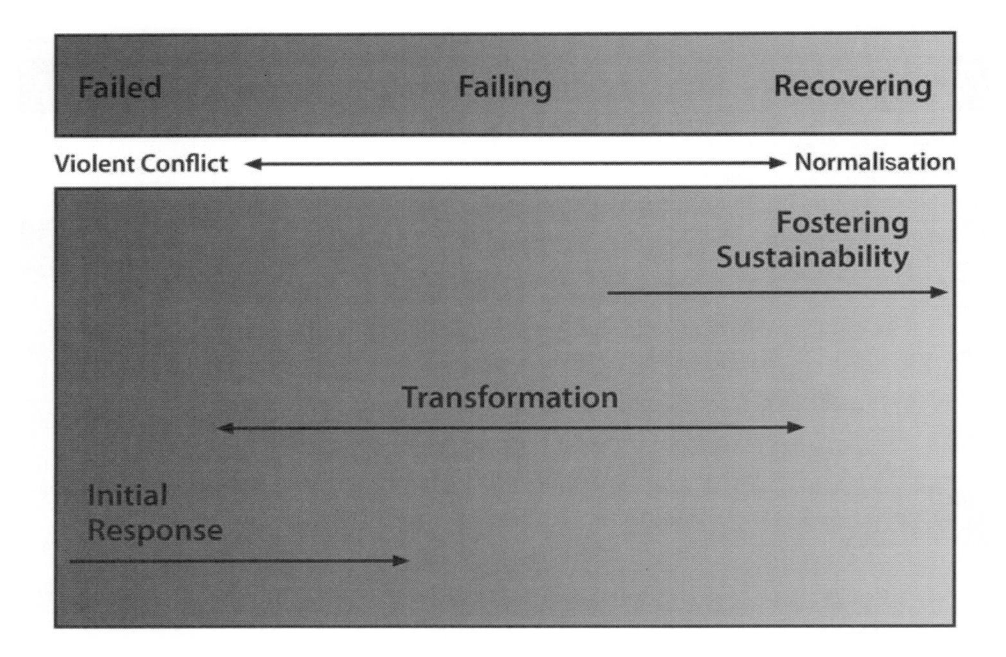

Figure 7.2 Responses across the spectrum of state failure
Source: (FM 3-07, 2–13)

- **Stable governance**, through encouraging accountable government, effective management of state resources, encouraging civic participation and empowerment, and supporting the provision of government services.
- A **sustainable economy**, through macro-economic stabilisation, control over the economy in face of threats from illicit activities, supporting a sustainable market economy, individual economic security, and creating employment.[29]

Since these end states self-evidently could not be delivered solely, or even mainly, by conventional military activity, stabilisation operations required efforts across five broad **sectors** of activity which covered the political, military, economic and social foundations for lasting peace.

- The first of these was **security** activities based on protecting the local population and territory: this was crucial to creating the space in which other activities could be conducted effectively.
- A second area of activity was **justice and reconciliation**: this embodies activities such as the creation of effective and accountable law enforcement; a transparent and accountable judicial system; and mechanisms for promoting human rights and reconciliation.
- A third focus was **humanitarian assistance and social well-being**, comprising some activities designed to ameliorate immediate suffering, such as providing emergency

aid, but also longer-term measures to prevent their re-occurrence, such as famine prevention.

- A fourth area of activity encompassed **governance and participation**, designed to help create processes and institutions of government that are legitimate and effective. The focus here would be on activities such as establishing rules, processes and norms of accountable and representative government; encouraging political participation; encouraging the local population to take increasing responsibility for their own affairs.
- Finally, **economic stabilisation and infrastructure** constitute another crucial sector of activity. This might require activities such as rebuilding essential infrastructure; securing key resources for the government; helping to effect economic restructuring; and helping to establish a regulatory framework conducive to effective economic development.[30]

To achieve goals across these sectors, FM 3-07 identified a range of **tasks** that the military need to be able to perform or support. These tasks were organised into five **primary stability tasks**:

- Establishing civil society in the host nation.
- Establishing civil control in the host nation.
- Restoring essential services.
- Supporting governance.
- Supporting economic and infrastructure development.

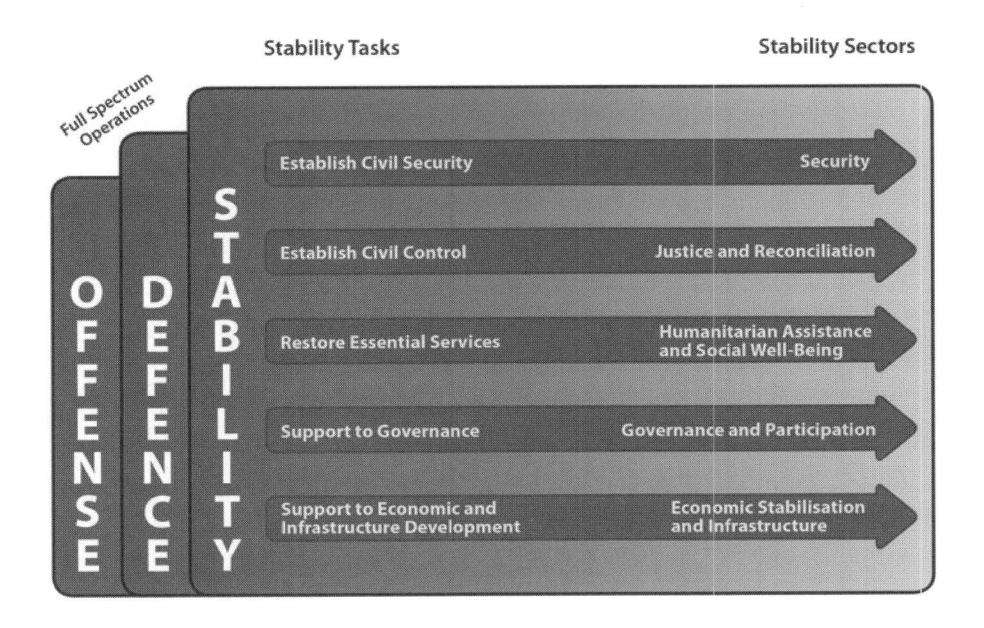

Figure 7.3 An integrated approach to stability operations

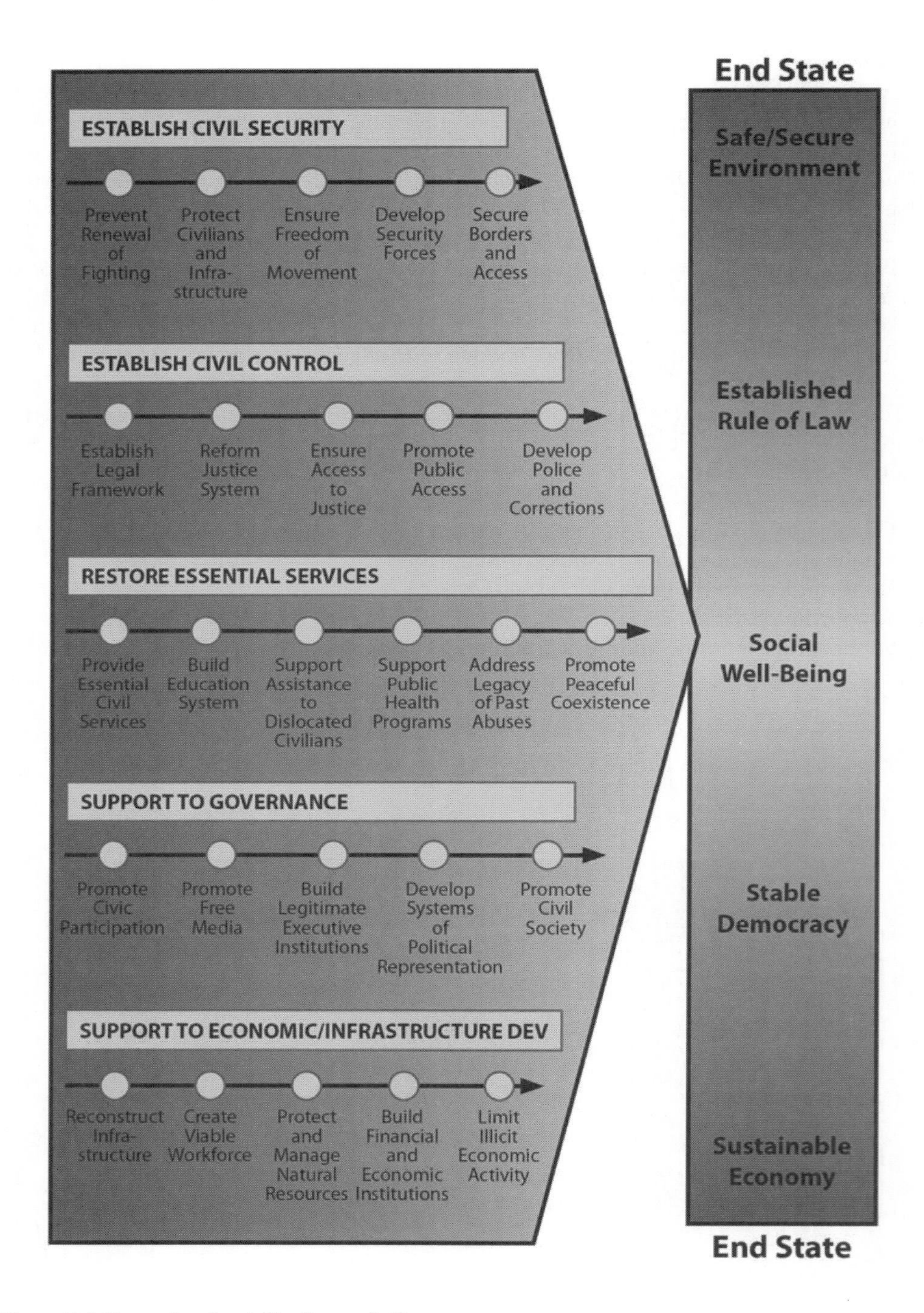

Figure 7.4 Example of stability lines of effort

These primary tasks are each associated with one of the stability sectors (see Figure 7.3). The kinds of more detailed tasks that each of these primary tasks might encompass are illustrated in Figure 7.4.

The relative importance of the military in performing these tasks is likely to vary. In the **initial response phase** (see Figure 7.1), the military might be one of the only organisations able to operate in the complex and possibly dangerous environment. In the **transformation and sustainability phases**, many tasks will be performed mainly or wholly by other actors. The nature of the tasks, the balance between them and the detailed implementation would depend on local conditions such as the capacities of the host-nation government, the presence of other actors such as NGOs, local actors and so forth.

What emerging stabilisation doctrine also reflected, however, was a set of **principles** that could guide effectively the performance of these tasks in pursuit of the objectives: whilst these were only implicit in FM 3-07, the US army's Peacekeeping and Stability Operations Institute identified an explicit set of principles based upon a wide survey of the existing stabilisation literature:

- **Host nation ownership and capacity**: peace can only be sustainable when it is the host nation that has the capacity to resolve conflict. Transforming conflict requires that responsibilities eventually be shifted from the international community to the host nation.
- **Political primacy**: the central focus in stabilisation efforts is the political settlement and all actions need to be considered in the light of the extent to which they advance or retard the creation of a lasting political agreement.
- **Legitimacy**: this is a function of broad-ranging acceptance and approval of an operation, reflected in acceptance of its mandate, consent of the host nation and agreement internationally.
- **Unity of effort**: this requires sustained co-operation between actors. This has a tactical dimension in relation to co-ordinating actors on the ground. It also has an operational and strategic dimension through the realisation of a 'whole-of-government' framework or a 'comprehensive approach' that ties together government departments at a national level, and also mechanisms to ensure international unity of effort.
- **Security**: this is a crucial enabler – without it, other tasks cannot be performed effectively.
- **Conflict transformation**: a guiding principle that should focus stabilisation efforts on reducing the causes of conflict and strengthening the host nation's capacity to mitigate and resolve conflict.
- **Regional engagement**: promoting co-operation with neighbours and other states key to the conflict.[31]

The importance accorded to stability operations reflected an explicit recognition that victory in modern contexts is a complex term. Stabilisation required that armies develop ever more sophisticated ways to harness their skills to tasks, often **highly political** in character, which could be peripheral to conventional military activities; requiring them to **integrate** their efforts ever more closely; to **subordinate** their objectives to those of other actors; and to deliver against complex, multi-dimensional, **long-term goals** in which traditional metrics of military success might be irrelevant or counter-productive. In stabilisation environments, what matters isn't military victory in the traditional sense, but '**strategic success**' – a concept that reduced the relative importance of the military component of victory in order to highlight the importance of other dimensions. As FM 3-07 argued, in contemporary conflict scenarios, 'the margin of victory will be measured in far different terms from the wars of our past';[32] instead, military force should aim 'to influence and deter spoilers working against the peace process or seeking to harm civilians; and not to seek their military defeat'.[33] Given the traumatic difficulties in Iraq and Afghanistan, these developments in doctrine would clearly seem to be an important advance for armies.

Assessing peace and stability operations

The idea that armies have a role to play outside of conventional warfare in helping to resolve conflict is now an established one, certainly amongst the military themselves. However, how important are such operations likely to be in the future?

What is success?

Peace operations, with their focus on impartial involvement, underwent a **crisis in the 1990s**. In the wake of problems in Somalia and Bosnia, and the dismal failure to intervene effectively in Rwanda, the early enthusiasm for peace operations waned. Between 1993 and 1996 the numbers deployed on UN peace operations shrank by more than two thirds, and between 1994 and 1999 only four new missions commenced, each of which was relatively small. Critics of peace operations could point to problems in identifying any unqualified successes. Three benchmarks can be used to assess peace operations since these are shared by almost all of these kinds of activity:

- **Violence abatement**: the extent to which the deployment reduced or eliminated armed violence.
- **Conflict containment**: the extent to which the deployment prevented the conflict from spreading.
- **Conflict settlement**: the extent to which the deployment resolved the issues and disputes between participants.[34]

It is certainly difficult to find examples of peace operations that provide more than partial successes. Some fail on all counts, such as the deployment in Somalia. Others succeed in some areas and not in others. Deployed in 1964 to separate Greek and Turkish Cypriots, UNFICYP, for example, helped to promote a reduction in violence and prevent the spread of the conflict; however, the UNFICYP deployment also, in a sense, **froze the dispute**. By limiting the costs of continuing to refuse to settle the conflict for the Greek and Turkish Cypriots, UNFICYP contributed to the lack of conflict settlement. UNPROFOR helped to prevent the conflict spreading; its contribution to violence abatement is complicated by claims that a quicker resolution to the violence might have been achieved without its deployment – that the deployment resulted instead in a '**slow-motion savagery**' that drew out the violence; likewise, whilst the agreement reached in 1996 constructed a settlement that has prevented a return to war, the federal Bosnia-Herzegovina state that has resulted has struggled to achieve true reconciliation between its Bosnian, Serb and Croat communities.

Contemporary utility

Despite these issues, peace operations have remained a feature of international conflict management efforts. For example, despite the crisis in confidence in the mid-1990s, four new missions were established in 1999:

- The UN Transitional Authority in East Timor (**UNTAET**), with 10,700 military and civilian personnel.
- The UN Organization Mission in the Democratic Republic of the Congo (**MONUC**), with an authorised strength of 6,000.
- The UN Mission in Sierra Leone (**UNAMSIL**), with 18,000 personnel.
- The UN Mission in Kosovo (**UNMIK**), with 1,000 civilian staff.

In 2007, what is now the African Union-UN Hybrid Mission in Darfur (**UNAMID**) was created, with nearly 15,000 military personnel and 5,000 police. In 2013, the UN established a stabilisation mission in Mali (MINUSMA) with nearly 5,000 troops and 800 police. Indeed, there are currently 15 UN peacekeeping operations (see Box 7.7) involving over 80,000 troops, 13,000 police and nearly 17,000 civilian personnel.[35] The renewed interest in peace operations has been the result of several factors:

- One was that some crises impinged directly on the **interests of Western states**: Australia took the lead in East Timor, for example, and managing the situation in Kosovo was of obvious interest to European states.
- A second reason was a shift in the conduct of peace operations in some cases away from the UN and towards **regional organisations**, which addressed the views of those who saw the problems with peace operations as stemming from the failings of the UN: ECOWAS took the lead in Liberia through the deployment of its ECOMOG (ECOWAS Monitoring Group) operation in 1990, a force that was also later deployed to Sierra Leone (1997), and Guinea-Bissau (1999); the Southern African Development Community intervened in Lesotho in 1998; in Sudan (2004) it was the African Union, until their operation was replaced by UNAMID in 2007; and NATO replaced the UN in Bosnia. It is worth noting, though, that experience has shown that regional organisations are not necessarily more effective than the UN and can often lack legitimacy, resources and experience.[36]
- Also, whatever their flaws, peace operations have survived, not least because they offer **a necessary bridge** between two alternatives: to do nothing, or to engage in full-scale military intervention. The former option risks conflicts destabilising surrounding countries, it is often domestically unsustainable given the impact of media attention, and it is often unpalatable to governments with an internationalist agenda. As Afghanistan and Iraq have demonstrated, full-scale military intervention brings its own difficulties in terms of mobilising the necessary political consensus and resources to make such an option feasible.

Box 7.7 UN peace operations (August 2013)

Africa

- United Nations Mission in the Republic of South Sudan (UNMISS)
- United Nations Interim Security Force for Abyei (UNISFA)
- UN Organization Stabilization Mission in the Democratic Republic of the Congo (MONUSCO)
- African Union-UN Hybrid Operation in Darfur (UNAMID)
- UN Operation in Côte d'Ivoire (UNOCI)
- UN Mission in Liberia (UNMIL)
- UN Mission for the Referendum in Western Sahara (MINURSO)
- UN Multidimensional Integrated Stabilization Mission in Mali (MINUSMA)

Americas

- UN Stabilization Mission in Haiti (MINUSTAH)

Asia and the Pacific

- UN Military Observer Group in India and Pakistan (UNMOGIP)

Europe

- UN Peacekeeping Force in Cyprus (UNFICYP)
- UN Interim Administration Mission in Kosovo (UNMIK)

Middle East

- UN Disengagement Observer Force (UNDOF)
- United Nations Interim Force in Lebanon (UNIFIL)
- UN Truce Supervision Organization (UNTSO)

Conclusions

Peace operations increasingly have become an important role for land forces. As the incidence of major conventional wars has declined, land forces have been required to undertake a **complex range of new tasks** in challenging and politically charged environments. Whilst some peace operations are still conducted in the areas between states, many are now intra-state operations contending with ambitious political mandates, severe humanitarian crises and awkward political dynamics. In performing roles associated with conflict prevention, conflict management and conflict resolution, traditional concepts of military victory have had to be revised. The military defeat of enemy fielded forces is no longer an adequate or often even a desirable objective. Victory is now associated much more with creating **legitimate political solutions** and **stable long-term peace**.

Reflecting the versatility of armies, doctrines of peace and stability operations illustrate the way in which the evolution of new techniques has allowed troops to adapt. A focus on such principles as consent, legitimacy and the constrained use of force has been accompanied by a focus on greater military co-ordination with non-military agencies and engagement across the political, economic and social spheres of activity. The thrust of this evolution in doctrine has been an attempt to make land power more flexible and more appropriate to the nuanced political goals laid out by policy.

Has all of this actually worked? In the next chapter we complete Part II of this book by examining the difficulties associated with using land forces for complex operations other than war. As Chapter 1 has noted, each of the advantages of land power carries with it potential disadvantages. History and recent experience suggest that despite an extraordinary amount of intelligent reflection by armies on the adaptation and innovation required for peace and stability operations, they remain difficult activities to perform, and there are difficult trade-offs that exist between performing these tasks and conducting traditional, large-scale, conventional warfare.

Notes

1 Paul F. Diehl, *Peace Operations* (Cambridge: Polity Press, 2008), 28–67.
2 *United Nations Peacekeeping Operations: Principles and Guidelines* (United Nations: Department of Peacekeeping Operations, 2008), 13.
3 Joint Publication (JP) 3-07.3, *Peace Operations*, 1 August 2012, x.
4 Ibid.
5 *United Nations Peacekeeping Operations*, 21.
6 *United Nations Peacekeeping Operations*, 31–34.
7 Opération des Nations Unies au Congo.
8 Trevor Findlay, *The Use of Force in UN Peace Operations* (Oxford: Oxford University Press, 2002), 51–89.
9 Alex J. Bellamy, Paul D. Williams and Stuart Griffin, *Understanding Peacekeeping* (Cambridge: Polity Press, 2010), 23–41, 94–97.
10 Bellamy *et al.*, *Understanding Peacekeeping*, 93–103.
11 William Flavin, 'U.S. Doctrine for Peace Operations', *International Peacekeeping* Vol. 15, No. 1 (February 2008), 40.
12 JP 3-07.3, vii.
13 Ibid., I-7.
14 Ibid., I-2.
15 Ibid., x.
16 *United Nations Peacekeeping Operations*, 18.
17 JP 3-07.3, x.
18 Ibid., x.
19 *United Nations Peacekeeping Operations*, 19–22.
20 JP 3-07.3, *Peace Operations*, 17 October 2007, I-6–I-12; *United Nations Peacekeeping Operations*, 36–39.
21 Field Manual (FM) 3-07, *Stability Operations*, October 2008, 1–19.
22 JP 3-07.3, I-3–I-12; *United Nations Peacekeeping Operations*, 36–39.
23 FM 3-07, 1–2.
24 William Flavin, 'U.S. Doctrine for Peace Operations', *International Peacekeeping* Vol. 15, No. 1 (February 2008), 42.
25 FM 3-07, 1–3.
26 JP 3-0, *Joint Doctrine*, 11 August 2011, V-5.
27 FM 3-07, 1–12.
28 Ibid., 1–5.
29 FM 3-07, 1-16–1-18.
30 FM 3-07, 2-7–2-9.

31 *Guiding Principles for Stabilization and Reconstruction* (Washington, DC: United States Institute for Peace Press, 2009), 3–12.
32 FM 3-07, vi.
33 *United Nations Peacekeeping Operations*, 35.
34 Paul F. Diehl and Daniel Druckman, *Evaluating Peace Operations* (Boulder, CO: Lynne Rienner, 2010), 29–48.
35 UN Peacekeeping Fact Sheet, www.un.org/en/peacekeeping/resources/statistics/factsheet.shtml (accessed 16/10/2013).
36 Alex J. Bellamy and Paul D. Williams, *Understanding Peacekeeping* (Cambridge: Polity, 2010), 121–29.

Suggested reading

Alex J. Bellamy, Paul D. Williams, and Stuart Griffin, *Understanding Peacekeeping* (Cambridge: Polity Press, 2010). An excellent and comprehensive examination of the development and future of peacekeeping operations.

Paul F. Diehl and Daniel Druckman, *Evaluating Peace Operations* (Boulder, CO: Lynne Rienner, 2010). Provides an insightful analysis of the sorts of criteria against which the success and failure of peace operations should be judged.

James Dobbins, John G. McGinn, Keith Crane, Seth G. Jones, Rollie Lal, Andrew Rathmell, Rachel M. Swanger and Anga R. Timilsina, *America's Role in Nation-Building: From Germany to Iraq* (Santa Monica, CA: RAND, 2003). A comparative analysis of US nation-building efforts from Germany and Japan in 1945 to the early phase of operations in Afghanistan.

William J. Durch (ed.), *The Evolution of U.N. Peacekeeping: Case Studies and Comparative Analysis* (Basingstoke: Macmillan, 1994). A classic analysis of the development of peacekeeping during the Cold War and during the transitional period that followed its demise.

Virginia Page Fortna, *Does Peacekeeping Work? Shaping Belligerents' Choices After War* (Princeton, NJ: Princeton University Press, 2008). Investigates the potential advantages and disadvantages of peacekeeping operations through the lens of two questions: does peacekeeping work and, if it does, how does it work?

Lise Morje Howard, *UN Peacekeeping in Civil Wars* (Cambridge: Cambridge University Press, 2008). An illuminating examination of the United Nations' performance in such difficult operations as Somalia, Rwanda and Bosnia, and the impact of these operations on UN peacekeeping approaches.

Jennifer Morrison Taw, *Mission Revolution: The U.S. Military and Stability Operations* (Chichester, NY: Columbia University Press, 2012). Examines the origins, implications and debates associated with the US military's shift to a focus on stability operations.

8 Peace and stability operations

Challenges and debates

Key points

- One of the foundation problems that face peace and stability operations is the contexts into which they are deployed: complex and variable intra-state conflicts with a plethora of political actors and often a high degree of risk.
- These operations are complicated by a lack of consensus over many of the key concepts and principles, making unity of effort more difficult to achieve.
- Even where the principles are well understood, it can be difficult in practice to make them work effectively.
- There are wider debates regarding whether it is even possible to achieve some of the more ambitious goals related to nation building and, even if it were possible to achieve them, whether the military is the right tool.

We have seen in the previous chapter that land power can be extraordinarily flexible. The period since 1945 has been marked by the development of increasingly complex definitions of 'victory', driven by the involvement of armies in an expanding array of tasks that have drifted well beyond those of combat against enemy military forces. Increasingly sophisticated doctrines of peace and stability operations have focused armies on performing alone, or in concert with other actors, an expanding range of political, economic, social and security tasks.

However, as armies in the modern period have found on numerous occasions, having an intelligent doctrine is in no sense a guarantee of success. Despite extensive experience, debate and reflection, peace and stability operations continue to be accompanied by serious challenges; these challenges have led some commentators to question the practical utility of land forces as tools of conflict prevention and management. This chapter focuses on those difficulties and explains the sorts of serious problems that routinely confront armies in the conduct of peace and stability operations. We will take a thematic approach to the subject, looking first at a foundation problem, the operational environment, before moving on to look at conceptual and practical difficulties. One might think that peace and stability operations should be simpler than conventional warfare, but experience suggests that often they are not.

The operational environment

One of the foundation challenges that has bedevilled peace and stability operations in the modern era has been the difficult operational context, in particular the growing significance of **intra-state conflicts** in the guise of civil wars and state failure. This context has had a paradoxical effect: on the one hand it has been a driver for the development of peace and stability operations because of the growing need to find some means to ameliorate and resolve these struggles; on the other, intra-state conflicts seem to carry with them certain inherent problems that make resolving them more difficult.[1]

The host nation

Intervening in an intra-state conflict puts a premium on the issue of **host nation capacity**. The scope of any peace and stability operation is defined in large part by the gaps that exist in the capacity of a host nation government to provide for the needs of its population. Almost by definition, intra-state conflicts tend to be associated with **weak or failed states**: in consequence, intervening forces are likely to have to perform more complex and wide-ranging tasks at the same time as they are less likely to be able to draw on the capacity or resources of the host nation government. Thus, the stronger the host nation government's capacity, the more restricted the tasks a peace and stability operation is likely to have to perform, and the more material support it will receive to perform those tasks. Conversely, operations such as those in Iraq, Afghanistan, Somalia and Haiti have been inherently problematic because of the profound gaps in host nation support, especially in the early stages.

Problems of political compromise

Another feature of these contexts is that the political issues surrounding them often tend towards **intractability**. One reason for this is that in intra-state conflicts, belligerents share the same political space: the issues at stake tend to be **less divisible** because they are about central issues regarding who holds political power in a state, so one group is likely to resist any peace settlement that empowers a rival. Another reason is that they tend often to have **'value-led' dimensions**: whilst the origins of these conflicts are often to do with *issues* associated with the division of power and resources, the conflicts become, often through the deliberate manipulation and distortion by such 'political entrepreneurs' as warlords or local government representatives, in many respects about *values* – about beliefs, ideologies, ethnicity and religion. Such conflicts tend to become **zero-sum**: because they hate, fear and/or mistrust the other side, political solutions can be viewed by a belligerent as legitimate only if they come at the expense of the other side.[2]

Political actors

A third feature of intra-state conflicts is that they tend to involve **multiple political actors**. More than 80% of interstate disputes involve only two primary parties. In intra-state conflicts, the belligerents often consist of a **multiplicity of parties, groups, militias, and warlords**. The greater the number of belligerents, the more difficult it is to reconcile their interests and to create consensus on an acceptable political framework. **Spoilers** (political actors who seek actively to undermine peace) become more likely in such conditions, and the peace operation may face difficult choices over whether particular conflict actors can legitimately be excluded from a peace process. Interventions in such disputes are also likely to feature **multiple interveners**, both in relation to numbers of states involved, and also in terms of institutions and organisations. This makes it problematic to define common goals and execute common strategies. Processes of globalisation have also strengthened other political actors which, in the context of humanitarian need, political gain, or financial incentives, may become important players in conflict. These can include actors such as **non-governmental organisations** (NGOs), **trans-national corporations** (TNCs), and **private military contractors** (PMCs). These actors can often play an important positive role in operations: NGOs are responsible for disbursing half of all humanitarian aid; PMCs can provide training, technical capabilities and additional manpower to augment peace and stability operation; TNCs can be crucial providers of resources and expertise for rebuilding critical infrastructure, as Halliburton was in Iraq. However, these actors can also complicate peace and stability operations. For example, TNCs can contribute to the 'conflict trade' in diamonds, light weapons, drugs, and have been an important source of conflict in places such as Afghanistan, Angola, Liberia and Sierra Leone by helping to provide the financial support for various local conflict actors. NGOs may unwittingly become complicit in ethnic cleansing by establishing refugee camps, as was the case in eastern Zaire in the wake of the crisis in Rwanda. PMCs create the problem of having military capabilities outside of the direct control of local military commanders.

Protraction

A fourth feature of intra-state conflicts is that as a consequence of the features identified above, they tend to be marked by their **longevity**. The longer conflicts continue, the more difficult they can become to resolve: one effect is that conflict becomes deeply embedded through the disintegration of normal society. For example, activities such as the widespread use of child soldiers, the use of rape as a weapon, and cycles of atrocity and counter-atrocity can result in deeply traumatised societies in which **conflict becomes 'normalised'**. Protracted internal wars develop a 'conflict history', in which **demonization** of the enemy and **polarisation** of political stances can result in an entrenched enmity towards other belligerents. These environments can also become marked by the emergence of **autonomous war dynamics**: wars may begin for one set of reasons, but as conflicts protract, they often develop new reasons for continuing unrelated to the original causes of armed conflict. One recurring difficulty in this respect is the problem of the political economy of violence: put simply, some participants benefit actively from war, especially economically. Warlords, for example, often have little incentive to co-operate in the establishment of peace and a strong central government. All of these contextual problems tend to be worsened by the fact that peace support operations are often a **last resort**: in other words, conflicts have often been underway for a considerable period before a peace support operation is actually deployed, often ensuring that these conflict dynamics are well entrenched. Indeed, many internal conflicts result from the breakdown of existing peace settlements that are unable to break well-established cycles of violence.[3]

Complexity

A fifth feature of contemporary conflicts is that they are often part of '**conflict complexes**': in other words, individual intra-state conflicts tend to be **connected to interstate conflicts** amongst neighbouring states. Thus, a peace and stability operation is unlikely to be able to achieve success solely by addressing issues in the state into which it has been deployed. Whether one is looking at the links between Liberia and Sierra Leone, Rwanda and Congo, or Iraq and Iran, many internal conflicts are also international conflicts as well. The conflict in Afghanistan, for example, is linked intimately to Pakistan's conflict with India. For Pakistan, the Afghan President Mohammed Karzai's government is viewed as pro-Indian; for India, the Taliban are seen as a potential Pakistan-sponsored catalyst for further conflict in Kashmir. There are a wide range of other regional actors linked by their own patterns of co-operation and conflict who have interests in the Afghan conflict and whose co-operation, or at least acquiescence, is required to resolve it, including Russia, China, Turkey and Iran.[4]

Variety

Finally, despite often similar generic difficulties of the sorts outlined above, intra-state conflicts are problematic because of their **variability**. The character of peace and stability operation will be affected profoundly by the particular attributes of the **context** into which it deploys, including:

- The degree of consent.
- The capacity of the host nation government.
- The nature of the regime.
- The state of local infrastructure.
- The shape of the local economy.
- The attitudes of the local population.
- The pattern of representation of NGOs.
- The extent of ethnic diversity or other potential political fault lines.
- The relative power and attitudes of local political entrepreneurs.
- Demographics.
- The local geography.
- The perspectives and capabilities of neighbouring countries.[5]

The point here is that the term 'intra-state conflict' covers contexts that vary wildly in their characteristics and requirements and which may require radically different approaches. Amongst other things this raises important question regarding the **transferability of lessons**: the extent to which one can import into one conflict the lessons learnt from another.

Conceptual challenges

Notwithstanding the practical problems associated with the contexts into which many peace and stability operations are deployed, there are recurring debates surrounding the **conceptual clarity** of the new doctrines designed to deal with them.

Varying definitions

First, given the importance attached in contemporary doctrine to unity of effort and common approaches to problems, peace and stability operations require a common understanding of doctrine from all participants, especially given the important multi-agency and multi-national dimensions of such operations. However, this **common understanding is often lacking**. For example, the United States and its allies have different conceptions of what stabilisation entails. Perhaps the most developed alternative to the United States' Field Manual (FM) 3-07 is the United Kingdom's doctrine for stabilisation: Joint Doctrine Publication (JDP) 3-40, *Security and Stabilisation: The Military Contribution*. JDP 3-40 is itself an innovative and intellectually stimulating document, but it conceives of stabilisation in a fundamentally different way. Stability operations are not a discrete type of task; instead, the term 'stability operations' is used in JDP 3-40 to describe **a specific kind of operation**: nation-building enterprises along the lines of Iraq and Afghanistan. From a British perspective, stabilisation is an operational level approach that seeks to build stronger states through simultaneous political, military and economic action in three spheres: building human security; fostering the capacity and legitimacy of the host nation government; and stimulating economic and infrastructure development. The role of the military in performing these tasks may vary (see Figure 8.1).[6]

Stability versus COIN and peace operations

These differences help to shape a second difficulty: the problem of **disentangling stabilisation from other military activities**. FM 3-07 notes that whilst there may be many different operational 'themes' in a campaign such as major combat operations, or irregular warfare, the overall mission is characterised by the nature of the stability tasks

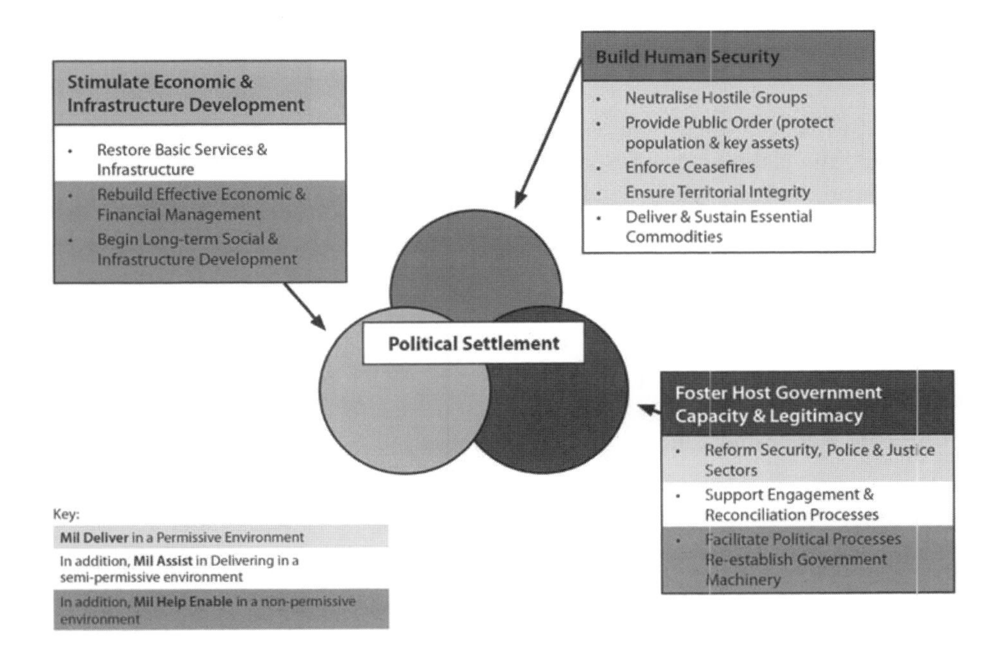

Figure 8.1 British approaches to stabilisation
Source: (Joint Doctrine Publication 3-40, *Security and Stabilisation: The Military Contribution* (Development, Doctrine and Concepts Centre, November 2009), 10–29)

performed. FM 3-07 asserts that **most military action is actually stabilisation**: it notes that the United States has only fought 11 conventional wars in its history – most conflicts, it asserts, from operations on the US–Indian frontier through to the war in Vietnam, have been stability operations. In the decade after the Cold War, FM 3-07 notes that the United States has undertaken 15 stability operations, including Haiti, Liberia, Somalia, the Balkans, Afghanistan and Iraq. Yet many do not take this view: in Afghanistan, for example, in 2008 the International Security Assistance Force (ISAF) commander, General David McKierney, asserted: 'The fact is that we are now at war in Afghanistan. It's not peacekeeping. It's not stability operations. It's not humanitarian assistance. It's war.'[7]

Stability operations versus counterinsurgency

So, what is the relationship between a stability operation and other types of operation? This question is unresolved. For example, if conflicts such as Vietnam that were previously termed counterinsurgency (COIN) campaigns are now 'stabilisation operations' then COIN has become a **sub-element of stabilisation** (indeed, this is the British view, which regards stabilisation as 'COIN plus'). However, some see stability operations through the lens of COIN. The strategist Colin Gray, for example, argues that 'stability operations need to be understood as integral to counterinsurgency strategy and doctrine'. Viewed in this way, stability operations are **subordinate to COIN**.[8]

Stability operations versus peace operations

The relationship with peace operations is also problematic. For some commentators, the United States has conducted peace operations throughout its history: defining peace operations in terms of tasks such as enforcing civil rights, demobilisation and disarmament, separating factions, overseeing formation of governments, training and advising local forces, human rights, educational reform, and providing a safe and secure environment, the claim is that US operations in places such as the Philippines, Cuba, China, Central America, Japan and Germany were peace operations. Since FM 3-07 defines such operations as stabilisation operations, can stabilisation operations be peace operations? The difficulty is that stabilisation operations on the lines of the Iraq or Afghanistan model cannot be peace operations since the latter are defined centrally by their **impartiality**, whereas armed nation building on the Iraq and Afghanistan model has been a decidedly partial activity. Overall, then, military training, education and doctrine have been left currently in an awkward situation: peace operations, COIN and stability operations doctrine exist in parallel but the relationship between them has been become increasingly difficult to conceptualise.

Operationalising principles

Peace and stability operations therefore operate in environments that are intrinsically difficult, and they comprise concepts that not everyone defines in the same way. Building on these observations, there is a third set of potential difficulties with these operations which derive from criticism of the principles of peace and stability operations. Essentially, and mirroring critiques of COIN doctrine outlined in Chapter 6, the debate here surrounds whether peace and stability operations doctrine provides an **effective set of practical tools**.

Unity of effort

One obvious example is provided by the challenge in realising the precept of **unity of effort**, expressed in concepts such as the 'comprehensive approach' or an 'integrated approach' or a 'whole-of-government' effort, which occupy a central position in peace and stability operations doctrine. For example, in July 2004, the United States created the Department of State Office of the Coordinator for Reconstruction and Stabilization, tasked with leading, co-ordinating and institutionalising civilian reconstruction and development capacity. A new Interagency Management System was established, and a new framework for planning across government departments. In the UK, the Foreign Office established a Stabilisation Unit (initially called the Post-Conflict Reconstruction Unit); Canada established a Stabilization and Reconstruction Task Force (START).[9]

Organisational problems

In practice, however, unity of effort is difficult to achieve: problems of **co-ordination and deconfliction** have continued to bedevil peace and stability operations. The central problem is that whilst structure and process-related issues can be contributory reasons why unity of effort can falter in an operation, they are often not the key causes: thus, creating new structures or procedures to improve co-ordination can mitigate some problems, but it will rarely solve them. Fundamentally, problems with co-ordinating different peace and stability operations actors stem from **deep-seated political and cultural reasons** that are embedded in the very **nature of human organisation**. For example, the seemingly perennial tensions between the military and humanitarian NGOs are built on fundamentally different **organisational perspectives**. NGOs, for example, tend to be decentralised and consensus-based; they have longer time lines because they are often deployed before and after the military elements of an operation. For NGOs, close co-operation with the military can be an obstacle to achieving their goals because it can compromise their neutrality and impartiality in the eyes of local communities.[10]

Differing perspectives

More broadly, some have argued that the military may have a tendency to see peace operations and stabilisation in relation to state building, whereas for the humanitarian and development communities the focus is on wider peace building: whilst these are related, the priorities and timelines can be different. Unity of effort initiatives by individual countries are also likely to be diluted by the **multi-national and multi-organisational character** of complex peace and stability operations. Different operations may involve different lead nations, participants, a variety of regional organisations, all with their own objectives and co-ordination difficulties. Even the United Nations (UN) is plagued by recurring problems of co-ordination between its various funds, agencies and programmes. Enduring national initiatives to improve unity of effort in peace and stability operations may therefore be swamped by the variable composition and objectives associated with evolving peace and stability operations.

The nature of peace and stability operations

Moreover, beyond the unity of effort challenges associated with the participants are those posed by the nature of peace and stability operations missions. Complex operations are **multi-dimensional**, and contemporary doctrine notes that activity across the five stability sectors are interrelated, with cross-cutting effects. However, these **cross-cutting effects may be negative** – in other words, success in one set of activities may undermine success in others. For example, military objectives may be achieved only at the expense of political objectives or vice versa. Afghanistan has exhibited numerous problems of this type: for example, pushing forwards on goals for social transformation such as greater gender equality creates difficulties politically given the resistance from Afghan traditionalists; drastically increasing the size of the Afghan security forces shows progress in building the capacity of the Afghan government, but it can come at a cost locally, where such 'security' forces are often perceived as partial, corrupt sources of insecurity. Efforts to eliminate corruption in the Afghan government provide another useful example: in 2010, 59% of Afghans saw corruption as their key concern (more, even, than security at 54%). At the same time, however, some writers have noted that corruption is also the 'bedrock of the Afghan government': without the capacity to use such inducements many governors would be unable to govern effectively.[11] In circumstances such as these, even if *actors* are co-ordinated, their *activities* may have consequences that cut across one another in ways that are not easy to resolve.

Perseverance

Perseverance is another key principle of contemporary peace and stability operations: it requires not just determination on the part of the military, but is related fundamentally to the **political will** of intervening states. Political will has been a perennial difficulty in peace and stability operations. In the past, this has manifested itself in the reluctance of member states to involve themselves in difficult and dangerous missions. In Rwanda in 1994, in the face of the emerging threat of genocide, the UN actually ran down its presence there, the UNAMIR operation (the UN Assistance Mission in Rwanda) being unable to intervene substantively in the face of a slaughter that would leave 800,000 dead. This difficulty has also manifested itself in a reluctance on the part of states to provide the necessary resources for ongoing operations. New operations are often poorly equipped and lack the physical and conceptual resources necessary for success. These problems are magnified in semi- and non-permissive environments: whether or not Western states are casualty sensitive, or just defeat sensitive, operations such as those in Somalia, Iraq and Afghanistan demonstrate that ambitious and costly operations can be impossible to divorce from the domestic politics of participants given that democracies tend to be sensitive to the costs of conflict. Since in peace and stability operations immediate **national interest** may be difficult to discern and/or progress may be difficult to prove unambiguously, perseverance may be a principle that is difficult to realise. Some of the scepticism with regards to peace building and stability operations therefore derives from the observation that successes like those in Germany and Japan after the Second World War required a massive and sustained physical, economic and political commitment that is difficult to replicate today.

End states

End states provide another example of the difficulty of operationalising peace and stability operations principles. The desirability of **clear and attainable end states** is a staple of military doctrines generally. One of the criticisms levelled routinely at peace and stability operations is that their mandates, the political authority that sets out their objectives and remit, are **too vague**. An example is UNOSOM II (United Nations Operation in Somalia, 1993–95). According to United Nations Security Council (UNSC) Resolution 794, UNOSOM II was tasked to 'Restore peace, stability and law and order with a view to facilitating the process of political settlement'. It was given latitude to use 'all necessary means to establish … a secure environment for the delivery of humanitarian relief'.[12] Critics argued that lacking any clear definition of how a condition of peace, stability and law and order should be defined in a Somalian context, without a clear political process in place, and given the latitude to use military means without restraint, UNOSOM II was primed for an unfocused escalation in the use of force.

Yet, in peace and stability operations, as in other military activities, end states and political objectives **tend usually to be ill-defined**. Often, governments find it difficult to define coherent grand strategic objectives which can then be translated into intermediate goals. In Iraq and Afghanistan, objectives cycled through toppling the existing regimes and withdrawing; nation building, surging to enhance security, then back to withdrawal, or some combination of these simultaneously. More importantly, governments may deliberately embrace vague end states because they are useful:

- Vague end states may actually be a **prerequisite for action** in circumstances where building a consensus is required with multi-national partners, or where domestic constituencies have to be reconciled to a peace or stability operation.
- Vague end states can be **operationally useful**: they allow flexibility to adapt to changing circumstances; they avoid creating benchmarks that might signal defeat; they make it less likely that opponents can manipulate aspects of the objectives, such as fixed exit dates. In Libya in 2011, for example, international intervention was justified under UNSC 1973 with a mandate to protect Libyan civilians 'by all necessary means'. The coalition was able to move from a use of airpower and remote attack against the advancing forces of Muammar al-Gaddafi to the use of such means in support of advancing rebels because of the ways in which the mandate could be reinterpreted without necessitating a second UNSC Resolution that might have been blocked by China or Russia.[13]

The effectiveness of peace and stability operations

The issues thus far have focused on the challenges of making peace and stability operations work. A fourth category of criticisms, those of **effectiveness**, attack the premise that nation building is a practicable activity in the first place. The common theme between these criticisms is that intervening forces lack the practical, cognitive or political wherewithal to effect the transformation of other societies, and that perhaps such operations need to be much less ambitious.

Historical evidence

One set of criticisms focuses on the **empirical evidence**. Here, the argument is that complex peace and stability operations are unlikely to work because the historical evidence suggests that operations that involve the military occupation of another country **fail more often than they succeed**. Since the post-Napoleonic Wars occupation of

France in 1815 through to the present, there have been 24 military occupations of other countries. Including the occupation of France in the wake of the Napoleonic Wars, only seven have been strong successes and the remaining six are clustered around the post-Second World War occupation and rehabilitation of former Axis powers (Germany, Japan, Italy, east and west Austria, and North Korea).[14] The root of the problem is that large-scale interventions in other countries are unlikely to succeed because they are large-scale interventions in someone else's country. Inevitably, local sentiment is likely to be suspicious of the methods and motives of intervening forces, almost irrespective of their conduct. As one commentator notes of the experience in Iraq, 'Most of the population disliked the U.S.-led coalition simply because it was the U.S-led coalition. Neither development projects not more flexible infantry tactics were going to turn that around'.[15]

The strongest **contributors to success** may be three factors external to the conduct of the intervention itself:

- A **perception of the need** for external intervention (for example, because of extensive war damage).
- A **credible commitment** by the intervening force to leave.
- A **common external enemy** that creates a sense of shared interests between the occupiers and key constituencies within the occupied country.[16]

Indeed, many factors and agencies that might contribute to long-term stability in a country are outside the gift of a peace or stability operation. Ultimately, these problems have led some to argue that effective nation building is beyond the capabilities of military intervention. Reflecting this view, Colin Gray argues that the United States needs to:

> ... forswear impossible tasks. America cannot build, or re-build, civil societies, let alone build 'nations', by act of will and determination, or by technical or even cultural skills in the conduct of stability operations and reconstruction.[17]

Ethno-centrism

A second perspective critical of the effectiveness of peace and stability operations focuses on Western culture and our **assumptions** regarding failed or failing states. Commentators who take this view argue that Western approaches to complex nation-building tasks are built on **ethno-centric analyses** of the nature of the failing state problem. Interventions have become a form of 'liberal imperialism', in which the problem is defined as 'bad governance' and solutions are imposed on local populations irrespective of their wishes. For example, we assumed that countries such as Afghanistan were 'ungoverned spaces'. In fact, Afghanistan was not an 'ungoverned space'; it was governed at a local level by traditional structures that provided justice and security for some, but this did not conform to Western assumptions of effective governance.

Taking this view, complex peace and stability operations are difficult not because of the 'planning school' focus on such features as strategy and metrics, or corruption, or lack of democracy, but rather because of Western culture and the associated bounded thinking about the nature of the peace and stability problem. Thus, critics note that Western approaches to complex peace and stability operations embodied in documents such as FM-3-07 are **guided by liberal assumptions** that conflict in failed or failing states is best addressed through promoting **rapid democratisation and market economics**.[18] In fact, this is a contested assertion and there are many who argue that liberalisation may promote internal conflict in transitional environments because the competition between competing groups cannot be mediated by well-established norms or democratic structures. For example, the World Bank and International Monetary Fund are often key to providing loans to enable development. However, critics argue that both are often too wedded to liberal orthodoxy and that initiatives such as structural adjustment programmes are often damaging. For example, by removing subsidies for the poor, market economics can promote instability, but liberal assumptions are often unconscious and so cannot be easily remedied. In consequence, there may be problems to which Western intervention can **provide no workable solution** and in these situations success may be almost impossible to achieve.

Lack of consensus

A third, and related, difficulty is the **lack of a consensus** surrounding a workable '**theory of victory**' for complex peace and stability operations: a set of well-understood, empirically proven strategies linking end states to the available means. FM 3-07 claims to be ' ... a roadmap from conflict to peace',[19] a statement that implies that it embodies an objectively validated process by which weak states can be turned into functioning liberal states. However, the concepts associated with democratisation are inherently contested. There is no general consensus, for example, on whether successful democratisation is best facilitated by focusing first on building strong host nation institutions and then moving towards political liberalisation ('institutionalisation before liberalisation'), or whether one should do the reverse (liberalisation before institutionalisation').

Neither is there a consensus over **how to measure progress** towards goals such as democratisation, normalisation, or social transformation. Under conditions in which political and military objectives coincide closely, as they may do in more total forms of war, military tangibles such as the control of ground or relative rates of attrition may be close enough approximations of progress. These tangibles are clearly much less relevant in peace and stability operations, but what should replace them? Metrics that are too general may be impossible to measure: the political scientist Robert Jervis notes trenchantly, for example, that as a measure of success, a criterion such as 'security' reads 'more like psychotherapy than international politics'.[20] More detailed criteria are still subject to debate. The US Government Accountability Office established five key metrics for Afghanistan:

- Creating a national Afghan army.
- Reconstituting the national police.
- Establishing a working judiciary.
- Combating illegal narcotics.
- Demobilising the Afghan militias.[21]

However, others have argued for different measures of success, including:

- The number of government officials who serve in a district and live in a district (as an indicator of the strength of local government).

- Transportation costs for goods (since this is a reasonable indicator of danger).
- Market activity (as an indicator of producer and consumer confidence).
- Reports of Improvised Explosive Devices (IEDs).
- School attendance.[22]

For one British provincial governor in Iraq, success was defined by his superior as a situation in which his province was: 'reasonably quiet and has not descended into anarchy and you are able to serve me some decent ice cream.'[23] For General Creighton Abrams in Vietnam the key metric for success was the answer to the question: 'who controls an area at night?'[24] Thus, even if military doctrines for peace and stability operations were not subject to debate, they are attached to broader political strategies that are.

The role of the military

A final set of controversies surround **the legitimacy of the military** as a tool for resolving conflict. Military doctrine on peace and stability operations is littered with the use of terms such as 'conflict transformation', 'justice' and 'reconciliation', appropriated from the broader literature on conflict resolution. This reflects the military's assumption that the military is a legitimate, even essential, tool for promoting peace and stability. For many, however, 'military conflict resolution' is an oxymoron, and real efforts at conflict resolution should seek to minimise the role of the military, and not develop doctrinal means to legitimise and expand it.

The basis for this perspective lies in the observation highlighted in Chapter 6 that the ethos and organisational predispositions of the military are towards war fighting. This is reflected in the way that peace and stability operations doctrine often continues to reflect the concepts and vocabulary of conventional warfare. Thus, the United States places stability operations within 'Full Spectrum Operations', defined as 'continuous, simultaneous combinations of offensive, defensive and stability tasks'. These are operations designed to seize, retain and exploit the initiative, 'accepting prudent risk to create opportunities for decisive results'.[25] In its reference to goals such as seizing the initiative, dictating terms and ensuring decisive results, FM 3-07 ties stability operations to a conventional warfare mindset and a tendency towards prioritising 'kinetic' ('violent') over 'non-kinetic' ('non-violent') means. This has a number of consequences. One is that critics argue that militaries **do not seem to have the requisite skills** to implement a development or aid programme, but there are two other interesting arguments.

A proclivity towards military escalation

First, critics argue that military organisations have a **tendency towards escalation**, because problems are seen in military terms. Thus, despite widespread recognition that metrics such as body counts are not wholly useful indicators of success in stability operations, military units often still measure the success or failure of their tours of operation in terms of relative casualties and participation in major operations.[26] In addition, the military **over-estimate the utility of force** in conflict resolution. For example, FM 3-07 argues that 'an inherent complementary relationship exists between lethal and non-lethal actions';[27] the US National Military Strategy asserts that 'winning decisively will require synchronising and integrating major combat operations, stability operations and significant post conflict interagency operations'.[28] Critics argue that this view underestimates the unintended second- and third-order consequences of military action. For example, air strikes in Afghanistan have proved to be an invaluable military tool, yet the political consequences of such attacks have been profoundly negative. Similarly, kinetic military activity designed to improve security in regions also erodes the neutral 'humanitarian space' required for NGOs to operate.

De-legitimising peace

Second, some argue that armies **misconstrue the nature of justice and legitimacy** in the long-term resolution of conflict. Violent means contaminate the pursuit of even just ends; the sort of moral compromises required to win wars make it difficult for the local population to see post-war political settlements as legitimate. In this respect, there is a paradox at the heart of military conflict resolution efforts: in their attempts to connect more intimately war fighting and peace building, modern military doctrine has achieved greater military efficiency at the expense of genuine political legitimacy. In consequence, it has made it more difficult, not less so, to realise the principle of *jus post bellum* (justice after war).

Conclusions

Land forces have a great deal of theoretical flexibility even in operations other than high-intensity conventional warfare, but this should not blind us to the fact that peace and stability operations have remained challenging. The operational environments in which they take place are complex and difficult; the key principles can be difficult to make work in practice; there are often variations in how different armies approach peace and stability tasks; and there are those who argue that the more ambitious nation-building tasks are simply beyond our abilities to achieve.

Indeed, definitional issues notwithstanding, the future of stability operations along the lines of Iraq and Afghanistan is uncertain. FM 3-07 asserts that 'America's future abroad is unlikely to resemble Afghanistan or Iraq, where we grapple with the burden of nation-building under fire'.[29] Yet FM 3-07 is a doctrine very much shaped by those experiences. Stability operations are a dependent variable – as Colin Gray notes: 'Stability operations, the demand for them and the provision of new capabilities to perform them well, are the downstream product of larger decisions on foreign policy and strategy.'[30] Iraq and Afghanistan have been so costly and so controversial, that it seems unlikely that there will be much appetite for similar enterprises in the short-to-medium term, as witnessed by the reluctance to intervene in the ongoing conflict in Syria.[31] In financially straitened times, this point raises real questions about the extent to which land forces are genuinely likely to invest in the long term to maintain the expertise and capabilities necessary for such ambitious operations. Thus, the question remains as to whether the current focus on stabilisation and reconstruction reflects a permanent shift in military focus, or simply a temporary one.[32] This specific issue leads us on to the more general subject of the problems of predicting the future of the land warfare environment. To consider this problem, we turn now to the third and last part of the book: future land warfare.

Notes

1 For an overview of the conflict resolution problems caused by intra-state environments see, for example, Roy Licklider (ed.), *Stopping the Killing: How Civil Wars End* (New York: New York University Press, 1993).
2 Mary L. Olson, 'From Interests to Identities: Towards a New Emphasis in Interactive Conflict Resolution', *Journal of Peace Research* Vol. 38, No. 3 (May 2001), 297.
3 *United Nations Peacekeeping Operations: Principles and Guidelines* (United Nations: Department of Peacekeeping Operations, 2008), 21.
4 Daryl Morini, 'A Diplomatic Surge in Afghanistan, 2011–14', *Strategic Studies Quarterly* (Winter 2010), 71–80, 89–90.
5 Paul F. Diehl and Daniel Druckman, *Evaluating Peace Operations* (Boulder, CO: Lynne Rienner, 2010), 157.
6 Stuart Griffin, 'Iraq, Afghanistan and the Future of British Military Doctrine: From Counterinsurgency to Stabilisation', *International Affairs* Vol. 87, No. 2 (2011), 323–27.
7 Rory Stewart and Gerald Knaus, *Can Intervention Work?* (New York: Norton, 2011), 52.
8 Stephanie Blair and Ann Fitzgerald, *Stabilisation and Stability Operations: A Literature Review* (30 June 2009), 8–9, dspace.lib.cranfield.ac.uk/bitstream/1826/4247/1/Stabilisation%20article_statebuilding_intervention_FitzGerald.pdf.
9 Ibid., 3.
10 Daniel L. Byman, 'Uncertain Partners: NGOs and the Military', *Survival* Vol. 43, No. 2 (Summer 2001), 105.
11 Rudra Chaudri and Theo Farrell, 'Campaign Disconnect: Operational Progress and Strategic Obstacles in Afghanistan, 2009–11', *International Affairs* Vol. 87, No. 2 (March 2011), 286–87.
12 United Nations Security Council Resolution 794, UN Doc. S/RES/794 (1992), www1.umn.edu/humanrts/peace/docs/scres794.html.
13 'Nato Rejects Russian Claims of Libya Mission Creep', *The Guardian*, 15 April 2011, www.theguardian.com/world/2011/apr/15/nato-libya-rasmussen-medvedev-criticism.
14 David Edelstein, 'Occupational Hazards: Why Military Occupations Succeed or Fail', *International Security* Vol. 29, No. 1 (Fall 2004), 49–91.
15 Stewart and Knaus, Can Occupation Work?, xxii.
16 See Edelstein, 'Occupational Hazards'.
17 Gray, 'Stability Operations in Strategic Perspective', 9.
18 Roland Paris, *At War's End: Building Peace After Civil Conflict* (Cambridge: Cambridge University Press, 2004), 19.
19 Field Manual (FM) 3-07, *Stability Operations* (Headquarters of the Army, 2008), 'Foreword'.
20 Robert Jervis, 'An Interim Assessment of September 11: What has Changed and What has Not', *Political Science Quarterly* Vol. 117, No. 1 (2002), 46.
21 Anthony H. Cordesman, *The Missing Metrics of 'Progress' in Afghanistan (and Pakistan)* (Center for Strategic and International Studies, 14 November 2007), 6–10, aix1.uottawa.ca/~rparis/Cordesman_paper.pdf (accessed 26/07/2011).
22 Antony Adoff, 'Measuring Success in Afghanistan', 18 March 2010, news.change.org/stories/measuring-success-in-afghanistan (accessed 26/07/2011).
23 Rory Stewart, *Occupational Hazards* (Basingstoke: Picador, 2006), 73.
24 Cordesman, *The Missing Metrics*, 24.
25 FM 3-07, 2-1.
26 Anthony King, 'Understanding the Helmand Campaign: British Military Operations in Afghanistan', *International Affairs* Vol. 86, No. 2 (2010), 325.
27 FM 3-07, 2-3.
28 FM 3-07, 1-11.
29 FM 3-07, 'Foreword'.
30 Gray, 'Stability Operations in Strategic Perspective', 4.
31 Ibid., 8.
32 William Flavin, 'U.S. Doctrine for Peace Operations', *International Peacekeeping* Vol. 15, No. 1 (February 2008), 49.

Suggested reading

Adekeye Adebajo, *UN Peacekeeping in Africa: From the Suez Crisis to the Sudan Conflicts* (Boulder, CO: Lynne Rienner, 2011). Provides both a set of Africa-related case studies of peacekeeping and an insightful analysis of the challenges that face peace operations in complex contexts.

Ian Johnstone, 'Managing Consent in Contemporary Peacekeeping Operations', *International Peacekeeping*, Vol. 18, No. 2 (2011), 168–82. Johnstone's examination of consent illustrates some of the key difficulties in realising the principles of peace operations.

Roland Paris, *At War's End: Building Peace After Civil Conflict* (Cambridge: Cambridge University Press, 2004). Paris highlights the lack of consensus on the most appropriate strategies for building peace.

Darya Pushkina, 'A Recipe for Success? Ingredients of a Successful Peacekeeping Mission', *International Peacekeeping* Vol. 13, No. 2 (2006), 133–49. An assessment of the foundations required for effective peace operations.

Rory Stewart and Gerald Knaus, *Can Intervention Work?* (New York: Norton, 2011). This provides two competing perspectives, each of which highlights important pros and cons associated with intervention operations.

Shahrbanou Tadjbakhsh (ed.), *Rethinking the Liberal Peace* (London: Routledge, 2011). The chapters in this book provide a comprehensive examination of the contested nature of the concept of liberal peace.

Part III

Future land warfare

Having examined the key concepts associated with modern land warfare, its development and the challenges of less conventional contexts, we turn now to the final part of this book and examine the debates on the future of land warfare. Generating land power carries with it inherent risks: since armies cannot be created overnight, choices need to be made now about what kind of land forces one needs for the future: equipment must be procured, personnel recruited and trained, doctrine devised, force structures assembled. The lead-in times for this can be long. Chapter 1 identified that armies are not infinitely flexible – no army can fight equally well in every context. So, the relevance of an army depends to an important extent on predicting correctly now the sorts of armed conflicts that one will fight in the future.

The final two chapters of this book explore this challenge of prediction in two ways. The next chapter, Chapter 9, looks at the most influential views on the future of land warfare, highlighting the lack of consensus on what future land warfare will look like. Chapter 10, the final chapter of this book, looks at the US army as a case study: what assumptions has the US army made about the demands of future land warfare; how, and with what success has it changed to meet these challenges; and what impact has this had on the land forces of other countries?

9 The future of warfare on land

Key points

- There is no consensus on what the future of land warfare will look like.
- For some, a Revolution in Military Affairs, founded upon precision munitions, new sensors and developed forms of networking, will result in a radical transformation in the armies of the future.
- For others, land warfare is, in a sense, going backwards: the future of land warfare will be irregular, local and decidedly low-technology.
- Still others see a future in which conventional and unconventional approaches to land warfare combine into a hybrid form of armed conflict.
- Since armies seem rarely to predict correctly the future form that land war will take, there are some commentators who argue that it is less useful to try and predict the future and more useful to focus on increasing an army's ability to adapt quickly to whatever form of land warfare emerges.

Theorists can debate endlessly what future warfare will look like, but governments, and their land forces, don't have that luxury. Major modern equipment systems, for example, can take decades to bring into service and so choices about what kind of land forces to create have to be made early; these choices will depend amongst other things on the sorts of conflicts that one thinks an army will have to fight. So what will the future bring for land warfare and what kinds of changes to land forces will be required to enable them to fight successfully?

Reflecting so much of what we have already encountered in our exploration of land warfare, there is no consensus amongst commentators on what the future holds. The Gulf War of 1990–91, and the wars surrounding the breakdown of Yugoslavia from 1992–96, in particular, proved to be important catalysts for new thinking on the future development of land warfare. However, the conclusions reached have varied, and fall into three general schools of thought.

The first embodies the notion that by the end of the twentieth century, technological developments were conducive to a revolutionary leap forwards in the conduct of warfare. This belief, couched in terms of the language of a **Revolution in Military Affairs (RMA)**, argued that warfare was going, or would go, through a paradigm shift as profound as the introduction of gunpowder. This technophile, military revolution perspective was countered by a second school, often termed the **'New Wars' school**, which argued that in key respects warfare was going backwards: in the future, high-intensity conventional warfare between states would become increasingly anachronistic. Instead, future warfare would be marked by bloody, local conflicts, often civil wars, in which the traditional military verities would be largely irrelevant. A third perspective, the **'Hybrid Warfare' school**, saw a future marked by blended regular and irregular warfare. Each of these perspectives had important ramifications for the relative importance and/or the conduct of land warfare in the future.

This chapter examines each of these schools of thought in turn, identifying the key arguments associated with each. The different perspectives span a range of future visions of land forces: dispersed, high-technology, conventional warfare involving small, heavily digitised forces; low-technology warfare featuring irregular, non-state actors in which conventional forces may or may not be involved; and blended wars in which future armies will face adversaries mixing both irregular and regular techniques of warfare. This chapter finishes by examining an alternative to the essentially predictive approaches of these three schools: a focus on **organisational adaptability**. In the end, trying to prepare for a single template of future warfare may be less useful than building into an army the capability to respond quickly to the actual conditions that emerge.

The Revolution in Military Affairs

Chapter 2 discussed Revolutions in Military Affairs as a general idea. We start this chapter by examining the debates surrounding *the* **Revolution in Military Affairs**: a label applied to a particular viewpoint on future warfare which emerged as a potent force in the wake of the Gulf War of 1990–91.

The Gulf War and the RMA

The outcome of the 1990–91 Gulf War stunned most analysts. Whilst the Iraqis were outnumbered (in the Kuwaiti Theatre of Operations there were some 340,000 Iraqis opposed by 500,000 Coalition troops), they had a range of advantages that made them a tough proposition in military terms:

- Coalition **force ratios** were well short of the 3:1 advantage normally desired for offensive operations.
- The Iraqis had had plenty of time to establish **prepared positions** which included multiple lines of entrenchments, barbed wire and minefields.
- The Iraqi army was also an **experienced army**, having been at war with Iran from 1980–88.

In consequence, many expected that the war would be a difficult and costly enterprise, with some estimating that the ground war might require two weeks and result in 30,000 US casualties.[1] Yet the war when it began on 17 January 1990 was an overwhelming allied military victory. The war commenced with an extended air campaign; the ground war when it began on the 24 February comprised a fixing attack to the front and a wide outflanking move, and lasted only 100 hours.[2] The Coalition destroyed more than 30 Iraqi divisions and took almost 90,000 prisoners. Yet Coalition battle deaths were a paltry 197: statistically, US males in the Iraq combat zone were safer than if they had been in the United States.[3] One of the key questions posed in the wake of the war was: why had the outcome been so one-sided?

The RMA

For one influential group of commentators, the answer lay in a **technological critical mass** that had been emerging since the 1970s and which built on the developments that had informed concepts such as AirLand Battle (see Chapter 4). Based on wider innovations in fields such as information technology, computing, cybernetics, microelectronics, miniaturisation, sensors and communications,[4] theorists in the 1980s began to speculate on the development of new military technologies in the spheres of **precision attack**, **sensors** and **networking** which, taken together, promised to produce revolutionary increases in military effectiveness:

- **New sensor technology** such as full spectrum sensors (optical, thermal, infra-red, laser and radar), and aerial and space-based platforms (reflected in developments such as Global Positioning System – GPS) promised to reduce dramatically, and some claimed even to eliminate, the fog of war.
- **New methods of attack**, such as stealth aircraft and precision-guided munitions (PGMs) would allow relatively small numbers of attacks to achieve unprecedented effects.
- **Intense networking** would magnify the effects of these technologies to a revolutionary level: connecting sensor systems and 'shooter' systems into a single system through deep integration would allow information to be passed within militaries at unprecedented rates. This would create what was termed 'Near-perfect Mission Assignment': an unprecedented rate and accuracy of attack. Admiral William Owens, Vice-Chairman of the Joint Chiefs of Staff, was one of the early proponents of this view, and he coined the term 'system of systems' to describe this emerging capability.[5]

Cumulatively, the technology presaged by the Gulf War promised the creation of **system of systems** structures, termed variously 'guided-munitions battle networks' or 'reconnaissance-strike complexes', with three basic components: sensors to find and track targets; platforms, weapons and munitions to conduct precision attacks, often at extended ranges; and a Command, Control, Communications and Intelligence (C3I) architecture capable of intense linking of the two.[6]

Thus, from an RMA perspective, the unexpectedly one-sided outcome of the Gulf War could be attributed to the fact that whilst the Iraqi army was a mass industrial-age military machine, the United States had fought them leveraging the fruits of the RMA: post-industrial, information-based societies now had the means to conduct post-industrial, information-based warfare based on networks. A new threshold in warfare appeared to have been crossed. Thus, for some, the Gulf War 'confirmed a major transformation in the nature of warfare'.[7] The Gulf War propelled the RMA concept into the frontline of debates on future warfare, but these debates were grouped around two rather different propositions which would each have different implications for land warfare. One view argued that the RMA demonstrated by the Gulf War was, centrally, an airpower revolution; the second view argued that the new RMA was a broader phenomenon.

The air-centric RMA

For one group of theorists, the technological advances associated with the 'system of systems' approach had **disproportionately advantaged airpower** at the expense of land power. The new RMA was therefore essentially an airpower revolution. The fruits of this thinking were expressed in a number of concepts, including nodal targeting, 'Shock and Awe' and 'effects-based operations', linked by the idea of viewing the enemy as a system that could be 'crashed' through precise attacks (essentially by airpower).

Nodal targeting

Colonel John A. Warden articulated the new RMA in terms of his '**Five Circles**' – a way of conceptualising targeting priorities. Warden argued that an enemy could be conceived as consisting of five 'circles' from a targeting perspective: the outer ring was its fielded forces; this was followed, moving inwards, by its population; then infrastructure; economic and industrial facilities; and finally, its leadership. Warden argued that the nearer the centre one targeted, the more decisive the outcomes would be: it was the centre-most targets that were the real enemy centres of gravity. The RMA enabled airpower to conduct what Warden termed '**inside-out war**'. In other words there was no longer any need to win wars by starting at the outside (the enemy's fielded forces) and moving in: one could 'leap the trenches' and instead put direct pressure on the enemy's **critical nodes** straight away. Moreover, new technology would allow one to do this in ways that would avoid the crude collateral damage of Second World War-style strategic bombing. The RMA promised, then, an ability to achieve decisive effects without requiring mass warfare.[8]

Shock and Awe

First articulated in 1997, Shock and Awe was in essence a kind of RMA 'hyperwar' built around the notion of '**rapid dominance**', inducing in the enemy at the outset of a war **systemic shock** of a scale that would paralyse them in the short term, and then subsequent attacks which would translate this shock into longer-term 'awe', embodying feelings of helplessness and impotence. The RMA origins of Shock and Awe were evident in the concept of rapid dominance which, advocates concluded, would require:

> a sophisticated, interconnected, and interoperable grid of netted intelligence, surveillance, reconnaissance, communications systems, data analysis, and real-time deliverable actionable information to the shooter. This network must provide total situational awareness and supporting nodal analysis that enables US forces to act inside the adversary's decision loop in a manner that on the high end produces Shock and Awe among the threat parties.[9]

Effects-based operations

Championed by US Air Force Colonel David Deptula, effects-based operations (EBO) similarly comprised a focus not on destroying the enemy, but on causing their systemic collapse by precise attacks. EBO, however, was focused very much on the need to consider **political effects**: it was a **systems-based approach to influencing the enemy** in which a focus on tactical destruction was to be replaced with a targeting concept that 'considers the full range of direct, indirect, and cascading effects, which may – with different degrees of probability – be achieved by the application of military, diplomatic, psychological, and economic instruments'.[10]

The Gulf War

For these analysts, the Gulf War provided unambiguous evidence of a **new air-centric RMA**, featuring:

- The extensive use of PGMs and stealth technology.
- The massive US advantage over the Iraqis in information and situational awareness.
- The decisive impact of airpower.

Whilst key aspects of the Coalition air plan focused on traditional roles such as securing control of the air, preparing the battlefield and providing close air support once the ground war started, the air plan contained, in particular, two important new elements:

- **Focus:** first, the air campaign would begin with attacks on the centre of Warden's Five Circles – the Iraqi leadership and Iraq's infrastructure.
- **Ambition:** second, it embodied an ambitious goal of 50% damage against the Iraqi ground forces – an unprecedented objective for an air campaign.

Both of these activities seemed in practice to be extremely effective. Precision attacks were able quickly to overwhelm Iraq's air defences; attacks were then launched on command, control and communication nodes, the electrical grid, bridges, key leadership targets, all of which were designed to render the Iraqi army ineffective by destroying Saddam Hussein's capacity both to understand what was going on and to exert meaningful control over his forces. Air attack against Iraq's fielded forces also seemed highly successful. Attacks against troops inflicted heavy casualties (according to Biddle, 48% of Iraq's tanks, 30% of its personnel carriers and 60% of its artillery were lost[11]); attacks against logistics targets undermined the Iraqi army's capacity to function; and the combination of both had a profoundly negative effective on Iraqi morale.

Warden attributed these successes to 'high technology, unprecedented accuracy, operational and strategic surprise through stealth, and the ability to bring all the enemy's key operational and strategic nodes under near simultaneous attack'.[12] The consequence, one author concluded, was that 'the world ha[d] just witnessed a new kind of warfare'.[13] Airpower was now the decisive element in warfare. Airpower, according to the Gulf War Air Power Survey, 'crossed some operational thresholds that … did suggest a transformation in war'.[14]

A developing RMA

Belief in the air-centric view drew sustenance from developments after the Gulf War. RMA-applicable technology continued to develop. In terms of **munitions**, for example, Joint Direct Attack Munitions (JDAM) were operational by 1997: these were essentially kits combining GPS and other equipment that could be attached to ordinary bombs to give them a precision-attack capability. At 15 miles, JDAM had a short range relative to cruise missiles but at US$16,000 each, they were also much cheaper. Satellite **guidance technology** also came on line: this allowed munitions to be launched beyond visual range and to avoid problems caused by bad weather. **Communications technologies** continued to develop, with better sensor technology and increasing digitisation.

Both sensor and attack capabilities were also enhanced by ever increasing deployment of **Unmanned Aerial Vehicles** (UAVs) capable of high-altitude surveillance and, latterly, attack.

The air-centric RMA seemed validated through further experience in conflicts in Kosovo in 1999 and in Afghanistan in 2001.

Kosovo

During the Kosovo crisis (see Box 9.1), airpower constituted the **North Atlantic Treaty Organization's (NATO) primary military instrument**, given that the President Clinton Administration had ruled out ground operations. Operation Allied Force, a 78-day air campaign, involved more than 37,000 sorties. The campaign featured the use of new munitions, platforms and intelligence technologies such as JDAM, Joint Surveillance Target Attack Radar System (JSTARS) and extensive use of UAVs. Whilst the air campaign encompassed traditional attacks against Serb air defences and fielded forces, latterly the campaign switched to a **focused assault** on targets such as Serbian power generation facilities, infrastructure and government media outlets. Moreover, despite the scale of the campaign, civilian casualties were relatively low at 528. **Low civilian casualties** and **low NATO losses** were instrumental in helping to manage the political difficulties surrounding the operation, which was a contentious multilateral enterprise. In the end, NATO achieved its objectives, an outcome which reinforced for some the decisive nature of airpower under the new conditions of warfare: the historian John Keegan, for example, argued that the Kosovo campaign was 'won by air power alone'.[15]

Box 9.1 The Kosovo conflict

Regarded by Serbs as an integral part of the Serbian state, Kosovo was populated predominantly by ethnic Albanians, who constituted around 90% of the population. Moves by the Serbian President Slobodan Milosevic to curb Kosovo's autonomy were met by a declaration of independence in 1990 by Albanian Kosovars. By the late 1990s a spiralling conflict between Serbian security forces and the Kosovo Liberation Army (KLA) resulted in the displacement of over 100,000 Albanians and fears in the international community of an emerging tide of ethnic cleansing. Negotiations with the Serbian government at Rambouillet in France broke down on 18 March 1999, and NATO embarked on 24 March on a 78-day air campaign. The campaign was controversial, not least because the United Nations Security Council (UNSC) Resolution 1199 of September 1998 had called for a negotiated settlement, but there was no specific resolution authorising war, leading some to question the legality of NATO action. The air campaign, anticipated as being two to four days long, was much more protracted, with critics arguing that the initial focus on attacking security targets in Kosovo should have been replaced much earlier with wider attacks on Serbia itself, especially on targets relating to the Serbian leadership. In essence, NATO decision makers seem to have assumed that Milosevic would capitulate as soon as NATO demonstrated its resolve to use force. On 27 May 1999, with air attacks focusing increasingly on Belgrade, and growing debate in NATO on the possibility of committing ground troops, Milosevic's key ally, Russia, recommended that he accept NATO's terms, which he did on 3 June 1999. An international peacekeeping force (KFOR) was deployed on 12 June 1999.

Afghanistan

In Afghanistan in 2001 (see Box 9.2), a traditional attack on the Taliban using conventional forces was judged impractical, given the constraints of geography and time. There were also worries that an extended campaign might carry political risks, especially if the campaign began to look at all like it might become bogged down. For these reasons, when Operation Enduring Freedom began on 7 October 2001, it had a variety of **innovative features**. Ground forces were constituted in the early stages largely by the **indigenous forces** provided by the Afghan Northern Alliance. The key US contribution came through a **mix of airpower and Special Operations Forces**. Airpower was initially used in traditional roles, attacking the limited Taliban air defence assets and fixed installations associated with command and control. Once Special Operations Forces were deployed, they could act in support of the Northern Alliance, using **laser designation technology** to expose Taliban troops to the full effects of US airpower. This initial campaign was very successful: with a US ground component of only 316 Special Forces and around 1,100 from the Central Intelligence Agency (CIA), the United States crushed the Taliban, taking Kabul on 7 December. Many commentators were convinced that the meshing of limited ground forces with lavish precision air attack constituted a new development in warfare. Indeed, President George W. Bush argued that the Afghanistan campaign in 2001 represented a 'revolution in our military … [that] promises to change the face of battle'.[16]

The trajectory of airpower effectiveness established by conflicts since 1991 seemed to demonstrate that **armies were an increasingly auxiliary element** in modern land warfare. Indeed, the lesson of the so-called '**Afghan model**' of warfare was that land forces needed to be configured in a different way: small, elite forces to support the application of firepower from the air, and perhaps larger forces designed primarily for constabulary operations, whose task it would be to secure ground. Whilst the experiences since 1991 had reinforced some of the developing verities of modern system land warfare, such as the importance of jointery, the kind of army required to service an airpower-centric RMA military of the twenty-first century was very different from that which had developed according to the modern system of warfare in the twentieth.

Box 9.2 Operation Enduring Freedom

Operation Enduring Freedom had its origins in the Al Qaeda attacks on the Twin Towers in New York on 11 September 2001 and the Taliban's subsequent refusal to hand Osama bin Laden over to US authorities. Preliminary action for Operation Enduring Freedom began in late September 2001 with the infiltration into Afghanistan of personnel from the CIA, tasked with establishing links with likely Afghan allies, especially the Northern Alliance, a coalition of forces drawn especially from Tajik, Uzbek and Hazara ethnic groups that opposed the Taliban. Air strikes began on 7 October 2001 and Special Forces began deploying into Afghanistan on 19 October.

Early air attacks were indecisive: the Taliban air defence capability was ramshackle and ineffective anyway, and, once this had been dispatched, attempts to provide air support for Northern Alliance forces proved problematic because of difficulties in command, control and communications. Once Special Operations Forces arrived, they were able to co-operate with Northern Alliance forces, allowing precision air strikes to be conducted in support of Northern Alliance ground attacks. In combination, the Special Operations Forces-Northern Alliance-airpower triumvirate crushed the Taliban: Mazar-i-Sharif was taken from the Taliban on 9 November and Kabul fell on 7 December.

The network-centric RMA

However, there was an alternative vision of RMA future warfare. This alternative perspective argued that land warfare would continue to be a crucial element in warfare generally, but that the RMA would invalidate many of the established principles through which operations generally would be conducted. From this viewpoint, the crucial issue with regard to the Gulf War was that the outcome reflected the **enormous predominance in information** possessed by the Coalition. The Gulf War, then, was not an airpower victory, but instead the **first information war**. What mattered in the Gulf War was the Coalition's real-time picture of the battlefield and its consequent ability to orchestrate the campaign through previously unknown levels of co-operation and jointery.[17] Similarly, Afghanistan in 2001 was portrayed from this perspective not so much as an airpower victory as an example of the potential of 'netwar'.[18] Perhaps the most dominant of ideas in this vein was that of Network-Centric Warfare (NCW), which emerged at the end of the 1990s.

Network-Centric Warfare

Network-Centric Warfare was essentially a development of the concept of the 'system of systems'. For its major advocates, NCW was the new military paradigm shift: it was a reflection of the broader and more fundamental transition from industrial to post-industrial, information-based societies. At the heart of NCW was the assertion that **information was increasingly becoming the central element in warfare**. As one commentator argued, 'information technology has become so important in defining military power that it overwhelms almost everything else'.[19] In such circumstances military victory was now founded upon **information dominance**:

> one's ability to collect, process, and disseminate an uninterrupted flow of information while exploiting or denying an enemy's ability to do the same.[20]

Information dominance would result in a decisive advantage over the enemy in **situational awareness**, and this would then cascade downwards into advantages in areas such as prioritisation of activity, speed and accuracy of decision making, precision in attack and so forth.

Achieving this dominance would rely on heavy investment in communications, sensors, computers and other information technologies: crucially, however, information dominance would require **highly integrated networking** so that information could be shared rapidly between the component parts of modern militaries. The 'system of systems' was, in NCW, a 'network of networks': linked grids of sensors, 'shooters' and information. The 'shooter grid', for example, would consist of geographically dispersed air, sea and ground-based assets. Intense networking would allow the benefits of information superiority to be enjoyed by the whole of the military system, which would confer a range of related benefits:

- '**Self-synchronisation**': with a near-perfect picture of the battlefield, friendly forces could adopt heavily decentralised approaches to achieving their objectives. Command would, in essence, be 'bottom-up' rather than 'top-down'.
- **Simultaneity**: information dominance would allow a move away from staged sequential battles towards near-simultaneous attacks by self-synchronising forces.
- **De-massification**: information dominance would allow mass effects to be delivered against critical targets by smaller, flexible, dispersed forces.
- **Tempo**: information dominance would enable a dramatic increase in the tempo of operations by removing the stultifying effects of an intermediate level of command; it would create much greater agility since friendly forces would be able more quickly to take advantage of local opportunities for deception and surprise.
- **Enemy 'lock-out'**: taken together, the preceding advantages would constrain decisively the enemy's options, locking them out of the Observe-Orient-Decide-Act (OODA) loop competition.[21]

If future land warfare essentially was to be net-centric, then the implications for armies were profound. Organisationally, the requirement would be for smaller, modular, combined arms manoeuvre teams; digitisation; a much greater focus on jointery. Conceptually, conventional warfare principles of mass, sequential operations and concentration would be replaced by a focus on agility, simultaneity and dispersal. A net-centric future is one that would strike fundamentally at the need for heavy armoured and mechanised forces.

The RMA debate

So, which of the two conceptions of the RMA outlined above is correct? Unfortunately for those tasked with generating future land power, no one can agree.

Challenges to an air-centric RMA

Critics of airpower-centric RMA visions of the future argue that this concept suffers from a range of compelling problems. History certainly demonstrates that air superiority is a crucial advantage. As the Gulf War and Afghanistan demonstrated, much of the increase in **firepower** that has driven tactical and operational changes in land warfare originates from developments in airpower. Moreover, air superiority manifests advantages in other spheres: for example, in the Gulf War it was an enabler in **intelligence gathering** and it helped mask allied troop movements, in particular the forces concentrated for General Norman Schwartzkopf's outflanking movement. However, the revolutionary nature of the employment of airpower in the Gulf War and after has been contested.

Conceptual flaws

First, critics attacked the **conceptual basis** of air-centric 'systemic shock' approaches. In relation to Warden's Five Circles and the concept of Shock and Awe, critics asked the question: what **centres of gravity** were there that would create successfully this overwhelming paralysis of the enemy? In 1991, Saddam Hussein had been subjected to a lengthy and overwhelming air assault, yet he had still refused to withdraw from Kuwait, and his forces still had to be ejected through a ground war.[22] In relation to EBO, critics highlighted three related problems.

- First, EBO was **not new**: wasn't the essence of all good strategy to consider intended effects first and then to choose the most appropriate tool to achieve them? That this had proven consistently difficult to do said more about the inherent difficulties in executing strategy than it did about the lack of a specific military doctrine on the topic.
- Second, EBO, like Shock and Awe, could not provide any effective answers to the question of **what effects would be generated** by attacking a given target. Didn't 'effect' depend to a significant degree on highly contingent factors such as the enemy response?
- Third, the attempts to remedy the previous two problems resulted in a concept that was **increasingly unwieldy** and, in many cases, baffling. Israel's attempts to turn EBO into a workable doctrine, entitled 'Systemic Operational Design', was dismissed by some as 'a pretentious post-modern approach'.[23]

Reflecting these criticisms, in August 2008 the head of the US Marine Corps, General James Mattis, announced that EBO would no longer be used as a planning tool: his argument was that EBO created dangerous predilections towards overestimating the predictability of war, as well as creating unrealistic expectations regarding the information and certainty required for decision making. In making this decision, one author notes trenchantly that Mattis's decision was 'the reaction of a military practitioner … confronted with a high-flying, half-baked, quasi-intellectual war-fighting concept'.[24]

Misinterpreting the evidence

Critics also drew attention to what was viewed as a **misinterpretation of the evidence**. Some noted the rather **conventional nature** of much of the technology employed during supposedly revolutionary military campaigns. In the Gulf War and Kosovo, for example, precision-guided munitions constituted only a tenth and a quarter, respectively, of the total bombs used. In the Gulf War, much of the material and psychological damage to the Iraqi army was inflicted by **massive conventional bombing**. Indeed, what post-Cold War campaigns demonstrated, so critics argued, was that **orthodox means of mitigating the effectiveness of airpower** were still applicable, despite new technology. In Kosovo, the Serbs proved able to keep elements of their air defence in operation, forcing allied aircraft to attack from higher altitude, so causing consequent difficulties in target identification. The Serbs also made extensive use of very traditional techniques, such as exploiting difficult terrain, camouflage, dispersion, proximity to civilians, and decoys to reduce significantly the effects of allied airpower.[25] In all the key 'airpower' conflicts, **ground forces played a central role**: Coalition forces in the Gulf War; the Kosovo Liberation Army in Kosovo; the Northern Alliance in Afghanistan. Overall, mitigation of the effects of airpower was least effective where friendly ground forces were strongest.

The importance of jointery

Nor did the methods employed in these campaigns fulfil the notion advanced by such theorists as Warden of airpower as an **independently decisive tool**. In the Gulf War, the initial phase of the air attacks on the most central of Warden's Five Circles, Iraq's leadership and infrastructure, did not induce the Iraqis to withdraw, nor did it succeed in neutralising Iraq's leadership. This, in combination with an attack on a command and control bunker that killed 400 civilians and the fact that the air force was running out of strategic targets, led Schwartzkopf to the conclusion, by the middle of February, that **strategic air attack had culminated** as an instrument in his campaign.[26] In Kosovo, commentators have argued that it was not airpower that broke Milosevic's will to continue, but his growing diplomatic isolation after Russia ceased to support him and the growing momentum in NATO circles towards a ground war.

Further, these campaigns seemed also to demonstrate that whatever the advantages given by control of the air, **controlling ground was still crucial** to achieving key objectives. In Kosovo, General Klaus Naumann commented in frustration that what NATO wanted was 'impossible, they want us to stop the individual murderer going with his knife from village to village and carving up some Kosovars; that you cannot do from the air'.[27] Indeed, it was noted that NATO air attacks actually increased the rate of attacks by the Serbs on the Kosovars. In Afghanistan, the lack of US troops on the ground contributed materially to the escape of Osama bin Laden from the Tora Bora mountains because the Northern Alliance troops who composed the bulk of the land forces focused more on the Taliban threat than on Al Qaeda.

For critics, then, there was a great deal of difference between asserting on the one hand that airpower played a crucial role in the campaigns of the post-Cold War world, which seemed indisputable, and of asserting on the other that it represented a paradigm shift that had created a new form of warfare that marginalised land forces.

Network-Centric Warfare?

However, if the air-centric vision of the future war RMA was controversial, there was no greater consensus surrounding its alternative: NCW. As a new paradigm in warfare, NCW has many critics. Many noted that, fundamentally, Network-Centric Warfare was **not new**, despite the claims of its advocates. Attempts to create a distinction between old platform-centric warfare and new NCW ignores the fact that co-operation and information sharing have been crucial components of warfare for a long time. However, criticism was also levelled at the concept because of its technological naivety and self-defeating nature.

Technological naivety

Critics pilloried NCW for its technophilia: as one commentator argued, 'all the complexities and uncertainties in a war are reduced to essentially no more than collecting, processing, and transmitting vast amounts of information, and to speed of command'.[28] For naysayers, Network-Centric Warfare underplays the limitations of technology:

- It underplays the **importance of people**, rather than networks, in war.
- It fails to note **human limitations** in the ability to assimilate and evaluate information.
- It does not consider the impact of **enemy action** such as the possibility that they might take counter-measures.
- It assumes that more information is better, yet many of the key aspects of war, such as political factors, are **intangible or subjective**, especially at the higher levels of war.

Self-defeating nature

Critics also argue that NCW may, in the chaos of war, be **self-defeating**. First, the more **complex a system is, the less likely it is that it will function effectively** in the test of war. In the Gulf War, there were many intelligence failures: 20% of US casualties were due to friendly fire; the unreliability of official communications networks forced many to use informal channels; there were problems of interoperability between national and international contingents. Indeed, General Walt Boomer noted that 'the intelligence stunk. I mean it was lousy. We didn't have all the pictures that we needed'.[29]

Second, the more integrated a system is, the more likely it is that shocks to one part of the system will be transferred quickly to other parts: **systemic shock may therefore be self-induced** if one part of the integrated system cannot function as required due to local conditions or enemy action. In this, we find an example of the often **paradoxical logic of military effectiveness** in which striving for 'efficiency' may actually result in less effective forces. Thus, the adoption of 'just in time' logistics practices derived from business, which rely on digitisation and the integration of the logistics organisation into one large, synchronised system, may be problematic for armies at war because it requires for its effective functioning a stable and reliable environment – in other words, one at odds with the nature of war. In land warfare, mass, either in terms of materiel or logistics, can be the foundation of adaptability and the ability to insulate oneself from the shocks of war.[30]

What's in a revolution?

Moreover, whatever the individual problems or air- or network-centric futures, they both also suffer from the **fundamental problems** with the concept of a 'military revolution' that we identified in Chapter 2.

The relevance of the 'modern system'

For example, how much change does a conflict need to demonstrate before it moves from being evolutionary to revolutionary? Dissenters argue that whilst post-Cold War conflicts all displayed elements of novelty, this novelty does not mean necessarily that a military paradigm shift is occurring: their core dynamics could still be **understood in traditional terms**. Examining the Gulf War and Afghanistan in 2001, Stephen Biddle concludes that the outcomes were shaped in military terms by the same factor that had shaped warfare since 1900: **force employment**. The one-sided nature of the Gulf War was not because the United States had a post-industrial military and Iraq had an industrial one. The Gulf War did feature what should have been hard fights for the Coalition; the Iraqis did spot Schwartzkopf's outflanking move and did move forces to block it; in the clashes between these forces and advancing US troops the Iraqis did have, despite Coalition air attack, the munitions and morale to fight; they were in prepared positions; and in the nine major engagements fought, Coalition forces never outnumbered the Iraqis by more than 50%, and in some were outnumbered by the same margin. Yet the Iraqis were crushed utterly, losing 350 armoured vehicles and hundreds dead, whilst the United States lost 13 dead, of which only one was killed by enemy action. The outcome was so one-sided because the Iraqis were so **poor at modern system warfare**: this meant that they were fully exposed to the effects of modern firepower. Similarly, in Afghanistan, whilst Biddle acknowledges innovation in the Special Forces/airpower relationship, victory came as the result of the skilfully orchestrated relationship between fire and manoeuvre, and effective joint warfare.[31] For Biddle, then, **relative skill at modern system land warfare** continues to be a key factor in explaining tactical and operational success in land warfare.

Relational factors

Critics of the RMA concept also raise another general point: have the wars since 1991 really been **robust enough tests** of the supposedly revolutionary technology? Conventional victories against Iraq in 1991 and 2003 were achieved against an army that was, according to one analyst, 'one of the worst led, worst fought forces in the history of warfare'.[32] For another commentator, the Iraqis were 'the perfect enemy'.[33] Whilst some argue that previous revolutions, such as *blitzkrieg*, have been executed against adversaries that embodied deep flaws, this returns us to the question raised in Chapter 2 of whether great military successes derive from the leveraging of a military revolution, or whether they are the result of less exciting but no less important factors such as differential quality in strategy, policy and generalship?

In the end, then, the notion that future war will be shaped by the characteristics of a 'system of systems' RMA is an idea that has been both influential but also highly controversial. However, the RMA is not the only vision of the future. What others are there, and is there more of a consensus surrounding them?

'New Wars'

At the polar opposite of the RMA debate is that surrounding the notion of '**New Wars**', a term coined by Mary Kaldor in her influential book *New and Old Wars: Organized Violence in a Global Era* (1998). Like the RMA concept, the New Wars idea is founded upon the assertion that a paradigm shift in warfare is taking place. Unlike the RMA, the New Wars paradigm sees a future not dominated by information-based societies, national armies and high-technology, but marked instead by **brutal, intra-state, communal struggles**, waged by **non-state actors**, in which the means used may be very basic indeed.

What's in a New War?

Just as the Gulf War of 1991 provided the basis of the explosion in the RMA debate, so the brutal conflicts in the **Former Yugoslavia** provided the catalyst for the emergence of the New Wars school. Despite the optimism surrounding the end of the Cold War, the 1990s saw the emergence of many vicious civil wars at the same time as it appeared that the incidence of interstate war was decreasing. On the basis of this phenomenon, many writers began to see the 1990s as a watershed in warfare in which future conflict seemed likely to become something very different from that of classical conflicts. Mary Kaldor's work was one of many charting the emergence of this brutal new future, including: William Lind *et al.*'s *The Changing Face of War* (1989); Martin van Creveld's *The Transformation of War* (1991); Herfried Munkler's *The New Wars* (2004); Rupert Smith's *The Utility of Force* (2005); and John Robb's *Brave New Wars* (2007).

The New Wars argument

At a general level, the New Wars argument has at its heart six related observations regarding the future of warfare:

- First, the **objectives of war are changing**: whereas old wars were about issues such as ideology and geo-politics, New Wars are motivated by the politics of identity, such as tribal or clan loyalties, linguistic differences and so forth.
- Second, New Wars are waged by **new participants**: states increasingly have lost the monopoly on the use of violence and force has now become decentralised, wielded increasingly by non-state actors such as militia groups, warlords and private security contractors.
- Third, New Wars involve **new strategies**. What used to be by-products of conflict, such as the use of violence against civilians and ethnic cleansing, have now become a deliberate and central part of the ways in which belligerents seek to achieve their objectives.
- Fourth, New Wars are conducted in a **globalised context**. This context, with its 'democratisation of economic power', has empowered non-traditional actors by giving them greater access to resources, weapons and the media.
- Fifth, and as a consequence of the previous points, New Wars **blur the distinctions between war, peace and criminality**. New Wars tend to be what Munkler terms 'autonomous': they are sustained struggles of see-sawing intensity, divorced increasingly from any rational cost-benefit analysis, sustained instead by circular motivations such as greed, hate and revenge.
- Last, many of the arguments in support of this New Wars thesis make explicit reference to Clausewitz in their analysis of the distinctions between new and old wars. In particular, they focus on Clausewitz's concept of the trinitarian nature of war (see Chapter 1). For New War advocates, **Old War is Clausewitzian war** (or post-trinitarian) in the sense that Clausewitz's trinity assumed that war comprised three elements: the people, the army and the state. New War is post-Clausewitzian because the state is no longer the key arbiter in war. Moreover, since the state is no longer a central actor in increasing numbers of conflicts, so Clausewitz's assertion that war is subordinate to rational policy also no longer holds true.

Whilst the many writers lumped into the New Wars school share some general assumptions regarding the current and future development of warfare, there are also **many variations**. Reflecting on the 'alphabet soup' of new ideas that emerged in the period after the Gulf War, the strategist Colin Gray notes the problems posed by what he called 'concepts du jour': transitory and contestable intellectual fashions that one after another seized the imaginations of sections within the defence community before falling into abeyance.[34] Two New Wars concepts, in particular, gained considerable traction: Fourth Generation Warfare and Asymmetric Warfare.

Fourth Generation Warfare

Fourth Generation Warfare (4GW) was a concept advanced by Colonel Thomas X. Hammes in his book *The Sling and the Stone: On War in the 21st Century* (2006). Hammes argued that the previous three generations of war (horse and musket, rifle and railway, *blitzkrieg* and manoeuvre warfare) had given way to a new form of war exemplified by the extended conflicts in Afghanistan and Iraq after 2001 and 2003, respectively. '4GW' was, for Hammes, **an evolution in insurgency**. Thanks to the dynamics of globalisation, non-state actors were now able to combat high-technology adversaries by leveraging political, economic and social networks to shift the locus of conflict from the technological battleground to the social and political arenas where Western states, in particular, were very vulnerable. In essence, for Hammes, 4GW represented a deliberate shift in strategy by adversaries who recognised that the West's military capabilities would overmatch them in the conventional sphere.

Asymmetric Warfare

The idea that New Wars involved elements of deliberate adaptation by adversaries also underpinned the concept of **Asymmetric Warfare**. The Asymmetric Warfare argument noted that this fact, allied with growing threats emerging from **failed state environments**, seemed to argue for actual military threats that resembled **insurgency and terrorism** rather than the Gulf War. Writers on Asymmetric Warfare pointed to the shortcomings of Western militaries in dealing with the complex conflicts in Somalia and Bosnia in the first half of the 1990s: their argument was that high-technology militaries (and the societies that created them) were **vulnerable to low-technology ripostes**, or to counter-attacks in the political, legal and economic dimensions of conflict. In essence, the Asymmetric Warfare argument was an expression of the paradoxical logic of strategy: that because RMA-type militaries were so effective at high-intensity conventional warfare, adversaries would always seek to situate armed conflicts in other arenas.[35]

The New Wars debate

There certainly seemed to be a **strong empirical basis** for the New Wars argument. The extended brutality of the conflict in **Bosnia** from 1992–95, the genocide in **Rwanda** in 1994, the prolonged suffering in places such as Somalia and **Congo** since 1991 – all seemed to reflect the themes that the New Wars school believed were becoming increasingly central to the future of warfare. These themes had important implications for land forces. Since high-intensity interstate war was becoming increasingly rare, armies that insisted on focusing on conventional warfare would be developing capabilities **increasingly irrelevant** to the likely conflict scenarios. Where military means were required, they were likely to be of the kind required in Bosnia: expeditionary capabilities with a focus on the sorts of structures, doctrine and training that had more in common with counterinsurgency and peace operations than traditional war fighting.

However, despite the fact that the New Wars paradigm was rooted in what seemed to be incontestable contemporary evidence, it has nevertheless remained controversial. The varied critiques of the New Wars thesis have highlighted a number of difficulties with the concept as a whole.

The persistence of interstate war

One counter argument is that **interstate war is still a feature** of the international system. The Gulf War, Kosovo, Afghanistan and the Iraq War of 2003 all had important interstate elements, even if, in the latter two cases, conflict transmuted into insurgency and communal violence. The weight attached to the existence of Old War varies in the literature on New Wars. Van Creveld's *Transformation of War* presents a stark case for the death of major conventional warfare. Kaldor, on the other hand, acknowledged that Old War might not die out everywhere and that both New and Old Wars might exist in parallel. Updating her book in 2006, Kaldor argued that Old War was evolving into what she called '**spectacle war**', designed primarily for public consumption, and that in conflicts such as Kosovo, Iraq and Afghanistan, the conflicts began as Old Wars and became New Wars.

Continuity in war

However, a second criticism is that **the characteristics of New Wars are not new**: in consequence, New War advocates have created a new category of war where none, in fact, exists. For example, it is not clear that the deliberate targeting of civilians or the use of fear to control populations is necessarily a new phenomenon: the Second World War contains many examples of the deliberate use of terror against civilians, whether this be the Holocaust, strategic bombing, anti-guerrilla operations, or the advance of the Red Army into Germany at the end of the war. History is littered with conflicts marked by protraction, proportionately high civilian casualties, brutality and a marked reluctance to engage in battle. The Thirty Years War (1618–48), for example, resulted in the loss of between 25% and 40% of the population of the German states. Equally, China's Taiping Rebellion (1850–64) resulted in up to 30 million dead, the majority being civilians. Nor is it clear that motivations have changed so radically: 'greed' as a motive is neither new, nor can it be separated out from other kinds of political motivation that have long been at the root cause of wars.[36]

Misinterpretation

A third criticism that has been levelled at the New Wars school is that it **misinterprets Old Wars**. At a theoretical level, it has been argued that the New Wars literature mangles Clausewitz's concept of the trinity of war. As writers such as Christopher Bassford have demonstrated, Clausewitz's trinity was composed in the first instance not by the people, the army and the government, but rather by hatred, passion and enmity, the play of chance and probability, and the subordination of war to rational policy.[37] These ideas transcend state-based warfare and, as the political scientist Colin Fleming notes, 'Clausewitz's assessment of the changing character of war throughout history illustrates his awareness that the character of war was constantly changing, often dramatically, from one age to the next'.[38] Thus, the distinction between Clausewitzian Old War and post-Clausewitzian New War is a mistaken one. Those things that defined the nature of war in the past continue to define it today and in the future.

Methodological problems

A fourth criticism is that the 'New Wars' argument is full of **methodological flaws**. For example, Fourth Generation Warfare has been critiqued for its biased selection of examples, and for misrepresenting the development of warfare through its 'generations' approach.[39] Similarly, Asymmetric Warfare has been criticised because the concept defies any easy definition: does asymmetry refer to asymmetries in the power of belligerents (strong versus weak), to the organisations involved (state versus non-state), to the methods (conventional warfare versus terrorism or insurgency), or to the normative aspects of the conflict (those adhering to international norms of behaviour versus those not doing so)? Moreover, how new are any of these ideas? War has an **inherently asymmetric dynamic**: few belligerents deliberately set about fighting a conflict in ways that magnify their opponent's strengths and most belligerents, given time, adapt to try to mitigate their own weaknesses and exploit those of their enemy. Even 'symmetrical wars' tend, therefore, to have inherent 'asymmetrical' aspects: in the Second World War, for example, British strategic bombing and German U-Boat warfare could be construed as 'asymmetric' responses to enemy strengths.

Overall, then, the New Wars debate has served an important purpose in **focusing contemporary thinking** on what might or might not be different about contemporary and future warfare. It also remains **an important counterpoint** to technology-centric RMA debates. However, there remains an important body of critics who argue that the New Wars perspective has overreached itself in trying to create a new category of warfare where none actually exists. New Wars, so critics argue, may actually be better understood either as low-intensity conflict, or it may be better to treat each conflict according to its specific context rather than trying to create a 'catch-all' category.

Hybrid Warfare

The third general category into which we can divide the future land warfare debate is that of '**Hybrid Warfare**'. This view of the future stands somewhere between the previous two perspectives, which focus on high- and low-intensity warfare, respectively, because it posits a future that is both **simultaneously** conventional and unconventional.

What is Hybrid Warfare?

Hybrid Warfare is in essence the belief that rather than a future marked by either conventional or unconventional warfare, we are likely to see a **convergence** occurring in modes of warfare: in effect, a **fusion of conventional and guerrilla warfare** into one hybrid threat. As Frank Hoffman, the key exponent of the hybrid war concept, notes:

> Instead of separate challengers with fundamentally different approaches (conventional, irregular, or terrorist), we can expect to face competitors who will employ all forms of war and tactics, perhaps simultaneously.[40]

Hoffman recognises that in general terms this blending of techniques is not new, but he distinguishes between old 'compound war' and the new 'hybrid war'.

Compound wars

In **compound wars** of the past, one can see separate conventional and unconventional forces co-ordinated strategically but fighting separately. So, for example, in the American War of Independence, American forces consisted of regular troops that fought regular battles, and irregular troops that were involved in guerrilla warfare. Likewise, in Vietnam, the North Vietnamese deployed regular forces trained for conventional operations, and the Viet Cong guerrilla forces used for irregular warfare.

Hybrid wars

In **hybrid wars** of the future, however, both conventional and irregular warfare will be applied by the same fighting force. There will be a high degree of **operational and tactical integration**, indeed a fusion of regular and irregular techniques by the same organisation. Pitting Israel against Hezbollah, the Lebanon War of 2006 is the key Hybrid Warfare example (see Box 9.3). In Lebanon, the Israeli Defence Forces (IDF) encountered an ostensibly irregular non-state actor, the Iranian-backed Shia group Hezbollah led by Hassan Nasrullah.

Box 9.3 The war in Lebanon, 2006

The Lebanon War of 2006 was a 34-day conflict between Israel and Hezbollah. The proximate cause of the war was a Hezbollah raid on 12 July 2006 which killed three IDF soldiers and captured two more. The underlying cause was the prolonged weakness of the Lebanese state that had allowed anti-Israeli actors, first the Palestine Liberation Organization (PLO) and then Hezbollah, to use southern Lebanon as a base for attacks against Israel. The IDF began the war on 12 July relying on remote attack, some artillery strikes but mainly air attacks, to crush Hezbollah and to induce the Lebanese government to move its troops into southern Lebanon and evict Hezbollah forces. Some 15,500 air sorties were launched against 7,000 targets. Ground operations were limited in nature, both to reduce the economic cost of the war but also to reduce the risk of casualties. Remote attack proved ineffective, since Hezbollah had invested in a large system of heavily protected bunkers, dispersal and camouflage, and also mobile rocket forces which it used to launch attacks on northern Israel. The IDF was forced to commit, at short notice, much larger numbers of ground forces, which required on 21 July the calling up of military reservists. The ground fighting dragged on and the IDF found itself ill-prepared for the tenacious Hezbollah defence, which combined features of irregular warfare with aspects of conventional armies including the use of high-technology equipment such as night-vision goggles and anti-tank guided weapons. On 14 August a ceasefire was finally signed. Hezbollah had lost between 650 and 750 fighters, although fewer than 190 of these had been killed in the ground fighting. The IDF lost 119 killed and the war was widely regarded as a victory for Hezbollah: it largely ignored the terms of the ceasefire, received an enormous political boost and consolidated its position in southern Lebanon.

Yet Hezbollah did not fight like a traditional irregular opponent: whilst it did exhibit some of the traits of a traditional insurgent, such as a cellular structure, a focus on mobility, evasion, ambush and long-range attack, it also embodied characteristics more often associated with regular forces as well: it was well disciplined, well trained, had effective command and control, was willing to contest ground and counter-attack, and utilised such modern technology as anti-tank guided weapons and night-vision goggles. The dangers posed by Hybrid Warfare were illustrated by the difficulties that the IDF experienced in combating Hezbollah. As one analyst noted:

> the Israeli Army, highly conditioned by its LIC [low-intensity conflict] experience, was initially confounded by an enemy that presented a high-intensity challenge that required joint combined arms fire and maneuver and a combat mindset different from that of Palestinian terrorists.[41]

For advocates of Hybrid Warfare, the Lebanon War was a classic example of the sort of conflict that land forces would have to fight in the future. For writers such as Hoffman, future warfare will be complex, dynamic and unpredictable because it will also be decisively hybrid.

The Hybrid Warfare debate

If the proponents of Hybrid Warfare are correct, then there are some obvious implications for land forces: the army of the future must be a balanced force capable of joint, combined-arms fire and manoeuvre; it must be a force flexible enough to meet the irregular aspects of the enemy threat, but it must also be sufficiently versed in the core competencies of conventional warfare. In many respects, it is not difficult to see why the Hybrid Warfare concept has been so popular. Amongst other things, it has tapped into a **post-Afghanistan military *zeitgeist***: the fear that the adaptations required to conduct stability operations in Iraq and Afghanistan have eroded decisively the conventional warfare capabilities of Western militaries. However, just as the RMA and New Wars visions are the subject of great contestation, so too is the value of the hybrid war theory. Critics of the concept highlight two general difficulties.

Hybrid Warfare is not a new category of war

First, some sceptics argue that hybrid war is **not a distinct category of warfare**. Here, the argument is that Hezbollah's effectiveness in 2006 was less to do with the fact that they were leveraging a new warfare paradigm, than the fact that they were just a tough military opponent:

Hezbollah strengths

First, Hezbollah were effective because they were **very competent not because they were unique**. Hezbollah was a potent insurgent force, well motivated and well armed. Since 2000 Hezbollah had recognised that a major clash with the IDF was likely and so it had had an extended period to analyse IDF strengths and weaknesses and to build up extensive defensive networks, acquire high-technology weapons, stockpile logistics, and develop a flexible defensive concept designed to stymie Israeli air attack and prolong the ground fight into an attritional struggle.

Israeli weaknesses

Second, we could argue equally that Hezbollah's perceived strength derived in part from the **IDF's weaknesses**. Critics have pointed to many flaws in Israel's 2006 campaign. It has been argued that Israel's **overall strategy was flawed**, with an overreliance on airpower and a failure to realise that confining the land operations initially to a narrow strip of 2–5 km along the border was never likely to be physically decisive. Israel suffered from poor intelligence: it failed to pick up on some important issues, such as the sophistication of Hezbollah's defences along the border, and there was a failure to pass information between organisations, leading to a failure to warn the army adequately about Hezbollah's missile firepower. Criticisms have also been made of **Israel's doctrine**, which in its focus on 'systemic operational design', reflected effects-based operations thinking. Moreover, critics point to **fundamental planning failures**, such as the neglect of contingency planning in case a major land operation *was* required, and an overambitious timetable for the operation. However, some problems with the IDF's campaign were conditioned by **structural factors** and were not easy to remedy: whilst perhaps in retrospect the only way to defeat Hezbollah was to have taken and held large swaths of southern Lebanon, this would have required mobilising far more than the original 15,000 troops, and would consequently have imposed heavy economic and political costs on the Israeli government.

Definitional problems

A second difficulty with Hybrid Warfare is the **lack of commonality in definitions**. Thus, whilst many writers have agreed with Hoffman on the importance of Hybrid Warfare, they have often defined it in very different ways, undermining fundamentally the concept's status as a workable paradigm.

Hybrid Warfare as stability operations

One view of Hybrid Warfare is strongly reminiscent of stability operations. For some writers, future war will be hybrid in the sense that it will be fought with **all the elements of national power** across of spectrum of activity ranging from full-scale armed combat through to complex stability operations.[42] For others, future war will be hybrid in that it will encompass **both physical and conceptual dimensions**: the former comprising the struggle against the enemy and the latter encompassing a political struggle for the support of local, domestic and international support. These views have two common elements: first, both seem indistinguishable from war as it is already fought; and second, both seem conditioned, not by the Lebanon War but by Western experiences in Iraq and Afghanistan. This concept of Hybrid Warfare is essentially **focused on stability operations**. Neither, critics would argue, provides a convincing definition of a new paradigm in warfare.

Hybrid Warfare and continuity

Even where writers have focused specifically in their concept of Hybrid Warfare on the blending of regular and irregular warfare, their definitions do not necessarily accord with that of Hoffman. Some define Hybrid Warfare as:

> conflict involving a combination of conventional military forces and irregulars (guerrillas, insurgents, and terrorist), which could include both state and non-state actors, aimed at achieving a common political purpose.[43]

This view **explicitly rejects the distinction between compound and hybrid wars** and, in consequence, the range of examples upon which this definition draws ranges from Roman campaigns in Germania, through to the Peninsular War of 1807–14, the Vietnam War and the 2006 Lebanon War. From this perspective, historically hybrid wars have probably been the norm and 'much of the present only echoes the past'.[44] Hybrid Warfare represents a point of continuity with the past and not a paradigm shift.

In conclusion, it can be said that the Hybrid Warfare concept is one which, whatever its political traction, is yet another **contested concept**. Biddle argues, for example, that even if there is a trend towards hybridity in the ways defined by Hoffman, 'It cannot yet be known how broad this may be, what its root causes are, or how far it will go'.[45] This doesn't mean that the concept itself is without merit: any concept that **promotes debate** is useful; the Hybrid Warfare debate raises important questions regarding the **role of non-state actors** in future warfare; and Hybrid Warfare is a concept that also **moves us beyond binary 'conventional versus unconventional' debates** on future land warfare. However, the concept is far from being a reflection of objective fact. Indeed, Hoffman himself argues that 'the utility of the concept has already been proven (but not the concept itself)'.[46]

Are predictive approaches useful?

We come now to the final part of this chapter and we finish on an interesting question. What each of the three views that we have looked at has in common is that they are **essentially predictive**: they reflect the assumption that it is possible and desirable to identify the character of the future land warfare environment. Preparing an army for the future is associated with **uncertainty** and this predicative approach tries to deal with this uncertainty through **pre-planning**. In other words, preparing land forces for future warfare requires two steps: deciding which of the three views of future warfare is the right one, and then generating land forces to fight that kind of warfare. Of course, the usefulness of this approach is determined by whether or not **one predicts correctly** the character of future land warfare. History suggests that this kind of prediction is invariably inaccurate. So, are there any alternatives to predictive approaches?

Military flexibility and adaptation

An alternative perspective argues that a predictive approach is largely self-defeating and that instead of trying to prepare for one future land warfare paradigm, armies should instead focus their efforts on **enhancing their inherent flexibility** as an organisation. This flexibility should allow them to adapt to whatever land warfare environment eventually does emerge. Adaptability is preferred over prediction as an army's foundation for future effectiveness. As one commentator notes, 'Those military organisations that have adapted to the actual conditions of combat are those that perform the best in war'.[47] The root of this argument in favour of flexibility over prediction lies in two observations.

The problems of prediction

First, historical evidence seems to suggest that, despite their best efforts, **armies usually fail to predict** the future character of warfare.[48] For this reason, Colin Gray argues that most wars are a race by belligerents to correct the mistaken assumptions with which they commenced the conflict. The roots of these failures of prediction are many and varied.

Inherent challenges

Predicting the future **is simply a difficult thing to do**. In relation to future war, one must make calculations based on very uncertain fundamentals:

- Uncertainty over timelines: how far forward must one predict?
- Uncertainty regarding opponents: who will one fight; how will they fight?
- Uncertainty over technology: will new means become available?
- Uncertainty in relation to the future political and economic context: will government policy change; will resources increase or decrease?

As a result of these uncertainties, full play is given to **the operation of human psychology** and the varied **decision pathologies** that shape human cognitive capacity: groupthink, organisational culture, personal psychohistory and so forth. For this reason, thinking on future warfare is often a bounded reflection of existing values and assumptions. As US General John R. Galvin notes, 'We arrange in our minds a war we can comprehend on our own terms, usually with an enemy who looks like us and acts like us'.[49] This foundation difficulty in compounded by other challenges:

- The intricacies of generating **strategic intelligence**, rooted in problems such as quantity and quality of available information and the politics of intelligence analysis and sharing.
- The difficulty in applying to the future **historical lessons and analogies** of the past.
- The **nature of war and strategy itself**: the fact that war is an art, not a science, and so lacks immutable principles upon which to base calculations; the crucial fact also that 'the enemy always has a vote'. It is simply difficult to replicate in peacetime the chaos, pressure and uncertainty of war that contributes to producing unexpected outcomes.

The problem of surprise

Second, the focus on preparing to meet a single assumed future warfare paradigm makes armies **vulnerable to 'technological and doctrinal surprise'**: armies that shape their force structures, equipment, doctrine and training to meet a narrowly defined future warfare model can find themselves vulnerable to shock and disorientation when the war they expect is not the one that they have to fight. The analyst Meir Finkel argues that 'the solution to technological and doctrinal surprise lies not in predicting the nature of the future battlefield or obtaining information about the enemy's preparations for the coming war, but in the ability to recuperate swiftly from the initial surprise'.[50]

Organisational flexibility as a solution

Because of these difficulties in predicting the future, some analysts argue that a focus on **organisational flexibility** is the best basis upon which to cope with the challenges of future warfare. Military flexibility of this kind has a number of components.[51]

First, flexibility has a **conceptual and doctrinal** component. Flexibility in this respect is supported when an army creates an organisational atmosphere that encourages officers to challenge existing conceptual and doctrinal assumptions, which encourages the asking of questions. This flexibility is also sustained by the creation of a doctrine that is balanced, which does not, for example, focus on offensive operations at the expense of the defence.

The second element in effective flexibility is **organisational and technological**. Flexible militaries require a balance of military capabilities: a balance between, for example, firepower and manoeuvre, between offensive and defensive capabilities. They also require organisational diversity: the army must embody a range of different technologies and weapons. Two other features also contribute to this dimension of flexibility: **redundancy**, describing a situation in which an army has spare capacity and resources; and **technological versatility and changeability**, an attribute that describes the ability of an army to think innovatively about how its technology can be used.

The third component of flexibility lies in the realm of **command and cognitive** skills. This component embodies traits such as mental flexibility, an attribute that stems from officers who have emerged from a creative and questioning military organisation. It requires also that an army has an effective concept of flexible command that embodies a system that encourages decentralisation and the display of initiative from even junior officers.

Fourth, true flexibility requires that an army has an effective mechanism for **rapid learning and the dissemination of lessons**. This capability is crucial if an army is to minimise technological and doctrinal surprise. To these four principles we can also add a fifth: **realistic peacetime preparation** for war. Training and exercises, for example, must reflect the challenging nature of wartime activity.

These five elements of flexibility are **interrelated** and **form a hierarchy**: conceptual and doctrinal flexibility, for example, underpins the other aspects of a flexible army.

The advantages of flexibility

A focus on **flexibility rather than prediction** focuses on the inevitability of surprise – that most armies do not get the war for which they have prepared. For this reason, preparing land forces adequately for future war is less about trying to predict the specific form that land warfare will take, which is inherently difficult to do, and more about ensuring that an army can adapt quickly to the actual conditions experienced. What matters, then, is **the distinction between flexible and inflexible armies**.[52]

- The **German army of 1941**, for example, was more resilient than the French army of 1940, because the former was far more flexible than the latter, judged by the five criteria outlined above. Thus, the German army was able to adapt in late 1941 to the need to shift from offensive to defensive operations because it had a balanced doctrine; it was able to neutralise quickly the technological disadvantage caused by the first encounters with the superior Soviet T34 tank through rapid tactical

experimentation and the dissemination of new techniques, each of which was facilitated by its open conceptual and doctrinal culture, its decentralisation of command, and its structures for the codification and distribution of improved methods.
- Conversely, **French defeat in 1940** was certainly not due to a deficiency in numbers, or even technological inferiority. Rather, it lay in an overly centralised system of command and control, tactical dogmatism, a one-dimensional defensive mindset, and a military culture that did not encourage free thinking. In essence, the French were surprised by the form of land warfare that they encountered, and were not sufficiently flexible to recover from the shock before it was too late.

The challenge of flexibility

Yet, despite the evident advantages of a focus on flexibility, it is **often difficult to realise** in real-world armies. Most armies would claim to be very adaptable; indeed, principles such as 'flexibility', 'adaptability' and 'agility' often feature heavily in military doctrine. Yet, in practice, most armies prove not to be so, illustrating that the obstacles to organisational flexibility often lie outside the conscious recognition of the military:[53]

- First, adaptability requires that a culture of flexibility be embedded deeply in an army. However, as Chapter 2 has identified in its discussions regarding the problems of military innovation and adaption, **the nature of military organisations** can itself pose difficult obstacles to the creation of a culture of flexibility – its routine procedures, social norms, hierarchical structure, internal politics and ethical standards. Creating a culture of flexibility may therefore be difficult because it requires a wholesale change in the existing military culture.
- Second, as Chapter 5 has noted in its observations about the problems of adopting the modern system of war, there are important **contextual limitations** on the ability of an army to adapt itself: these may be political, industrial, cultural or organisational. Flexibility is not just a tactical concept: it has operational and strategic dimensions as well. As Germany's experience in the Second World War demonstrates, strategic political inflexibility can undermine fundamentally the value of an army's tactical or operational adaptation.
- Third, **war itself provides inherent challenges** to the flexibility of a military organisation. War is a difficult environment within which to adapt: it embodies immense psychological and political pressures; it can be difficult to derive and disseminate appropriate lessons.
- Fourth, **balance is difficult to sustain** in an army. As Chapter 1 has argued, no army can be fully effective in every military environment, and the wider an army spreads its expertise, the more likely it is that its capabilities will be inadequate individually for each specific context.
- Finally, flexibility is also about **understanding the enemy** if adaptation is to be relevant. However, understanding a potential enemy requires an understanding that extends beyond an opponent's military organisation and requires the comprehension of features such as complex cultural contexts.

Cumulatively these problems explain the repeated phenomenon that armies often tend to try to fit the lessons of actual war into existing pre-war concepts rather than discard them in favour of wholly new approaches.

Conclusions

There exists a wide array of cogently argued visions for the future of land warfare. For some, this future is marked by the consequences of the **Revolution in Military Affairs**: a military revolution based on the development of a 'system of systems', drawing together advances in sensors, precision firepower and networking technologies to a produce a military paradigm shift. This RMA might be one that privileges airpower, or it might be one that, through NCW, redefines the principles and structures through which armies will fight. In these futures, armies must transform themselves radically, rejecting the traditional focus on mass and heavy forces, and embracing further complex decentralisation and disaggregation.

For others, the future of warfare will be marked by the demands of **New Wars**, which will see a move away from high-intensity, conventional conflicts and towards brutal, messy, internecine struggles in which the methods used are almost neo-medieval. In this future, conventional forces, if they are involved at all, will confront 'super-empowered' non-state actors such as demagogues, warlords, militias and criminals who are leveraging the opportunities afforded them by globalisation. In this future, armies must embrace the needs for extended low-intensity conflict, building on the skills required for counterinsurgency and stability operations.

Alternatively, the future might be a complex, blended **Hybrid Warfare** in which conventional and unconventional techniques are used simultaneously by belligerents to produce a challenging new form. In this future, armies must maintain a full range of capabilities and expertise, enabling them to adapt to the techniques adopted by flexible opponents.

Or, none of these paradigms might be correct. It might be that what land forces require for future warfare is a focus on **organisational flexibility**: an approach that avoids the problem of trying to predict the future of warfare and which focuses instead on nurturing the adaptability of the armies of the future – on creating an external and internal context conducive to rapid adaptation once the nature of the future land warfare challenge becomes clear.

The difficulty facing armies is that whilst each of these visions of the future is contested, states do not have the luxury of perpetual analysis: choices have to be made; equipment procured; force structures developed; doctrines created. In the final chapter of this book, we look at a case study in future land warfare: the US army and its attempts to adapt itself to the demands of the future. The challenges that have been faced by the US army provide a stark example of the problems faced by land forces both in predicting future warfare and in adaptation to unexpected conditions.

Notes

1 Edward C. Mann, III, *Thunder and Lightning: Desert Storm and the Airpower Debates* (Maxwell, AL: Air University Press, 1995), 20.
2 See Michael R. Gordon and General Bernard E. Trainor, *The Generals' War: The Inside Story of the Gulf War* (New York, NY: Little, Brown, 1995), 122–58.
3 Grant T. Hammond, 'Myths of the Gulf War: Some "Lessons" Not to Learn', *Airpower Journal* (Fall 1998), 6.
4 Stephen J. Cimbala, 'Transformation in Concept and Policy', *Joint Force Quarterly* Vol. 38 (July 2005), 28.
5 See Admiral William A. Owens, 'The Emerging U.S. System-of-Systems', *National Defence University Strategic Forum* Vol. 63 (February 1996).

6 Keith L. Shimko, *The Iraq Wars and America's Military Revolution* (Cambridge: Cambridge University Press, 2010), 93.
7 Richard P. Hallion, *Storm Over Iraq: Air Power and the Gulf War* (Washington, DC: Smithsonian Books, 1997), 1.
8 Shimko, *The Iraq Wars*, 110.
9 Harlan K. Ullman and James P. Wade, *Shock and Awe: Achieving Rapid Dominance* (Washington, DC: National Defence University, 1996), 45.
10 Torgeir E. Saeveraas, 'Effects-based Operations: Origins, Implementation in US Military Doctrine and Practical Usage', in Karl Erik Haug and Ole Jorgen Maao (eds) *Conceptualising Modern War* (London: Hurst and Company, 2011), 191.
11 Stephen Biddle, 'Victory Misunderstood: What the Gulf War Tells us About the Future of Conflict', *International Security* Vol. 21, No. 2 (Autumn 1996), 149.
12 Shimko, *The Iraq Wars*, 79.
13 Frederick W. Kagan, *Finding the Target: The Transformation of American Military Policy* (New York, NY: Encounter, 2006), 161.
14 *Gulf War Air Power Survey* (Washington, DC: US Government Printing Office, 1993), 298–99.
15 Hallion, *Storm Over Iraq*, ix.
16 Eliot A. Cohen, 'Stephen Biddle on Military Power', *The Journal of Strategic Studies* Vol. 28, No. 3 (June 2005), 421–22.
17 Kagan, *Finding the Target*, 305–6.
18 Max Boot, *War Made New* (New York: Gotham, 2012), 328.
19 Bruce D. Berkowitz, *The New Face of War: How War Will Be Fought in the 21st Century* (New York, NY: Simon and Schuster, 2003), 2.
20 Milan Vego, *Joint Operational Warfare: Theory and Practice* (Newport, RI: US Naval War College, 2009), XIII-4.
21 Vego, *Joint Operational Warfare,* XIII-3–XIII-10.
22 Shimko, *The Iraq Wars*, 113.
23 See Avi Kober, 'The Israeli Defense Forces in the Second Lebanon War: Why the Poor Performance?' *Journal of Strategic Studies* Vol. 31, No. 1 (February 2008), 3–40.
24 Saeveraas, 'Effects-based Operations', 203.
25 Barry R. Posen, 'The War for Kosovo: Serbia's Political Military Strategy', *International Security* Vol. 24, No. 4 (Spring 2000), 51–52.
26 Gordon and Trainor, *The Generals' War*, 324–26.
27 Dag Henriksen, *NATO's Gamble: Combining Diplomacy and Airpower in the Kosovo Crisis, 1998–1999* (Annapolis, MD: Naval Institute Press, 2007), 192.
28 Vego, *Joint Operational Warfare*, XIII-3.
29 Shimko, *The Iraq Wars*, 84.
30 See Chris C. Demchak, 'Creating the Enemy: Global Diffusion of the Information-Technology Based Military Model', in Emily O. Goldman and Leslie Eliason (eds) *The Diffusion of Military Technology and Ideas* (Stanford: Stanford University Press, 2003), 307–47.
31 Stephen Biddle, *Afghanistan and the Future of Warfare: Implications for Army and Defense Policy* (Carlisle, PA: Strategic Studies Institute, 2002), 31–46.
32 Adrian R. Lewis, *The American Culture of War: A History of U.S. Military Force from World War II to Operation Enduring Freedom* (Abingdon: Routledge, 2007), 321–25.
33 John Mueller, 'The Perfect Enemy: Assessing the Gulf War', *Security Studies* Vol. 5, No. 1 (Autumn 1995), 79, 111.
34 Colin S. Gray, *Strategy and History: Essays on Theory and Practice* (Abingdon: Routledge, 2006), 114.
35 See Steven Metz and Douglas V. Johnson, II, *Asymmetry and U.S. Military Strategy: Definition, Background, and Strategic Concepts* (Carlisle, PA: Strategic Studies Institute, 2001).
36 Ole Jorgen Maao, 'Mary Kaldor's New Wars: A Critique', in Karl Erik Haug and Ole Jorgen Maao (eds) *Conceptualising Modern War* (London: Hurst and Company, 2011), 79.
37 Christopher Bassford, 'The Primacy of Policy and the "Trinity" in Clauswitz's Mature Thought', in Hew Strachan and Andreas Herberg-Rothe (eds) *Clausewitz in the Twenty-First Century* (Oxford: Oxford University Press, 2007), 74–90.
38 Colin M. Fleming, 'New or Old Wars? Debating a Clausewitzian Future', *The Journal of Strategic Studies* Vol. 32, No. 2 (April 2009), 233.

39 Antulio J. Echevarria, II, *Fourth Generation War and Other Myths* (Carlisle, PA: Strategic Studies Institute), 6–9.
40 Frank G. Hoffman, 'Hybrid Warfare and Challenges', *Joint Force Quarterly* Vol. 52 (1st Quarter 2009), 35.
41 David E. Johnson, *Military Capabilities for Hybrid Warfare: Insights from the Israeli Defense Forces in Lebanon and Gaza* (Santa Monica, CA: RAND, 2010), 3.
42 See Margaret S. Bond, *Hybrid War: A New Paradigm for Stability Operations in a Failing State*, Strategy Research Project Paper (Carlisle, PA: US Army War College, 2007).
43 Peter R. Mansoor, 'Introduction: Hybrid Warfare in History', in Williamson Murray and Peter Mansoor (eds) *Hybrid Warfare: Fighting Complex Opponents from the Ancient World to the Present* (Cambridge: Cambridge University Press, 2012), 2.
44 Williamson Murray, 'Conclusions', in Murray and Mansoor (eds) *Hybrid Warfare*, 291.
45 Stephen Biddle and Jeffrey A. Friedman, *The 2006 Lebanon Campaign and the Future of Warfare: Implications for Army and Defense Policy* (Carlisle, PA: Strategic Studies Institute, 2008), xvii.
46 Frank G. Hoffman, 'Further Thoughts on Hybrid Threats', *Small Wars Journal* (March 2009), 2, smallwarsjournal.com/blog/journal/docs-temp/189-hoffman.pdf?q=mag/docs-temp/189-hoffman.pdf (accessed 10/07/2012).
47 Williamson Murray, *Military Adaptation in War: With Fear of Change* (Cambridge: Cambridge University Press, 2011), 38.
48 Michael Howard, 'The Use and Abuse of Military History', *RUSI Journal* Vol. 138, No. 1 (1993), 28.
49 Brian McAllister Linn, *The Echo of Battle: The Army's Way of War* (Harvard, MA: Harvard University Press, 2009), 4.
50 Meir Finkel, *On Flexibility: Recovery from Technological and Doctrinal Surprise on the Battlefield* (Stanford, CA: Stanford University Press, 2011), 2.
51 Ibid., 2–4.
52 See Finkel, *On Flexibility*, 69–72, 138–47.
53 See Murray, *Military Adaptation in War*, 7–15, 327–28.

Suggested reading

Meir Finkel, *On Flexibility: Recovery from Technological and Doctrinal Surprise on the Battlefield* (Stanford, CA: Stanford University Press, 2011). This is a stimulating book on the weaknesses of predictive approaches to the future and the advantages of organisational flexibility.

Thomas X. Hammes, *The Sling and the Stone: On War in the 21st Century* (St Paul, MN: Zenith, 2004). An influential, but controversial work that sees future adversaries increasingly in terms of evolved insurgencies.

Frank Hoffman, *Conflict in the 21st Century: The Rise of Hybrid Wars* (Arlington, VA: Potomac Institute for Policy Studies, 2007). The classic expression of the hybrid war argument.

Mary Kaldor, *New and Old Wars: Organized Violence in a Global Era*, second edn (Stanford, CA: Stanford University Press, 2007 [1998]). The key text on the 'New Wars' debate, updated here to include an analysis of the war in Iraq.

Bill Owens with Ed Offley, *Lifting the Fog of War* (Baltimore, MD: Johns Hopkins Press, 2001). An enthusiastic hurrah for the possibilities inherent in the RMA.

Christopher M. Schnaubelt, 'Whither the RMA?' *Parameters* (Autumn 2007), 95–107. Schnaubelt provides an incisive overview and critique of the RMA concept.

Keith L. Shimko, *The Iraq Wars and America's Military Revolution* (Cambridge: Cambridge University Press, 2010). An excellent analysis of the RMA concept and its application to the wars in Iraq.

Rod Thornton, *Asymmetric Warfare: Threat and Response in the 21st Century* (Cambridge: Polity, 2007). Another vision of the future threat environment: here, Western conventional military superiority will drive adversaries to seek new ways to target our weaknesses.

10 The paradigm army

Key points

- For much of the period since the end of the Cold War, the US army has approached its development for future warfare through the lens of transformation, an essentially Revolution in Military Affairs (RMA) perspective on the demands of future armed conflict.
- Transformation has been an ambitious, and ultimately problematic, concept and many of its key tenets have been eroded over time. There is now much less certainty surrounding what future land warfare might look like.

What should the army of the future look like? Should it reflect an essentially evolutionary development of Cold War theory and practice? Alternatively, should it reflect the embrace of revolutionary new technology and concepts? The final chapter of this book examines contemporary land warfare through the lens of the US army in the post-Cold War world. As a case study, the US army is interesting for two reasons. First, where the US army has led, others have followed. Political and military elites buy into common perspectives on the most legitimate military paradigm. In essence, for many other armies, the US army is a model for emulation: the concepts that have underpinned its development since the 1990s have had a powerful influence on the armies of other states, shaping their view of the demands of current and future warfare. Second, the development of the US army since the 1990s has been accompanied by debates that resonate with the themes developed in this book: the precepts of modern systems of warfare; the political and cultural dynamics of military change; the contested nature of the debates on future warfare.

In order to organise the discussion, this chapter looks at developments in the US army in relation to **transformation**, a theme that has been both pervasive and controversial. Transformation is a term that dominated defence reform debates in the 1990s and the first decade of the twenty-first century. It is a radical vision of the demands of future warfare and it embodies themes that promise to create a very different kind of US army: smaller, modular, rapidly deployable and highly networked. Transformation has also had a much wider impact, shaping other countries' views on what constitutes an effective modern army.

As this chapter shows, despite the prominent role that transformation has played in the development of the post-Cold War US army, it has remained controversial, with many arguing that at best it has been irrelevant and at worst it has actually undermined the army's effectiveness in conflicts such as those in Afghanistan from 2001 and Iraq from 2003. These controversies speak to themes recurrent in this book: the difficulties posed by the uncertain demands of the future, and the often difficult trade-offs between the different choices available for armies in coping with this uncertainty.

Transformation

As a US defence concept, 'military transformation' had its origins in a number of developments:[1]

- **The influence of the RMA concept**: as the previous chapter has discussed, the tremendous US victory in the 1990–91 Gulf War was taken in many quarters as vindication of the view that the United States stood on the threshold of a new military revolution. Many felt that further developments in technology, doctrine and organisation would be necessary to embed and sustain the US lead in the new RMA.
- **The end of the Cold War**: the collapse of the Soviet Union in 1991 created a challenging new environment for the US army – cuts in military spending, as domestic political pressures mounted for a 'peace dividend', combined with wider uncertainty regarding the necessity of maintaining the sorts of heavy forces that had been required to match the Soviet Union.
- **Strategic uncertainty**: the predictability of strategic competition with the Soviet Union was replaced by an unpredictable international environment and a rash of politically controversial and practically difficult operations, including deployments in Somalia, Bosnia, Haiti and Kosovo. As one commentator noted, 'the Army of the 1990s simply had no idea about what contingencies it would have to prepare for'.[2]
- **Political support**: military transformation was popular politically; key individuals including President George W. Bush and his Secretary of Defense Donald Rumsfeld were keen to push the idea forward.
- **Institutional survival**: the RMA seemed to benefit the US air force the most. Unless the army developed its own vision of future warfare that would match the core themes of the RMA, it was likely to suffer heavily in successive reviews of defence.

Combined, these factors created both perceived problems and perceived opportunities. The former included the fact that the US army seemed in 1991 to be a legacy force equipped to face an adversary that no longer existed; moreover, in conditions where cuts in the size of land forces seemed inevitable, maintaining its relevance seemed to dictate far-reaching reforms. The perceived opportunities lay in the developments associated with the RMA, which held out the promise of US forces that could be smaller but might actually be even more effective. The process of leveraging the fruits of the RMA to reform US land forces took place under the label of 'military transformation'. The consequence of these factors was a demand for army reform that coalesced around a number of desirable attributes:[3]

- An ability to respond rapidly.
- An ability to operate in 'austere' conditions (i.e. where the local infrastructure was underdeveloped and without established bases).

- The capacity to conduct longer-term deployments.
- The facility to conduct globally dispersed operations simultaneously.
- The requirement to be able to create flexible task groups tailored to each specific contingency.
- The ability to win across the widest possible range of conflicts (see Figure 10.1).

What was transformation?

The US Department of Defense has defined transformation as:

> A process that shapes the changing nature of military competition and cooperation through new combinations of concepts, capabilities, people and organisations that exploit our nation's advantages and protect against our asymmetrical vulnerabilities to sustain our strategic position, which helps underpin peace and stability in the world.[4]

Military transformation was a vision not just of the requirements of contemporary military operations, but also of what would be required to fight in the future. The label 'transformation' to describe a programme for reforming the US military was first used during the Bill Clinton Administration (1993–2001), and in practice the concept was an RMA-type view of the developing character of warfare featuring an emphasis on such features as Network-Centric Warfare (NCW), information dominance and stand-off precision firepower. The concept received renewed emphasis under the Presidency of George W. Bush. The Bush Administration embraced enthusiastically the idea of the RMA, seeing it, amongst other things, as a vision of warfare that, as Bush noted, 'perfectly matches the strengths of our country – the skill of our people and the superiority of our technology'.[5] Viewing the Pentagon itself as one of the obstacles to radical transformation of the US armed forces, Bush's Secretary of Defense, Donald Rumsfeld, argued that **new technology** alone would not suffice: **new concepts** and **organisational change** would also be required. Rumsfeld's commitment to transformation was reflected in his creation in 2002 of a specific Office of Force Transformation. Each of the US armed services was required to draw up a programme for transformation to meet the challenges of contemporary and future warfare.

The dimensions of army transformation

Beginning in 1991 with General George R. Sullivan, each successive Chief of Staff of the US army had their own vision of the character and pace of transformation, but the general concept was nevertheless embraced enthusiastically. Departing from the prescriptive doctrine and force structures of the Cold War, the overarching aim was to leverage the revolutionary potential of the RMA to create an army that could win decisively at any scale and in any type of armed conflict. The transformed army would be **rapidly deployable**: able to deploy a brigade globally in 96 hours, a division in 120 hours and five divisions within 90 days. It would be an army that was **more agile**; it would replace mass with precision effect; it would be **more lethal** even though its forces would be **smaller**; it would be more **versatile**; more dispersed; more sustainable; more connected; less rigidly hierarchical. To attain these goals, the transformation of the US army had a number of core themes, including those of technology, force structures, doctrine and jointery.

Technology

Two kinds of technology in particular underpinned army transformation. First, and reflecting the RMA assumptions underpinning military transformation as a concept, emphasis was placed on **information systems**, especially digitisation. For example, the army benefited from the Department of Defense's development of SIPRNET (Secret Internet Protocol Router Network), effectively a military version of the civilian internet which provided a secure communication network. More directly, however, under the 'Force XXI' programme begun in 1993, the US army sought to realise Network-Centric Warfare (see Chapter 9) through the creation of a digitally linked force. Here, for example, the army connected units through the creation of the Army Battle Command System (ABCS). Perhaps the most well-known element of ABCS is Blue Force Tracking (BFT), a system that displays the geographical position on the battlefield of friendly forces, but ABCS also includes other elements that integrate, for example, aspects of artillery, logistics and air defence into a common system.[6]

A second important technological aspect of army transformation was the development of **alternative weapons systems**. The army that fought and won the Gulf War was equipped with platforms, such as the Abrams Main Battle Tank and the Bradley Infantry Fighting Vehicle, that were procured during the Cold War. These legacy systems were regarded as an obstacle to the creation of an agile, rapidly deployable army. Under the 'Objective Force' programme outlined in 1999, the US army aimed to

Figure 10.1 Operating environments and missions of US forces
Source: (Force XXI Operations, TRADOC Pamphlet 525-5, 1 August 1994, 1–4)

develop a Future Combat System (FCS): a mix of manned and unmanned vehicles, unmanned aerial vehicles and specialist munitions, linked by an advanced digital network. In the interim, the army would procure a wheeled Infantry Fighting Vehicle, the Stryker: relatively light, mobile and digitally equipped. The aim was to create a force that was much more deployable than the Cold War legacy forces, but which was more lethal.

Force structures

Organisationally, transformation's key theme was **modularisation**, 'a force design methodology that establishes a means to provide interchangeable, expandable, and tailorable forces'.[7] Divisions were replaced as the key US army formation by the brigade. Overall, the army would comprise 40 infantry brigades, eight Stryker brigades and eight heavy (armoured) brigades. Moreover, these brigades would now become combined arms Brigade Combat Teams (BCTs), absorbing into their order of battle a slice of the headquarters, and supporting elements that had previously been held at a divisional level. BCTs would consist of a standardised organisation, allowing brigades to become functionally interchangeable. BCTs would also be more capable than their Cold War equivalents: they would have a larger logistics element, allowing them to sustain themselves more effectively in austere environments; they would have a larger headquarters, allowing them to plan and conduct more complex operations; they would also have a higher proportion of officers, and organic reconnaissance, artillery and brigade special troops.

Two other themes also emerged over time. One of these was the **increase in size** of the army under the 'Grow the Army' initiative. Whilst the total size of the army was cut in the immediate aftermath of the end of the Cold War, between 2003 and 2011 it was increased by over 50,000, from 1,062,042 (regular and reserve), to 1,112,119. Another was a **rebalancing** of the army: increasing the number of infantry battalions and psychological operations companies, for example, and reducing the number of chemical warfare brigades; shifting manpower from the administrative element to the 'operating army'; and placing additional emphasis on reserve forces.[8]

Doctrine

Developments in technology and structures would mean nothing, however, without an appropriate doctrine to guide their use. Doctrinal change was founded on a shift from threat-based to capabilities-based planning: of focusing not on the requirements to defeat a specific enemy but instead on the likely sorts of capabilities that might be required in the future. A process of reflection begun by Training and Doctrine Command (TRADOC) in 1990 culminated in 1993 in a new version of Field Manual (FM) 100-105, *Operations*. The new doctrine embodied many changes, not least embracing jointery, combined operations, and the importance of the relationship between the operational and strategic levels of war. The new doctrine also made reference to 'Full-Dimensional Operations': joint and combined operations that might span, sometimes simultaneously, the full spectrum of conflict. Reflecting this concern, FM 100-105 gave consideration to Operations Other Than War (OOTW) and even to conflict termination, recognising that 'Success on the battlefield does not always lead to success in war'.[9] In 2001, FM 100-105 was replaced by a new doctrine, FM 3-0, *Operations*, and the concept of Full-Dimensional Operations was developed into **Full-Spectrum Operations**. Effectiveness across this spectrum would require an appropriate mix of capabilities for combinations of offensive, defensive, stability and support operations (see Figure 10.2). FM 3-0 2001 highlighted the centrality to future army effectiveness of strategic responsiveness, joint operations and multi-national operations.[10] Army doctrine also embraced effects-based operations (EBO, see Chapter 9), instituting a formal system to relate the impact of violent ('kinetic') and non-violent ('non-kinetic') activity against targeted military, political, economic, social, informational and infrastructure 'systems'.

Jointery

US army transformation took place within a wider context of US defence efforts to improve joint co-operation. For example, the Goldwater-Nichols Act of 1986 had

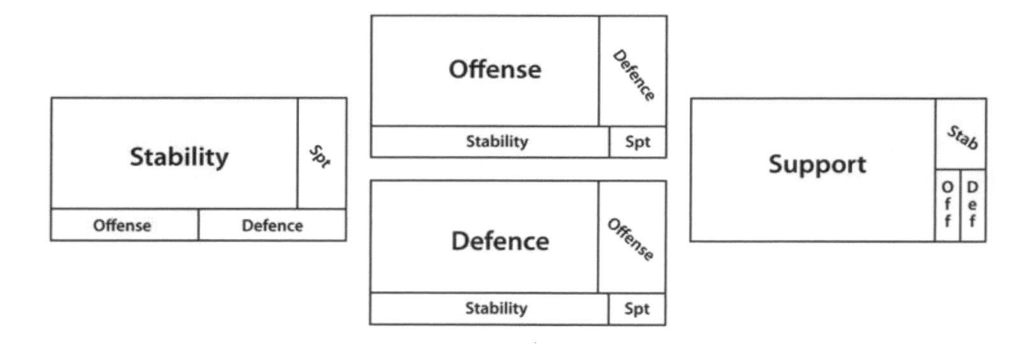

Figure 10.2 Full-Spectrum Operations
Source: (FM 3-0, *Operations* (Headquarters Department of the Army, June 2001) 1-14–1-16)
Full spectrum operations embodied the idea that the US army could be effective across the whole spectrum of war, from high-intensity conventional conflict to support for the civil authority, by focusing on developing its expertise in four areas and them employing these, in different combinations, as dictated by local conditions. Offense: destroy or defeat an opponent to impose US will and achieve decisive victory. Defence: defeat, delay, economise in forces. Creating conditions for later success. Stability: developmental, co-operative and coercive activities in response to a crisis. Support: assist foreign or domestic civil authorities to mitigate crises and relive suffering.

already ended operational control of single-service military forces by their own service chiefs. In the 1990s this process of facilitating jointery continued. In 1993, for example, US Atlantic Command (ACOM), which became Joint Forces Command (JFCOM) in 1999, was made responsible for training forces from all of the services for joint operations. Regional Combatant Commands (RCCs) were developed into meaningful joint headquarters organisations. The navy and air force also made investments in **equipment** that supported army capabilities: US navy purchases of fast sea-lift vessels, for example, and air force investment in C-17 Globemaster III heavy transports facilitated an increase in the strategic mobility of the army. To improve its capacity for joint operations, the army established more rigorous joint components in its **training regime**: for example, it included a joint element in its Battle Command Training Program (BCTP) from 1992.[11] At the same time, the whole focus on digitisation and modularity was designed to promote more effective integration between army units and those of their sister services.

The transformation debate

Military transformation was intended to produce a radically more effective US army – an army that was at the vanguard of new approaches to land warfare in the twenty-first century. Moreover, the US concept of transformation had a much **wider influence**. The scale of US success during the Gulf War of 1990–91, combined with the breadth, enthusiasm and optimism of the transformation vision, swept other militaries along. Perhaps unsurprisingly, such close US allies as the United Kingdom developed similar, if more modest, programmes of change. However, the status of the US military as the cutting edge of military effectiveness at the dawn of the twenty-first century gave military transformation a great deal of power as a model for reform. In an official report in 1997 by the US Government Accountability Office, it was noted that more than 100 states were embarking on processes of military modernisation. Despite the diversity of the states involved, which ranged from Ukraine to Malaysia, the vision of modernisation pursued by them was remarkably similar and reflected the core themes of the US transformation concept, including a desire to promote greater networking, better strategic mobility, tailored and modular forces, better joint and multi-national connectivity, and versatility across the spectrum of warfare.[12] However, as Chapters 5 and 9 have noted, first, there is no automatic reason why military reform need necessarily increase military effectiveness, and second, the RMA model upon which transformation is based is a contested one.

Not surprisingly, therefore, military transformation proved to be a controversial concept. Debates on the utility of transformation as a model highlighted a number of issues.

The rigour of transformation as a concept

Paralleling the difficulties in defining military revolutions and hybrid warfare outlined in the previous chapter, transformation has remained for many an **opaque concept** without much rigour. Transformation has been defined in many different ways (see Box 10.1), making it more difficult to discern how extensive or radical transformation was, and whether it constituted a continuous process or had a defined end state.

Box 10.1 Alternative definition of transformation

In war, transformational change means conserving equipment and operational methods that are still relevant while incorporating new technologies, tactics, and organisations that chart a more realistic path to future victory.

(Douglas A. MacGregor, *Transformation Under Fire: Revolutionizing How America Fights* (Westport, CT: Praeger, 2003), 8)

Without clarity on these issues, it has proven difficult to assess the extent to which transformation has or has not embodied elements of continuity with the past or the extent to which it extends to non-military agencies. Consequently, it also became difficult to assess the success or limitations of transformation. For some, transformation resulted in no substantial change in the crucial period of the 1990s because much of the US army remained wedded to legacy systems, but whether this is a fair assessment depends to an extent on how ambitious a definition of transformation one uses.[13]

Transformation and the reality of warfare

Reflecting the themes developed in Chapters 1 and 9, another source of debate was the extent to which the tenets of transformation would actually work in practice when faced with the realities of the nature of warfare, such as the inevitability of friction, and the limitations of RMA approaches.

Defending transformation

Advocates of transformation could point to two factors.[14] First, in Afghanistan in 2001–02 and in Iraq in 2003, the US forces won **startling successes**. Moreover, evidence from the battlefields seemed to suggest that, certainly at the tactical level, aspects of transformation, such as modularity, seemed to reduce military risk because the modular BCTs increased the capabilities available to local commanders and improved combined arms co-operation because the forces providing those diverse capabilities had trained together.

Second, these successes were won by an army that was **transforming but not transformed**. By 2003, for example, the US army had only a single division (the 4th Infantry Division) that was fully digitised, and the first Stryker brigade was not available for combat operations in Iraq until December of that year. Thus, if elements of both the Iraq and Afghan campaigns appeared very conventional, this was because many of the forces involved continued to rely on legacy systems.

Critiquing transformation

For transformation sceptics, however, the early stages of the wars in Afghanistan and Iraq provided little concrete evidence that the United States was leveraging a radically more effective land warfare paradigm. One source of criticism was that under the pressure of combat many aspects of **transformation simply did not work well** (see Box 10.2). Both wars demonstrated limitations in battle command systems, ground-air co-operation, knowledge of enemy forces and logistics. Moreover, it also became evident that even in more conventional military operations greater precision could not always substitute for mass, as Osama bin Laden's escape from the Tora Bora mountains illustrated.[15]

Box 10.2 The war in Iraq

Operation Iraqi Freedom (OIF) began on 19 March 2003 and was concluded formally on 1 May 2003, when President Bush announced the end of major combat operations. The campaign involved nearly 200,000 Coalition ground troops, of which the United States provided three-quarters. Whilst the campaign featured some heavy fighting in such places as Nasiriyah, Najaf and Karbala, Coalition success was overwhelming: Iraqi conventional forces disintegrated rapidly under the pressure of the Coalition's swift war of manoeuvre and exploitation. From a US perspective, many aspects of the campaign seemed to vindicate military transformation:

- Victory was rapid, with the key elements of the conventional campaign completed in three weeks.
- Coalition losses were low, at fewer than 200 killed.
- There was an unprecedented degree of air-ground co-ordination and integration.
- There was an unprecedented degree of integration between conventional and Special Operations Forces, including the placing of conventional forces under the control of Special Operations Forces.
- The networking of improved intelligence, surveillance and reconnaissance systems gave commanders at the divisional level and above an effective view of the battlespace.

Despite the scale of the Coalition victory, there were still many aspects of the campaign that proved problematic:

- Technical difficulties, including software and bandwidth problems, caused routine crashes of network systems.
- At the tactical level, commanders often lacked information on the position of enemy forces.
- The rapidity of the Coalition advance left large uncontrolled areas behind the advancing forces in which Iraqi irregulars were able to operate.
- The speed of advance combined with 'just in time' automated logistic management techniques resulted in a supply system that 'functioned barely above subsistence level'. Front-line troops suffered shortages of food, fuel and medical supplies.

(See Fontenot *et al.*, *On Point*, 408)

Compounding this argument, and reflecting earlier assessments of the Gulf War, some argued that in 2001 and 2003 the fundamental source of the United States' dramatic success lay in the **weakness of its opponents**. For example, whatever the actual strength of the Iraqi army (with estimates varying between 150,000 to 400,000), it is clear that they were much less capable than in 1991, being largely demoralised and poorly led. The Taliban, too, operated under key disadvantages, including their unpopularity with many Afghans and their lack of an effective air defence capability.[16]

More broadly, some argued that elements of the 'transformative' aspects of the two wars were actually rather **evolutionary in character**. With respect to modularity, for example, the creation of separate brigades for specific tasks had a long history, as did the idea of task organisation – of tailoring forces to meet specific local conditions.[17] Moreover, enemy adaptation often shifted the terms of the conflict to quite traditional forms of warfare. For example, Taliban adaptation to the effects of precision airpower required the coalition to return to traditional methods of close combat.[18]

Thus for some, the wars in Afghanistan and Iraq provided little concrete proof of the veracity or otherwise of the transformation concept.

The relevance of transformation

A second group of observations surrounded the relevance of a transformed military across the whole spectrum of warfare. Here, the fundamental criticism was that even if transformation lay at the heart of US success in 2001 and 2003, these successes constituted only partial victories. In each case, as Chapters 7 and 8 have discussed, the initial narrow military victory gave way to increasingly problematic stability operations in which it seemed fundamentally difficult to tie tactical and operational level military activity on the ground to measurable strategic level progress. Transformation, after all, was tied to the concept of Full-Spectrum Operations and the idea that US land forces would be effective across a wide span of contingencies including those at the lower end. Indeed, the importance of these sorts of operations was recognised in official publications (for example, the 1997 *Quadrennial Defense Review* argued that Operations Other Than War (OOTW) would become increasingly prevalent in the future), and in practical action, given the involvement of US ground forces in Somalia in 1993, Haiti in 1994 and Bosnia in 1995.[19]

Defending transformation

In defence of transformation, supporters argued two points. First, the difficulties in stabilising Iraq and Afghanistan were not linked to military transformation but to **broader mistakes** in Coalition strategy and policy, such as the lack of a clear strategy, or specific examples of poor policy decisions. For example, in Iraq the decision by the Coalition Provisional Authority to abolish the Iraqi army and to exclude Baath Party members from government employment created economic hardship and worsened the growing security problems.[20]

Second, the transformation concept itself was **adapted** to the world after the terrorist attacks of 11 September 2001 (9/11), under a process that some termed 'transforming transformation'. As Chapters 6 and 7 have noted, US approaches to counterinsurgency (COIN) and stability operations underwent considerable revision as codified in the 2006 *National Security Strategy* and 2006 *Quadrennial Defense Review*. New doctrines were accompanied by new equipment to cope with such threats as Improvised Explosive Devices (IEDs), and an increase in the number of troops on the ground.[21] Overall, 'transforming transformation' seemed to shift the transformation concept away from an RMA paradigm and towards a focus on OOTW.[22]

Critiques of transformation

However, others were less forgiving, arguing that the evidence of the stability phases of the wars in Afghanistan and Iraq highlighted several real problems with the transformation concept.

First, the problems of stabilisation in Afghanistan and Iraq seemed more than casually related to the **transformation agenda**. The problem seemed to be that belief in the power of a transformed military to achieve decisive military results convinced policy makers such as Rumsfeld that a light footprint would be sufficient to secure ambitious political objectives. As the previous chapter has noted, this focus on the defeat of the enemy military forces missed the point that securing the peace would require more troops on the ground, not fewer, and that the problems of stabilisation were less amenable to technological solutions than those of conventional warfare. Despite the rhetoric of Full-Spectrum Operations, transformation had delivered land forces predicated on the assumption that, through maximum force, rapid success, and a clear and early exit, war winning would lead to peace winning. Emblematic of this problem was the debate over the numbers of troops required for the Iraq War: General Eric Shinseki argued that, with the example of Bosnia, as many as 480,000 troops might be required; Rumsfeld wanted far fewer – perhaps as few as 60,000. Whilst Rumsfeld recognised that more would be required in the immediate aftermath of the war, it was assumed both that allies would provide many of them, and that the demands of Phase IV operations would be more limited because Iraq would quickly be stabilised.[23]

A second criticism made was that the whole idea of 'transforming transformation' represented such a **significant shift from transformation's original meaning** that it had, in effect, illustrated the bankruptcy of transformation as a concept. Even if the revised concept had many thematic similarities with the original, for example in terms of a continued focus on digitisation, it was argued that transforming transformation was simply a process of dedicating US ground forces to the needs of counterinsurgency in Iraq and Afghanistan. In essence, whereas transformation had been shaping an army for conventional warfare, transforming transformation shaped it for COIN.[24]

Adaptation: the case of China

A further set of criticisms of transformation as an idea related to the inherently adversarial nature of armed conflict discussed in Chapter 1. Intrinsic to the transformation agenda was the assumption that it was a mechanism to sustain and entrench US military superiority. However, as the previous chapter has discussed, there is considerable debate about whether, even if RMAs actually exist, they can translate into a long-term advantage for only one state. The US military's status as a paradigm army, established through its victories in 1991, 2001 and 2003, almost inevitably makes it the standard against which other armies are likely to measure themselves. Thus, such processes as imitation and adaption outlined in Chapter 2 might make transformation less revolutionary in its impact than enthusiasts supposed.

One of the most obvious advocates of this position is the People's Republic of China (PRC), a state which constitutes perhaps the most obvious of the United States' strategic competitors. The approach to post-Cold War military change by China's People's Liberation Army (PLA) has been shaped by many factors:[25]

- **Economic growth**: reforms begun in the late 1970s delivered high rates of economic growth, averaging around 10% of gross domestic product (GDP) in the 1990s. This, combined with the modernisation and diversification of China's industry, provided the material means for new military investment.
- **International relations**: the demise of a common adversary (the Soviet Union), a gradual re-focusing of the United States towards the Asia-Pacific, which China sees as threatening, and what the United States sees as an increasingly muscular Chinese foreign policy have raised periodic tensions in Sino-US relations.
- **Military conflicts**: US successes in the Gulf War, Kosovo, Afghanistan and Iraq highlighted the effectiveness of US forces and created great interest in the ways in which these wars were fought.
- **Strategic culture**: for some, the PLA's interest in the RMA debate was shaped by a traditional approach to war characterised by a focus on the importance of superior knowledge, information and the manipulation of perceptions.
- **External debates**: China has closely followed Western debates (and Soviet thinking before 1991) on military change and its implications.

In consequence of these factors, the PLA has been an enthusiastic advocate of the concept of an emerging *xin junshi geming*, or 'new military revolution'. This concept has much in common with transformation. Notably, the PLA has emphasised the importance in future warfare of information as a result of wider social and economic changes in society. Increasingly, information superiority over an adversary has become a vital component for military success. Future war, according to the PLA, will be *xinxi hua*: 'informationalised war'. As one Chinese academic has noted, 'war in the information age will emphasize the asymmetrical contest of information that is silent and invisible. This trend is hastening the birth of a brand-new form of war'.[26] In the early years of the twenty-first century, the PLA outlined a doctrine of the 'RMA with Chinese Characteristics'. This approach emphasised many themes that mirror transformation efforts. These include:[27]

- A reduction in the size of the PLA to allow a focus on quality over quantity.
- The creation of elite reaction units.
- The importance of joint, integrated military operations.
- Investment in advanced technology equipment.
- Investment in Command, Control, Communications, Computers, Intelligence, Surveillance and Reconnaissance (C4ISR) to link platforms in four dimensions.
- The need to acquire capabilities to conduct information warfare.
- Improving the training and education of personnel.

However, whereas transformation was built on the idea of leveraging change to enhance US military superiority, the PLA believe that informationalised warfare will have the reverse effect: it will strengthen the Chinese military disproportionately. In particular, the PLA has highlighted three factors:

- The globalisation of information will allow new technology and ideas to flow into China at an accelerating rate, helping the PLA to leap forward in its capabilities.
- Evidence from conflicts suggests that the United States has exploitable political weaknesses, especially in relation to political resolve.

- A reliance by adversaries on information superiority and networks creates targetable weaknesses. Cyber warfare, anti-satellite technology and precision attack could be used to threaten enemy command and control, sensors, communications networks and remotely operated platforms, and inflict, or threaten to inflict, crippling damage to an enemy's military systems.[28]

Developments in the PLA since the end of the Cold War therefore create interesting issues for the debates on the utility of transformation as a model of future land warfare. On the one hand, the PLA's enthusiasm for informationalised warfare provides some support for the idea that such features as information superiority, networking, jointery and high-technology equipment might form important features for future military effectiveness. Yet the PLA example provides two important caveats: first, does transformation open up new points of vulnerability for future armies, such as their reliance on digital networks, that can be exploited by adaptive adversaries? Second, might the technology and doctrines of transformation actually allow potential US adversaries to close the gap in military effectiveness? In either case, it might prove difficult to use transformation as a mechanism to entrench US military superiority.

The transformation gulf

A final issue for debate is the extent to which the US model of transformation is likely to become the future norm for advanced land forces. The fact that many armies globally have been influenced by the US transformation agenda might lead to the assumption that the transformed army model will diffuse inevitably throughout other militaries.

At one level, there would seem to be concrete evidence for this, especially given the model of modernisation adopted in North Atlantic Treaty Organization (NATO) countries. The United Kingdom provides a useful example of this phenomenon. In the early 2000s the model of modernisation adopted by Britain prioritised three themes, each of which reflected core elements of US military transformation:[29]

- **Networking**: the United Kingdom pursued its own version of NCW, entitled Network-Enabled Capability (NEC). NEC was defined as 'the coherent integration of sensors, decision-makers, weapon systems and support capabilities'.[30] For the British army, NEC was a 'key enabler', linking together other key features of modernisation. Technologically, it was supported by acquisitions that included a Defence Information Infrastructure (DII), a new satellite communications service and a new secure tactical communications system.
- **Doctrine**: Drawing on the US EBO, Britain adopted an effects-based approach to operations (EBAO), shifting the focus of military thinking from destruction to a consideration of often non-violent effects and the effective integration of military and non-military activity.
- **Force structures**: increasing emphasis was placed on creating forces that were more joint, expeditionary and modular in structure. This was reflected in the creation of a Joint Rapid Reaction Force and the move later to create more deployable brigades and battlegroups. In particular, the army sought to increase its medium-weight capability through the acquisition of the Future Rapid Effects System (FRES), a

family of vehicles designed to give the army a flexible, mobile platform for future warfare.

There were several 'push' and 'pull' factors that led the United Kingdom to adopt this model, including changes in the threat environment and resource constraints. However, emulation of the United States was another factor: acquiring transformation-type capabilities was seen as a signal of Britain's commitment to partnership with the United States, and it was hoped that it would also improve interoperability.[31] Despite the apparent similarities between the US and British approaches to modernisation, however, in practice there have been many differences: in essence, the British approach to transformation was less ambitious, less technophile and much more incremental. There were several reasons for this:

- Overall, the UK's **lack of resources** mandated a more incremental approach to change. FRES, for example, was never intended to be the digitised tank substitute that the US Future Combat System was supposed to be.
- NEC was a much less ambitious take on the US NCW, reflecting the UK's **reading of experiences** in Iraq and Afghanistan. For the UK, both conflicts demonstrated the limitations of NCW and the importance of the traditional war-fighting verities, such as tempo and skill.
- British **command philosophy** also played a role, with the army suspicious of command and control systems that might undermine 'mission command' and encourage senior officers to meddle in tactical actions.
- Another contributory factor was **technological scepticism** – a belief that there were limits to how far technology genuinely could substitute for properly trained personnel.[32]

This rather incremental take on transformation was reflected in the British concept of an EBAO. EBAO was a rejection of the process-led EBO approach, and was instead a more nebulous concept that gave priority to a 'way of thinking' about the relationship between military activity and political effect, and which prioritised also the need for multi-agency approaches to operations. However, even this less ambitious approach proved problematic. Having rejected EBO as too rigid, EBAO proved too vague and too difficult to institutionalise: a 2006 report noted that most officers deploying into Iraq did not understand the concept. Nor did other departments of the British government seem keen to embrace it. In the end, EBAO was dropped, although some of its concepts, such as the importance of 'influence' in military operations, survived in doctrine. Even the technological aspects of British transformation proved problematic. The new secure tactical communications system, Bowman, was criticised for being too costly, too heavy and too unreliable. The core element of FRES, the Utility variant, was cancelled in 2008, and the programme was described in 2010 by one British official as 'dead in the water'.[33]

In practice, then, whilst the general tenets of British and US transformation were similar, in practice there was a great deal of difference. This problem extended to other NATO allies such as France, Poland and Germany: if the language of transformation used by these various states was similar, the substance was in reality often very different.[34] Whilst this is not surprising, given the discussion in Chapter 5, nevertheless this creates potentially serious difficulties for coalition warfare in which, paradoxically, the pursuit of the same modernisation themes might widen the disparities between the United States and its allies. If multi-national integration is indeed a crucial element in future land warfare, then the US pursuit of its own transformation agenda may make this integration more, not less, problematic.

Assessing the impact of transformation on the effectiveness of the US army is challenging, since it is difficult to draw incontestable causal relationships between transformation and the army's performance in Iraq and Afghanistan. Whether or not the successes and failures in these campaigns were the result of the equipment, doctrines or force structures created by transformation, or whether they can be attributed to other factors such as strategy or enemy action, is difficult to say. Certainly, however, under the Barack Obama Administration, transformation has lost some of its impetus as a concept. The term 'transformation' largely fell out of use and the Office of Force Transformation had already been disbanded in 2006. Moreover, other key aspects of the programme have run into difficulties. As noted in the previous chapter, effects-based operations was dropped as a concept: its complexity and flaws had already led many officers to ignore it. The Future Combat Systems programme was cancelled in 2009, being branded 'a very expensive failure'.[35]

Overall, it would seem fair to say that military transformation has not wrought the epoch-making changes envisaged for it in the 1990s. Whilst the US army post-Afghanistan will seek to develop its capacity to support the capabilities of foreign militaries through forward presence, it is also seeking to return to a focus on some key pre-9/11 themes, not least the ability of ground forces to inflict decisive conventional defeat on invading forces or to undercut illegitimate regimes. Nevertheless, the ripples from the transformation agenda continue to have an impact: many of its tenets remain important, not least modularity and networking, and its impact on US army force structures and doctrine is likely to continue to influence current and future operations.

Beyond transformation

Looking forward, the US army faces many challenges in preparing itself to meet the demands of future warfare. The US military has set about trying to analyse the nature of the future threat environment and the sorts of forces required to meet that threat. Looking at the future, the Pentagon argues that it will be marked by continued conflicts: such factors as demographic changes, the variable effects of globalisation, resource shortages, competition with rising or revisionist powers, the erosion of state authority and the empowerment of non-state actors will all result in the continued need to deter threats or employ the use of military forces. Indeed, in this analysis, the Pentagon is not alone (see Box 10.3).

Box 10.3 The future character of conflict

British thinking focuses on a future of growing instability caused by state failure, strategic competition, struggles for access to resources, climate change, the effects of globalisation and extremist political ideologies. The operating environment for military forces will be characterised by five attributes:

- Congested: the battlespace is likely to be densely populated by civilians and/or by the vessels and aircraft of neutral or friendly powers.

- Cluttered: adversaries will be able to exploit the environment to conceal themselves, exploiting better local knowledge and the constraints imposed on us by the need for legitimacy and discrimination.
- Contested: exploiting the proliferation in sophisticated technology, enemies will be able to broaden the dimensions of conflict, exploiting the political, social, cyber and information spheres.
- Connected: future conflict will focus increasingly on the key nodes that connect together the various geographic, logistic, technological and communications networks.
- Constrained: legal and moral requirements will demand greater discrimination in the use of force.

(From *Strategic Trends Programme: Future Character of Conflict* (Development, Concepts and Doctrine Centre, Shrivenham, February 2010), 1–6, 21–25)

These future conflicts are likely to pose difficult challenges:

- The **diversity** of future threats will require a capability to succeed across a wide spectrum of conflict against a mix of state and non-state actors.
- The growing importance of the **cyber and space dimensions** will create potential new vulnerabilities for technologically networked forces.
- The **diffusion** globally of sophisticated technology is likely to erode the qualitative advantage in hardware possessed traditionally by the US military.
- The **competitive logic** of war is likely to drive adversaries to focus on US weaknesses and critical enabling capabilities such as computer and communications networks.
- With an increasing proportion of US forces based in the United States itself, the army, especially, may be **far from where it is needed**, opening up opportunities for adversaries to interfere with the movement or sustainment of forces, or leaving US forces without bases of operation.
- The potential proliferation of nuclear, biological and chemical capabilities may **deter the United States** from using military force.
- Media scrutiny, the importance of the 'battle of the narratives' against adversaries, as well as the demands of effective strategy may require even **more discrimination** in the use of force and even **more understanding** of the local context.[36]

Globally Integrated Operations

In consequence of its process of reflection, the Pentagon has defined nine key missions for the US armed forces of the future (see Box 10.4). In order to achieve these goals the Pentagon has identified the need for a concept of Globally Integrated Operations: 'a globally postured Joint Force to quickly combine capabilities with itself and mission partners across domains, echelons, geographic boundaries, and organizational affiliations.'[37] Unsurprisingly, the thrust of this new concept is to push further the boundaries of jointery, advocating, for example, the development of common understandings of mission command; alternatives to mass formations, including cyber and global strike; developing new, more flexible approaches to joint organisation that move beyond just functional or geographic boundaries; and putting more focus on partnering with other government agencies, or allies. The US armed forces needed to become, according to the Pentagon's new concept, '**pervasively interoperable**'.

Box 10.4 Future missions of the US military

- Counterterrorism and irregular warfare.
- Deter and defeat aggression.
- Counter weapons of mass destruction.
- Defend the homeland and provide support to civil authorities.
- Project power despite anti-access/area denial challenges.
- Operate effectively in cyberspace.
- Operate effectively in space.
- Maintain a safe, secure and effective nuclear deterrent.
- Provide a stabilising presence.
- Conduct stability and counterinsurgency operations.
- Conduct humanitarian assistance, disaster relief and other operations.

(From *Sustaining U.S. Global Leadership: Priorities for 21st Century Defense* (Department of Defense, January 2012), 4–6)

One of the especially interesting elements of the concept of Globally Integrated Operations was its recognition of the need to prepare for operations in '**degraded environments**': in other words, it recognised that future adversaries might have the capability to attack and disrupt the United States effectively in the cyber and space spheres, creating the need to develop resilience enough to survive this.

The US army and the future

In its *Army Strategic Planning Guidance* document of 2013, the US army laid out the parameters of its future development. This document was high on references to adaptability and flexibility in relation to the complexity and uncertainty of the future threat environment. Under the title of 'Army Modernization', this was a less ambitious vision of future development than that of military transformation. The two key concepts were those of 'regionally aligned' forces and 'mission-tailored' forces, the regionally aligned forces being those units shaped to meet the needs of specific regional commands, and the mission-tailored forces being those developed to meet the three key US army missions – counterterrorism and irregular warfare, deterring or defeating aggression, and defending the homeland. The US army will seek over the coming years to enhance its practice of mission command; to modernise its equipment; to invest particularly in improving the capabilities of forces below the brigade level; to improve its resilience against cyber and space attack; to improve its capabilities for joint theatre entry scenarios; to develop its regional and cultural expertise; and to try to hold on to the experience gained so painfully in Iraq and Afghanistan.

How effective will this new approach to the future be? Though less ambitious than the programmes of the 1990s, there remain serious challenges to the US army's ability to position itself successfully to meet future threats.

Financial constraints

The impact of the **global financial crisis** beginning in 2008 has led to cuts in US defence spending. The army is likely to have to undergo cuts of US$170 billion over ten years from 2013. In consequence, the size of the active duty army will fall by 80,000 and the number of BCTs may be cut by as many as 13. Modernisation will have to take place over a much longer period of time, in circumstances of continued fiscal uncertainty.

US grand strategy

The **foreign policy 'pivot'** towards the Asia-Pacific that has taken place under the Administration of President Obama has shifted US defence policy focus towards a region in which air and maritime forces seem to be pre-eminent instruments of US power. The US response to conflicts in Libya in 2011 and in Syria (beginning in 2011 and ongoing at time of writing) also illustrate the effects of the wars in Iraq and Afghanistan, not least a reluctance to countenance in the immediate future the significant deployment of ground forces.

A continued lack of consensus

Even amongst supporters of army investment, many continue to **challenge the value of post-Cold War assumptions** regarding the desirable attributes of future land forces. For some, for example, the scope for US army expeditionary operations is likely to be curtailed by the development by likely enemies of improved anti-access capabilities, such as air defence systems, mines and missiles, and the focus should instead be on long-term forward deployments and a greater investment in technology such as missiles. For others, Iraq and Afghanistan demonstrate the continued value of heavy armoured forces not just in terms of conventional operations against potential adversaries such as Iran, North Korea and China, but in unconventional operations as well, where they provide flexibility, accurate firepower and protected mobility.[38]

Doctrinal debates

Even the **doctrinal basis** of the future US army is controversial. In October 2011, the army revised its doctrine under the new title *Unified Land Operations.* The term Full-Spectrum Operations was dropped and replaced by the term 'Decisive Action', used to describe simultaneous offensive, defensive, stability and defence support for civil authority.[39] It also introduced two new ideas: 'Combined Arms Maneuver', the focusing of combat power to defeat enemy ground forces; and 'Wide Area Security', with a focus on protection and the consolidation of control.[40] However, not all were impressed by the new doctrine, some characterising it as 'mumbo jumbo' and others as an assemblage of 'reasonable but timid generalities'.[41]

Forgetting the past

The US army intends to move from a focus on counterinsurgency to a focus now on a full range of operations, declaring the need to 'reinvigorate capabilities that have declined'. However, historical precedent suggests it may be difficult to retain the experience of Iraq and Afghanistan. One of the many interesting consequences of the Vietnam War was the way in which the army chose to respond to defeat, **refusing to learn** from it.[42] Instead, the US army embraced an explanation for defeat that focused on strategic level political causes and renewed its focus on conventional warfare. The foundation for the renewal of the army in the 1970s came, then, from a deliberate focus on conventional warfare, a form of war more consonant with the values of the army.

Foundation difficulties

Perhaps the most compelling challenges to the US army's future plans, however, derive from some of the key themes that have been evident again and again throughout the preceding chapters of this book. One is the fact that it is inherently **difficult to predict** the needs of future war and, as previous chapters have shown, land forces aren't generally very successful at guessing. Acknowledging that the future is complex and uncertain is one thing; being able to make an effective practical response is another. As the strategist Colin Gray notes, 'we know nothing, literally zero, for certain about the wars of the future, even in the near-term'.[43]

Another recurrent problem has been the evident **difficulty in realising** such commonsense ideas as greater integration and jointery between armed services, or greater flexibility and capacity for adaptation. The Pentagon itself recognised that the concept of Globally Integrated Operations carried with it risks. Some of these related to the availability of key enablers, such as the right technology or willing partners.

However, other potential risks lie in the **tensions and trade-offs** that are inherent to land warfare. Might an over-emphasis on decentralisation lead to a loss of focus in operations? Might removing some of the redundancies in existing methods and organisation undermine the resilience of the US military? Might a focus on flexible integration prevent the development of trust and understanding between individuals and organisations that comes with long-standing co-operation?

Conclusion

By virtue of its dominant role in such alliances as NATO and because of its battlefield success from the Gulf War onwards, US perspectives on what constitute an effective land force for current and future warfare have gained wide currency. Since the end of the Cold War, a focus on 'transformation', an RMA-based vision of the needs of future warfare, has led the US army to focus its development on four themes: new technology, revised force structures, doctrinal development, and a renewed focus on joint warfare. Transformation tapped into the belief that 'there is … reason to think that a major change – call it transformation or not – in warfare has occurred … the RMA is here to stay'.[44] Shaped by this belief, the US army embarked on a process of creating a modularised, digitised and networked force, focusing on principles such as effects, deployability and agility in opposition to a traditional focus on linearity, weight and mass. At the same time, the army continued to highlight a formalised version of operational art and the operational level of war – an approach that was adopted by its allies.

It is testament to the sorts of uncertainties explored in the previous chapters of this book that despite the extraordinary intellectual and practical efforts made by the US army since the end of the Cold War, transformation has remained the subject of considerable scepticism. Critics have argued that US battlefield victories from 1991 onwards have relied more on fighting sub-standard opponents than on leveraging a new paradigm of war. Moreover, despite a commitment to effectiveness across the full spectrum of warfare, operations in Iraq and Afghanistan challenged the premise that it was possible to create doctrine, equipment and force structures that would allow one army to win decisively whatever the scale of conflict. Thus, even if the US army was leveraging an RMA, it is possible that this RMA did not, in fact, apply to every aspect of warfare. Even if transformation had improved some US capabilities in land warfare, there remained difficulties, not least compensatory adaptation by strategic competitors, and the growing gulf in capabilities between the United States and its allies.

The debates on the rights and wrongs of these issues continue: it is testament to the difficulties in shaping an army to meet the challenges of future conflicts that there are few issues that are not to some degree contested. The historian Michael Howard commented: 'No matter how clearly one thinks, it is impossible to anticipate precisely the character of future conflict. The key is to be not so far off the mark that it becomes impossible to adjust once that character is revealed.'[45] The difficulty that continues to face the US army is that it simply isn't clear even where that mark should be.

Notes

1 Theo Farrell, 'The Dynamics of British Military Transformation', *International Affairs* Vol. 84, No. 4 (2008), 778; Keith L. Shimko, *The Iraq Wars and America's Military Revolution* (Cambridge: Cambridge University Press, 2010), 106.
2 Frederick W. Kagan, *Finding the Target: The Transformation of American Military Policy* (New York, NY: Encounter, 2006), 204.
3 Stuart E. Johnson, John E. Peters, Karin E. Kitchens, Aaron Martin and Jordan R. Fischbach, *A Review of the Army's Modular Force Structure* (Santa Monica, CA: RAND, 2012), 16.
4 Olof Kronvall, 'Transformation: The Key to Victory?' in Karl Erik Haug and Ole Jorgen Maao (eds) *Conceptualising Modern War* (London: Hurst and Company, 2011), 260.
5 Shimko, *The Iraq Wars*, 132.
6 Gregory Fontenot, E.J. Degen and Tommy Franks, *On Point: The United States Army in Operation Iraqi Freedom* (Annapolis, MD: Naval Institute Press, 2005), xiii–xiv, 11.
7 *Force XXI Operations, A Concept for the Development of Full-Dimensional Operations for the Strategic Army of the Twenty-First Century* (TRADOC Pamphlet 525–5, 1 August 1994), Glossary-5.
8 Johnson *et al.*, *A Review of the Army's Modular Force Structure*, 18–21.
9 Fontenot *et al.*, *On Point*, 9.
10 Ibid., 20–21.
11 Fontenot *et al.*, *On Point*, xiv, 17–19.
12 See Deborah Sanders, 'Ukraine's Military Reform: Building a Paradigm Army', *The Journal of Slavic Military Studies* Vol. 21, No. 4 (2008), 599–614.
13 Shimko, *The Iraq Wars*, 108.
14 Johnson *et al.*, *A Review of the Army's Modular Force Structure*, 38; Fontenot *et al.*, *On Point*, 1.
15 Elinor Sloan, *Military Transformation and Modern Warfare: A Reference Handbook* (Westport, CT and London: Praeger Security International, 2008), 20–21, 123–24.
16 Kagan, *Finding the Target*, 294.
17 Johnson *et al.*, *A Review of the Army's Modular Force Structure*, 49–52.
18 See Steven Biddle, *Afghanistan and the Future of Warfare: Implications for Army and Defense Policy* (Carlisle, PA: Strategic Studies Institute, 2002).

19 David Jablonsky, 'Army Transformation: A Tale of Two Doctrines', *Parameters* Vol. 31, No. 3 (Autumn 2001), 44.
20 Michael R. Gordon and Bernard E. Trainor, *Cobra II: The Inside Story on the Invasion and Occupation of Iraq* (London: Atlantic, 2006), 484.
21 Kronvall, 'Transformation', 265.
22 Ibid., 264.
23 Anthony H. Cordesman, *The Iraq War: Strategy, Tactics, and Military Lessons* (Washington, DC: Center for Strategic and International Studies, 2006), 494–506.
24 Kronvall, 'Transformation', 266, 283.
25 Jacqueline Newmyer, 'The Revolution in Military Affairs with Chinese Characteristics', *The Journal of Strategic Studies* Vol. 33, No. 4 (August 2010), 491–94.
26 Ibid., 495.
27 Ibid., 496–99.
28 Ibid., 483–89.
29 Farrell, 'The Dynamics of British Military Transformation', 784–85, 800.
30 *Delivering Security in a Changing World: Future Capabilities*, 2003 Defence White Paper (Ministry of Defence, HMSO, 2004), 5.
31 Farrell, 'The Dynamics of British Military Transformation', 786.
32 Ibid., 787–89.
33 '£16bn Future Rapid Effects System Faces Axe in Defence Cuts', *The Daily Telegraph*, 9 October 2010, www.telegraph.co.uk/news/uknews/defence/8052782/16bn-Future-Rapid-Effects-System-faces-axe-in-defence-cuts.html (accessed 07/08/2013).
34 Terry Terriff and Frans Osinga, 'Conclusion: The Diffusion of Transformation to European Militaries', in Terry Terriff, Frans Osinga and Theo Farrell (eds) *A Transformation Gap: American Innovations and European Military Change* (Stanford, CA: Stanford University Press, 2010), 191.
35 Andrew Krepinevich, quoted in BreakingDefense.com, 'Total Cost to Close Out Cancelled Army FCS Could Top $1 billion', 19 June 2012, breakingdefense.com/2012/06/19/total-cost-to-close-out-cancelled-army-fcs-could-top-1-billion/ (accessed 12/08/2013).
36 See *The Joint Operating Environment 2010*, US Joint Forces Command (February 2010).
37 *Capstone Concept for Joint Operations: Joint Force 2020* (10 September 2012), 4.
38 See Jim Thomas, 'Why the Army Needs Missiles: A New Mission to Save the Service', *Foreign Affairs* Vol. 92, No. 3 (May 2013), 137–44; Chris McKinney, Mark Elfendahl and H.R. McMaster, 'Why the U.S. Army Needs Armor: The Case for a Balanced Force', *Foreign Affairs* Vol. 92, No. 3 (May 2013), 129–36. See also *Statement by General Raymond T. Odierno*, First Session, 113th Congress, 26 February 2013, docs.house.gov/meetings/AS/AS00/20130918/101291/HHRG-113-AS00-Wstate-OdiernoUSAR-20130918.pdf (accessed 12/08/2013).
39 Army Doctrine Publication (ADP), *Army Doctrine Update 1–12*, US Army Combined Arms Center, 16/12/2011, 8.
40 ADP 3-0, *Unified Land Operations* (October 2011), Glossary 1.
41 Major J.P. Clark, 'The Missed Opportunity: A Critique of ADP-3-0, Unified Land Operations', *Military Review* (July–August 2012), 48; Andrew B. Nocks, 'More Mumbo-Jumbo: The Clutter and Confusion Within the Army's Operational Concept of Unified Land Operations', *Small Wars Journal*, 8 June 2012, smallwarsjournal.com/jrnl/art/more-mumbo-jumbo-the-clutter-and-confusion-within-the-army%E2%80%99s-operational-concept-of-unified (accessed 12/09/2013).
42 Hew Strachan, 'Operational Art and Britain, 1909–2009', in John Andreas Olsen and Martin van Creveld (eds) *The Evolution of Operational Art: From Napoleon to the Present* (Oxford: Oxford University Press, 2011), 119.
43 Colin S. Gray, 'War – Continuity in Change, and Change in Continuity', *Parameters* (Summer 2010), 5.
44 Eliot A. Cohen, 'Change and Transformation in Military Affairs', *Journal of Strategic Studies* Vol. 27, No. 3 (September 2004), 407.
45 Quoted in *Strategic Trends Programme: Future Character of Conflict* (Development, Concepts and Doctrine Centre, Shrivenham, February 2010), 1.

Suggested reading

Thomas K. Adams, *The Army After Next: The First Postindustrial Army* (Stanford, CA: Stanford University Press, 2008). Adams presents a critical view of transformation and the impact that it has had on the US army's ability to fight in Iraq and Afghanistan.

Colin S. Gray, 'War – Continuity in Change, and Change in Continuity', *Parameters* (Summer 2010), 5–13. A voice of studied realism on the problems of trying to conceptualise changes in war.

Walter E. Kretchik, *U.S. Army Doctrine: From the American Revolution to the War on Terror* (Lawrence, KS: University of Kansas Press, 2011). A readable and comprehensive overview of US army doctrinal development from 1776 to 2008.

Douglas A. MacGregor, *Breaking the Phalanx: A New Design for Landpower in the 21st Century* (Westport, CT: Praeger, 1997). Outdated now, but worth reading for the 1990s take on the future challenges of land warfare and the adaptation required by the US army in response.

Frans Osinga and Theo Farrell (eds), *A Transformation Gap: American Innovations and European Military Change* (Stanford, CA: Stanford University Press, 2010). Provides a comprehensive overview of US and European transformation efforts.

Robert H. Scales, Jr, *Yellow Smoke: The Future of Land Warfare for America's Military* (Lanham, MD: Rowman and Littlefield, 2006). An RMA-centric view of the kind of US army required for the future.

Keith L. Shimko, *The Iraq Wars and America's Military Revolution* (Cambridge: Cambridge University Press, 2010). Shimko provides a thoughtful and balanced analysis of the performance of transformation in the light of the Iraq wars.

David Talbot, 'How Technology Failed in Iraq', *MIT Technology Review* (1 November 2004). www.technologyreview.com/featuredstory/403319/how-technology-failed-in-iraq/. Provides a fascinating insight into the technological limitations of the transformed army in Iraq.

Conclusion

Over the ten chapters of this book, we have investigated the development of land warfare in its different forms since 1900 and the associated debates and controversies. Whilst it is a challenge to conclude a subject as broad as modern land warfare, cumulatively, the chapters of this book have introduced and reinforced a number of themes that are important to grasp for a proper understanding of modern land warfare.

Land warfare is important

Understanding land warfare matters because land has a crucial political significance. Whilst the air and maritime environments can be vital media before and during armed conflicts, only land forces can take and hold ground. Land power, the ability to influence and coerce on or from the ground, therefore carries with it the capacity for decision in warfare: control of ground facilitates control of people and key points, and so provides the basis for achieving broader political goals. Land forces can achieve those goals not just through fighting high-intensity conventional wars, but also through counterinsurgency and peace and stability operations. There is a flexibility inherent in land power that derives from its focus on the soldier and the soldier's ability to interact with other human beings on an individual level.

Land warfare is complex

Land warfare is not a simple activity. If it were, it would be much easier to do well. Instead, land warfare is defined by a medium, land itself, that creates both problems and opportunities. By using the terrain effectively, land forces can mitigate the terrible effects of modern firepower, but in order to utilise the terrain effectively land forces must be complex and adaptive organisations. Making the best use of ground generates difficult tensions and trade-offs that have shaped the conduct of armies and the outcomes of land campaigns. The advantages and disadvantages of such competing trade-offs as fire and movement, mass and dispersal, centralisation and decentralisation create difficult choices for land forces. Innovation in doctrine can harness advantages and minimise disadvantages, but poor tactical and operational methods can open up land forces to crushing defeat against more competent foes.

Land warfare is a dependent activity

Land warfare cannot be approached in isolation if it is to be waged successfully. Tactically and operationally, land warfare must be conducted in relation to inter-service interaction (joint warfare) and multi-national operations (combined warfare). Joint warfare is crucial if the services are to mitigate their own weaknesses and amplify their strengths. Land warfare conducted without proper regard for the other armed services is unlikely to be successful except in circumstances of huge advantages in other spheres. Equally, conducting land warfare without reference to the demands of strategy and policy is a recipe for a disaster, likely to lead to uncoordinated campaigns and purposeless military activity. There is no point in winning battles if the cost is losing wars. All of this means that an understanding of land warfare is a necessary, but not sufficient, condition for success.

Land warfare is difficult to do well

There are no easy routes when it comes to prosecuting land warfare effectively: there are no templates that can be applied that will ensure success; there are no 'silver bullet' solutions. The nature of land warfare has not changed over time, and those things that continue to define its nature, such as friction, its human component and interaction with the adversary, make land warfare relentlessly difficult to do well. Moreover, the fact that the character of war changes between, even within, each conflict means that the context for land warfare changes constantly. In consequence, history has shown how difficult it is to translate even plausible-looking military theory into effective military practice. Principles of war, principles of counterinsurgency, principles for peace and stability operations – these have developed as the result of an accumulation over time of often hard-won experience. However, the history of land warfare is littered with armies that have triumphed in one conflict only to lose in the next, or of armies that have failed in one conflict only to repeat their failings in another. Even where armies have had extensive and relevant doctrine for a conflict, there is no guarantee that this doctrine will be applied appropriately, or, indeed, at all.

Modern land warfare has been an evolutionary activity

At the core of this book has been Stephen Biddle's idea of the 'modern system of warfare': a tactical and operational method of warfare based on such interlocking themes as dispersion, fire and movement, depth, initiative, and reserves. Modern system tactical warfare emerged during the First World War; modern system operational level warfare emerged during the Second World War. Whilst each of these two developments constituted in some respects an important break with the past, they were still themselves examples, perhaps, of 'ratcheted change': each reached backwards as well as forwards. Some aspects of the modern system tactical approach were already evident in the pre-war debates on the implications of modern firepower. At the operational level, the continuity was even more pronounced, with operational art arising from the application of tactical level themes drawn from the First World War but on a different scale.

Evolution has been even more pronounced since 1945. Much that has been identified as 'revolutionary' has been seen by other writers as manifestations of evolution. So, for example, the concept of networking, which has been accorded by some writers a prominent role in the Revolution in Military Affairs (RMA), has been seen by others as an example of military 'business as usual'. Networking, far from being a revolutionary theme, has instead been central to the whole idea of military communications and the linking of sub-elements of land forces with one another and with other services. After all, the recognised benefits of networking were at the heart of the push of rapid efforts to utilise new communications technology such as the field telephone and wireless radio. Even in counterinsurgency and peace and stability operations, the heart of modern doctrine is rooted in ideas and concepts with a long historical provenance.

Land warfare is not a decisively technological activity

Technology has played an important role in driving forward developments in the theory and practice of land warfare. Aircraft, new means of communication, and the increasing range and rate of fire of artillery and small arms have created new problems and opportunities for armies since 1900. However, land warfare is not decisively a technological activity. Superior technology hasn't in the past guaranteed success in land warfare and this is unlikely to change in the future. In part, and reflecting points made earlier, this is because there are other important dimensions to warfare of any kind, including effective strategy, generalship and psychological factors. Reflecting this book's focus on the modern system, however, there are also important conceptual dimensions to land warfare reflected in formal and informal doctrine and in the phenomenon of force employment. If technology matters, it is often because of how it is used; equally, new methods can negate the effects of new technology. This has been the crucial function of modern approaches to land warfare, without which offensive action would collapse under the weight of attrition inflicted by modern firepower.

Land warfare is heterogeneous

There is no single model of land warfare. Whilst, all things being equal, the modern system of warfare might constitute the most effective method by which land warfare can be fought, it doesn't necessarily follow that all land forces fight in the same way. Context is everything. For some armies, the modern system of land warfare is simply too difficult to implement effectively; for others, the trade-offs associated with implementing it are too significant. For other armies, issues associated with organisational culture or command style might require the modern system to be modified in its application. For many states, the modern system of land warfare might be irrelevant because the primary roles for their land forces are internal. Just as importantly, land warfare, like all war, is an adversarial activity: how an army fights will be conditioned amongst other things by what an adversary does. Since war is a competitive activity, it is always likely that an adversary will try to shift the terms of conflict onto ground in which they hold the advantage. For that reason, being acknowledged as an effective exponent of modern system land warfare may actually encourage adversaries to find alternative ways of contesting victory. In the end, land warfare has since 1900 been conducted in many different ways and this is likely to continue in the future.

Land warfare is problematic to predict

There is no shortage of predictions on the future of land warfare. From the shiny, techno-centric RMA 'hyper war', to the dirty and brutal visions of 'war amongst the people', the future of land warfare has been the subject of extensive analysis and forecasting. However, history suggests that we are actually very poor at predicting the future of land warfare. As the circumstances prior to the First World War illustrate, this difficulty isn't necessarily the result of closed minds or blinkered thinking; it is just that predicting the future is simply inherently difficult, given the uncertainties surrounding future technological developments, strategic and political contexts, and the characteristics of future adversaries. Whilst one answer might be to focus on increasing the inherent flexibility and adaptability of land forces, the historic difficulties experienced by armies in performing counterinsurgency and peace and stability operations suggests that some of the barriers to military innovation and adaptation are difficult to overcome and lie in the very fabric of human organisation.

What we can say with some certainty is that future land warfare will not look exactly like the immediate Western experiences in Iraq or Afghanistan. First, all wars are different. Second, for most armies of the world, Iraq and Afghanistan are not even the present, let alone the future.

If there is perhaps one theme that underlies everything that *Understanding Land Warfare* has tried to achieve, it is the idea that we know much less about the topic of modern land warfare than we often suppose. It doesn't require much of a survey of the literature on land warfare to determine that many of the really important issues are still subject to debate, and that there are fewer objective facts than we might be comfortable with. Inevitably, then, this volume has raised more questions than it answers. But that doesn't matter. The Introduction to this book argued that one of the purposes of *Understanding Land Warfare* was to provide a foundation of understanding of warfare on land that would better enable the reader to explore the issues further. It is hoped that in highlighting the debates and issues, this book has achieved that goal and readers will find themselves better equipped to navigate their way through further engagement with the subject.

Select bibliography

Adams, Thomas K., *The Army After Next: The First Postindustrial Army* (Stanford, CA: Stanford University Press, 2008).

Bellamy, Alex J., Williams, Paul D. and Griffin, Stuart, *Understanding Peacekeeping* (Cambridge: Polity Press, 2010).

Biddle, Stephen, *Military Power: Explaining Victory and Defeat in Modern Battle* (Princeton, NJ: Princeton University Press, 2004).

Black, Jeremy, *War Since 1945* (London: Reaktion, 2004).

Black, Jeremy and MacRaild, Donald M., *Studying History* (Basingstoke: Palgrave, 2000).

Boot, Max, *War Made New* (New York: Gotham, 2012).

——*Invisible Armies: An Epic History of Guerrilla Warfare from Ancient Times to the Present* (New York, NY: Liveright, 2013).

Citino, Robert M., *Quest for Decisive Victory: From Stalemate to Blitzkrieg in Europe, 1898–1940* (Lawrence, KS: Kansas University Press, 2002).

Corum, James S., *The Roots of Blitzkrieg: Hans von Seekt and German Military Reform* (Lawrence, KS: University Press of Kansas, 1992).

Diehl, Paul F. and Druckman, Daniel, *Evaluating Peace Operations* (Boulder, CO: Lynne Rienner, 2010).

Dobbins, James, McGinn, John G., Crane, Keith, Jones, Seth G., Lal, Rollie, Rathmell, Andrew, Swanger, Rachel M. and Rimilsina, Anga R., *America's Role in Nation-Building: From Germany to Iraq* (Santa Monica, CA: RAND, 2003).

Edelstein, David, 'Occupational Hazards: Why Military Occupations Succeed or Fail', *International Security*, Vol. 29, No.1 (Fall 2004), 49–91.

Ellis, John, *The Sharp End: The Fighting Man in World War II* (London: Aurum Press, 2009).

English, John A. and Gudmundsson, Bruce I., *On Infantry* (Westport, CT: Praeger, 1994).

Farrell, Theo, *The Norms of War: Cultural Beliefs and Modern Conflict* (Boulder, CO: Lynne Rienner, 2005).

Farrell, Theo and Terriff, Terry (eds), *The Sources of Military Change: Culture, Politics, Technology* (Boulder, CO: Lynne Rienner, 2002).

Finkel, Meir, *On Flexibility: Recovery from Technological and Doctrinal Surprise on the Battlefield* (Stanford, CA: Stanford University Press, 2011).

Frieser, Karl-Heinz, *The Blitzkrieg Legend: The 1940 Campaign in the West* (Annapolis, MD: Naval Institute Press, 2005).

Gat, Azar, *A History of Military Thought: From the Enlightenment to the Cold War* (Oxford: Oxford University Press, 2001).

Glantz, David M., *Soviet Military Operational Art: In Pursuit of Deep Battle* (London: Frank Cass, 1991).

Goldman, Emily O. and Eliason, Leslie C., *The Diffusion of Military Technology and Ideas* (Stanford, CA: Stanford University Press, 2003).

Gooch, John (ed.), *The Origins of Contemporary Doctrine*, The Occasional No. 30 (Camberley: Strategic and Combat Studies Institute, September 1997).

Gordon, Michael R. and Trainor, Bernard E., *Cobra II: The Inside Story on the Invasion and Occupation of Iraq* (London: Atlantic, 2006).

Gray, Colin S., *Modern Strategy* (Oxford: Oxford University Press, 1999).

Griffin, Stuart, *Joint Operations: A Short History* (Training Specialist Services HQ, 2005).

Gudmundsson, Bruce I., *Stormtroop Tactics: Innovation in the German Army 1914–1918* (Westport, CT: Praeger, 1995).

Heuser, Beatrice, *The Evolution of Strategy: Thinking War from Antiquity to the Present* (Cambridge: Cambridge University Press, 2010).

Hoffman, Frank, *Conflict in the 21st Century: The Rise of Hybrid Wars* (Arlington, VA: Potomac Institute for Policy Studies, 2007).

Hoiback, Harald, *Understanding Military Doctrine: A Multidisciplinary Approach* (London: Routledge, 2013).

Holmes, Richard, *Acts of War: The Behaviour of Men in Battle* (London: Cassell, 2004).

House, Jonathan M., *Combined Arms Warfare in the Twentieth Century* (Lawrence, KS: University Press of Kansas, 2001).

Johnson, William, *Redefining Land Power for the 21st Century* (Carlisle, PA: Strategic Studies Institute, 1991).

Kagan, Frederick W., *Finding the Target: The Transformation of American Military Policy* (New York, NY: Encounter, 2006).

Kelly, Justin and Brennan, Mike, *Alien: How Operational Art Devoured Strategy* (Carlisle, PA: Strategic Studies Institute, September 2009).

Kilcullen, David J., 'Countering Global Insurgency', *The Journal of Strategic Studies*, Vol. 28, No. 4 (August 2005), 597–617.

Knox, MacGregor and Murray, Williamson, *The Dynamics of Military Revolution, 1300–2050* (Cambridge: Cambridge University Press, 2003).

Latawski, Paul, *The Inherent Tensions in Military Doctrine* (Sandhurst Occasional Papers No. 5, 2011).

Lewis, Adrian R., *The American Culture of War: A History of U.S. Military Force from World War II to Operation Enduring Freedom* (Abingdon: Routledge, 2007).

Linn, Brian McAllister, *The Echo of Battle: The Army's Way of War* (Cambridge, MA: Harvard University Press, 2007).

Lind, William S., *Maneuver Warfare Handbook* (Boulder, CO: Westview, 1985).

Linn, Brian McAllister, *The Echo of Battle: The Army's Way of War* (Harvard, MA: Harvard University Press, 2009).

Lupfer, Timothy T., *The Dynamics of Doctrine: The Changes in German Tactical Doctrine During the First World War* (Fort Leavenworth, KS: Combat Studies Institute, July 1981).

Luttwak, Edward N., *Strategy: The Logic of War and Peace* (Cambridge, MA: Cambridge University Press, 2001).

MacGregor, Douglas A., *Breaking the Phalanx: A New Design for Landpower in the 21st Century* (Westport, CT: Praeger, 1997).

Mackinlay, John, *The Insurgent Archipelago: From Mao to Bin Laden* (London: C. Hurst and Co., 2009).

Marston, Daniel and Malkesian, Carter (eds), *Counterinsurgency in Modern Warfare* (Oxford: Osprey, 2010).

McInnes, Colin, *Men, Machines and the Emergence of Modern Warfare, 1914–1945* (Camberley: Strategic and Combat Studies Institute, 1992).

McInnes, Colin and Sheffield, G.D., *Warfare in the Twentieth Century: Theory and Practice* (London: Unwin Hyman, 1988).

Metz, Steven and Johnson, Douglas V., II, *Asymmetry and U.S. Military Strategy: Definition, Background, and Strategic Concepts* (Carlisle, PA: Strategic Studies Institute, 2001).

Milevski, Lukas, 'Fortissimus Inter Pares: The Utility of Landpower in Grand Strategy', *Parameters* (Summer 2012), 6–15.

Millett, Allan R. and Murray, Williamson (eds), *Military Effectiveness* (Cambridge: Cambridge University Press, 2010).

Mockaitis, Thomas R., *Resolving Insurgencies* (Carlisle, PA: Strategic Studies Institute, 2011).

Murray, Williamson, *Military Adaptation in War: With Fear of Change* (Cambridge: Cambridge University Press, 2011).

Murray, Williamson and Mansoor, Peter (eds), *Hybrid Warfare: Fighting Complex Opponents from the Ancient World to the Present* (Cambridge: Cambridge University Press, 2012).

Nagl, John A. and Burton, Brian M., 'Thinking Globally and Acting Locally: Counter-insurgency Lessons from Modern Wars – A Reply to Smith and Jones', *Journal of Strategic Studies*, Vol. 33, No. 1 (February 2010), 123–38.

Neiberg, Michael S., *Warfare in World History* (London: Routledge, 2001).

Newmyer, Jacqueline, 'The Revolution in Military Affairs with Chinese. Characteristics', *The Journal of Strategic Studies*, Vol. 33, No. 4 (August 2010), 483–504.

Olsen, John Andreas and van Creveld, Martin (eds), *The Evolution of Operational Art: From Napoleon to the Present* (Oxford: Oxford University Press, 2011).

O'Neill, Bard, *Insurgency and Terrorism* (Washington, DC: Potomac Books, 2005).

Osinga, Frans and Farrell, Theo (eds), *A Transformation Gap: American Innovations and European Military Change* (Stanford, CA: Stanford University Press, 2010).

O'Sullivan, Patrick, *Terrain and Tactics* (London: Greenwood, 1991).

Paret, Peter (ed.), *Makers of Modern Strategy: From Machiavelli to the Nuclear Age* (Princeton, NJ: Princeton University Press, 1986).

Paris, Roland, *At War's End: Building Peace After Civil Conflict* (Cambridge: Cambridge University Press, 2004).

Parker, Geoffrey, *The Military Revolution: Military Innovation and the Rise of the West* (Cambridge: Cambridge University Press 1996).

——(ed.), *The Cambridge History of Warfare* (Cambridge: Cambridge University Press, 2005).

Pollack, Kenneth M., *Arabs at War, 1948–1991* (Lincoln, NE: University of Nebraska, 2004).

Posen, Barry R., *The Sources of Military Doctrine: France, Britain, and Germany Between the World Wars* (Ithaca, NY: Cornell University Press, 1984).

Rogers, Clifford (ed.), *The Military Revolution Debate: Readings on the Military Transformation of Early Modern Europe* (Boulder, CO: Westview Press, 1995).

Ryan, Mark A., Finkelstein, David M. and McDevitt, Michael A. (eds), *Chinese Warfighting: The PLA Experience Since 1949* (London: M.E. Sharpe, 2003).

Samuels, Martin, *Doctrine and Dogma: German and British Infantry Tactics in the First World War* (London: Greenwood Press, 1992).

Shambaugh, David, *Modernizing China's Military: Progress, Problems, and Prospects* (Berkeley, CA: University of California, 2004).

Sheffield, Gary, *Forgotten Victory: The First World War – Myths and Realities* (London: Headline, 2001).

Shimko, Keith L., *The Iraq Wars and America's Military Revolution* (Cambridge: Cambridge University Press, 2010).

Stewart, Rory and Knaus, Gerald, *Can Intervention Work?* (New York: Norton, 2011).

Strachan, Hew, 'Strategy or Alibi: Obama, McChrystal, and the Operational Level of War', *Survival*, Vol. 52, No. 5 (October–November 2010), 157–82.

Strachan, Hew and Herberg-Rothe, Andreas (eds), *Clausewitz in the Twenty-First Century* (Oxford: Oxford University Press, 2007).

Taw, Jennifer Morrison, *Mission Revolution: The U.S. Military and Stability Operations* (Chichester, NY: Columbia University Press, 2012).

Toffler, Alvin and Toffler, Heidi, *War and Anti-War: Survival at the Dawn of the 21st Century* (London: Little, Brown and Co., 1993).

Townshend, Charles (ed.), *The Oxford Illustrated History of Modern War* (Oxford: Oxford University Press, 1997).

Ucko, David H., *The New Counterinsurgency Era: Transforming the U.S. Military for Modern Wars* (Washington, DC: Georgetown University Press, 2009).

Ullman, Harlan K. and Wade, James P., *Shock and Awe: Achieving Rapid Dominance* (Washington, DC: National Defence University, 1996).

van Creveld, Martin, *Command in War* (Cambridge, MA: Harvard University Press, 1985).

Vego, Milan, *Joint Operational Warfare: Theory and Practice* (Newport, RI: US Naval War College, 2009).

von Clausewitz, Carl (edited and translated by Michael Howard and Peter Paret), *On War* (Princeton, NJ: Princeton University Press, 1976).

Index